FRESHWATER POLLUTION AND AQUATIC ECOSYSTEMS

*Environmental Impact and
Sustainable Management*

FRESHWATER POLLUTION AND AQUATIC ECOSYSTEMS

Environmental Impact and Sustainable Management

Edited by

Gowhar Hamid Dar, PhD
Khalid Rehman Hakeem, PhD
Mohammad Aneesul Mehmood, PhD
Humaira Qadri, PhD

AAP | APPLE ACADEMIC PRESS

First edition published 2022

Apple Academic Press Inc.
1265 Goldenrod Circle, NE,
Palm Bay, FL 32905 USA

4164 Lakeshore Road, Burlington,
ON, L7L 1A4 Canada

CRC Press
6000 Broken Sound Parkway NW,
Suite 300, Boca Raton, FL 33487-2742 USA

2 Park Square, Milton Park,
Abingdon, Oxon, OX14 4RN UK

Library and Archives Canada Cataloguing in Publication

Title: Freshwater pollution and aquatic ecosystems : environmental impact and sustainable management / edited by Gowhar Hamid Dar, PhD, Khalid Rehman Hakeem, PhD, Mohammad Aneesul Mehmood, PhD, Humaira Qadri, PhD.

Names: Dar, Gowhar Hamid, editor. | Hakeem, Khalid Rehman, editor. | Mehmood, Mohammad Aneesul, editor. | Qadri, Humaira, editor.

Description: First edition. | Includes bibliographical references and index.

Identifiers: Canadiana (print) 20210244143 | Canadiana (ebook) 20210244321 | ISBN 9781771889582 (hardcover) | ISBN 9781774638835 (softcover) | ISBN 9781003130116 (ebook)

Subjects: LCSH: Water—Pollution. | LCSH: Aquatic ecology.

Classification: LCC TD420 .F74 2022 | DDC 628.1/68—dc23

Library of Congress Cataloging-in-Publication Data

Names: Dar, Gowhar Hamid, editor. | Hakeem, Khalid Rehman, editor. | Mehmood, Mohammad Aneesul, editor. | Qadri, Humaira, editor.

Title: Freshwater pollution and aquatic ecosystems : environmental impact and sustainable management / edited by Gowhar Hamid Dar, Khalid Rehman Hakeem, Mohammad Aneesul Mehmood, Humaira Qadri.

Description: First edition. | Palm Bay, FL : Apple Academic Press, [2022] | Includes bibliographical references and index. | Summary: "This new volume addresses the environmental impacts of pollution on freshwater aquatic ecosystems and presents sustainable management and remediation practices and advanced technology help to address the different types of pollutants. Freshwater Pollution and Aquatic Ecosystems: Environmental Impact and Sustainable Management considers the need for sustainable, efficient, and cost-effective tools and technologies to assess, monitor, and properly manage the increasing issues of aquatic pollution. It provides detailed accounts of the phenomena and mechanisms related to aquatic pollution and highlights the problems and threats associated with pollution contamination in freshwater. It provides useful insight into the sustainable and advanced pollution remediation technology adopted by different countries for the monitoring, assessment, and sustainable management of pollution. The chapters in the volume evaluate the sources of harmful pollutants, which include industrial effluents, sewage, and runoff from agricultural industries, which result in toxic microbes, organic waste, oils, and high load of nutrients. Unsustainable management practices of domestic sewage and indiscriminate use of chemical pesticides lead to the technological disturbance of aquatic biota. In addition to harming aquatic biota, these pollutants find their way into the human body through inhalation, ingestion, or absorption and finally tend to bio-accumulate in trophic levels of the food chain, which poses a major risk to human beings. This book will be a valuable resource for ecologists, environmentalists, scientists, and many others for their work in understanding and management of aquatic pollutants in freshwater biospheres"-- Provided by publisher.

Identifiers: LCCN 2021028757 (print) | LCCN 2021028758 (ebook) | ISBN 9781771889582 (hardback) | ISBN 9781774638835 (paperback) | ISBN 9781003130116 (ebook)

Subjects: LCSH: Water--Pollution--Environmental aspects. | Ecosystem management. | Sustainable development. | Aquatic organisms--Effect of water pollution on.

Classification: LCC QH545.W3 F74 2022 (print) | LCC QH545.W3 (ebook) | DDC 577.6/27--dc23

LC record available at https://lccn.loc.gov/2021028757

LC ebook record available at https://lccn.loc.gov/2021028758

ISBN: 978-1-77188-958-2 (hbk)
ISBN: 978-1-77463-883-5 (pbk)
ISBN: 978-1-00313-011-6 (ebk)

Dedicated to the scientific community.

About the Editors

Gowhar Hamid Dar, PhD
Assistant Professor, Environmental Science,
Sri Pratap College, Cluster University Srinagar,
Department of Higher Education,
Jammu and Kashmir, India

Gowhar Hamid Dar, PhD, is an Assistant Professor in Environmental Science, Sri Pratap College, Cluster University Srinagar, Department of Higher Education (Jammu and Kashmir), India, where he has been teaching for many years. He has a PhD in Environmental Science with a specialization in Environmental Microbiology (fish microbiology, fish pathology, industrial microbiology, taxonomy and limnology). He has published more than 40 papers in international journals of repute and a number of books with international publishers. He is guiding a number of students for their master's theses. He has been working on the isolation, identification, and characterization of microbes; their pathogenic behavior; and impact of pollution on development of diseases in fish fauna for the last several years. He has received many awards and appreciations for his services toward science and development. In addition, he also acts as a member of various research and academic committees.

Khalid Rehman Hakeem, PhD
Professor, King Abdulaziz University,
Jeddah, Saudi Arabia

Khalid Rehman Hakeem, PhD, is a Professor at King Abdulaziz University, Jeddah, Saudi Arabia. After completing his doctorate (botany; specialization in plant eco-physiology and molecular biology) from Jamia Hamdard, New Delhi, India, he worked as a lecturer at the University of Kashmir, Srinagar, India. At Universiti Putra Malaysia, Selangor, Malaysia, he was a Postdoctorate Fellow and Fellow Researcher (Associate Professor) for several years. Dr. Hakeem has more than 10 years of teaching and research experience in plant eco-physiology, biotechnology and molecular biology, medicinal plant research, plant-microbe-soil interactions, as well as in environmental studies. He is the recipient of several fellowships at both national and international levels. He has served as a visiting scientist at Jinan University, Guangzhou, China. Currently, he is involved with a number of international research projects with different government organizations. To date, Dr. Hakeem has authored and edited more than 35 books with international publishers. He also has to his credit more than 80 research publications in peer-reviewed international journals and 55 book chapters in edited volumes with international publishers. At present, Dr. Hakeem serves as an editorial board member and reviewer for several high-impact international scientific journals. He is included in the advisory board of Cambridge Scholars Publishing, UK. He is also a fellow of the Plantae group of the American Society of Plant Biologists, member of the World Academy of Sciences, member of the International Society for Development and Sustainability, Japan, and member of the Asian Federation of Biotechnology, Korea.

Mohammad Aneesul Mehmood, PhD
Department of Environmental Science,
School of Sciences, Sri Pratap College Campus,
Cluster University Srinagar, Jammu and Kashmir,
India

Mohammad Aneesul Mehmood, PhD, specializes in limnology and environmental toxicology. He has been teaching graduate and postgraduate students for the past two years in the Department of Environmental Science, School of Sciences, Sri Pratap College Campus, Cluster University Srinagar, Jammu and Kashmir, India. He has been supervising many students for their MSc projects. In addition, he has published a number of papers in international journals of repute and a number of books with international publishers. He completed his doctorate at the Division of Environmental Science, Sher-e-Kashmir University of Agricultural Sciences and Technology of Kashmir, with a meritorious certificate from the university. He received the Dr. Mumtaz Ahmad Khan Gold Medal and the Shri Bhushan Memorial Gold Medal during his master's program. He has qualified various national competitive tests in the discipline of environmental science. He was also awarded with an INSPIRE Merit Fellowship (JRF & SRF) by the Department of Science and Technology, GoI, during his doctoral program.

Humaira Qadri, PhD
Head, Department of Environment and
Water Management, Cluster University Srinagar,
Sri Pratap College Campus, Jammu and Kashmir,
India

Humaira Qadri, PhD, has been actively involved in teaching postgraduate students of environmental science for the past ten years at the Sri Pratap College Campus of Cluster University Srinagar, Jammu and Kashmir, India, where she also heads the Department of Environment and Water Management. She has published scores of papers in international journals and has more than ten books with national and international publishers. She is also a reviewer for various international journals and is the principal investigator of some major projects on phytoremediation. She is guiding a number of research students for PhD programs and has supervised more than 60 master's dissertations. She also has been on the scientific boards of various international conferences and holds life memberships of various international organizations. With a number of national scientific events to her credit, she is an active participant at national and international scientific events and has organized a number of national conferences on science. A gold medalist at the master's level, she earned a number of awards and certificates of merit. Her specialization is in limnology, nutrient dynamics, and phytoremediation.

Contents

Contributors

Ajaz Ahmad
Department of Clinical Pharmacy, College of Pharmacy, King Saud University, Riyadh–11451, Saudi Arabia

Rukhsana Akhtar
Department of Clinical Biochemistry, Govt. Degree College (Baramulla), Khawaja Bagh, Baramulla, Jammu and Kashmir, India

Aarif Ali
Department of Biochemistry, University of Kashmir, Jammu and Kashmir, India

Md. Niamat Ali
Cytogenetics and Molecular Biology Laboratory, Center of Research for Development, University of Kashmir, Hazratbal, Srinagar–190006, Jammu and Kashmir, India

Sadaf Ali
Department of Biochemistry, Government Medical College (GMC), Karan Nagar, Srinagar–190010, Jammu and Kashmir, India

Shafat Ali
Department of Biochemistry, Government Medical College (GMC), Karan Nagar, Srinagar–190010, Jammu and Kashmir, India
Centre of Research for Development (CORD), University of Kashmir, Srinagar–190006, Hazratbal, Jammu and Kashmir, India

Henna Amin
Department of Pharmaceutical Sciences, Faculty of Applied Sciences, University of Kashmir, Srinagar–190006, Jammu and Kashmir, India

Insha Amin
Division of Veterinary Biochemistry, Faculty of Veterinary Science and Animal Husbandry, SKUAST-Kashmir, Alustang, Shuhama, Srinagar–190006, Jammu and Kashmir, India

Shiekh Amir
Department of Forensic Medicine & Toxicology, Government Medical College (GMC), Karan Nagar, Srinagar–190010, Jammu and Kashmir, India

Mohammad Yasir Arafat
Fish Biology and Limnology Research Lab., Department of Zoology, University of Kashmir, Hazratbal, Srinagar–190006, Jammu and Kashmir, India

Rezwana Assad
Department of Botany, University of Kashmir, Srinagar–190006, Jammu and Kashmir, India, E-mail: rezumir@gmail.com

Yahya Bakhtiyar
Fish Biology and Limnology Research Lab., Department of Zoology, University of Kashmir, Hazratbal, Srinagar–190006, Jammu and Kashmir, India, E-mail: yahya.bakhtiyar@gmail.com

Iqra Bashir
Department of Botany, University of Kashmir, Srinagar–190006, Jammu and Kashmir, India

Kulsum Ahmad Bhat
Wildlife Laboratory, Department of Zoology/Phytochemistry Laboratory, Center of Research for Development, University of Kashmir, Hazratbal, Srinagar–190006, Jammu and Kashmir, India

Showkat Ahmad Bhat
Department of Biochemistry, Government Medical College (GMC), Karan Nagar, Srinagar–190010, Jammu and Kashmir, India

Sheikh Bilal
Division of Veterinary Biochemistry, FVSc & AH, SKUAST-K, Shuhama, Jammu and Kashmir, India

Yusra Al Dhaher
Department of Biology, College of Science, UAE University, P.O. Box–17551, United Arab Emirates University, Al-Ain–15551, United Arab Emirate

Rafiqa Echikoti
Department of Biochemistry, Government Medical College (GMC), Karan Nagar, Srinagar–190010, Jammu and Kashmir, India

Adil Farooq
RAK College of Pharmaceutical Sciences, RAK Medical and Health Sciences University, Ras Al Khaimah–11172, United Arab Emirates

Sana Farooq
Department of Biochemistry, Government Medical College (GMC), Karan Nagar, Srinagar–190010, Jammu and Kashmir, India

Aimen Firdous
Department of Processing Technology, Kerala University of Fisheries and Ocean Studies (KUFOS), Panangad, Kerala–682506, India

Hilal Ahmad Ganaie
Department of Zoology, Government Degree College (Boys), Pulwama–192301, Jammu and Kashmir, India, E-mail: hilalganie@hotmail.com

Showkat Ahmad Ganie
Department of Biochemistry, University of Kashmir, Jammu and Kashmir, India

Henna Hamadani
Mountain Livestock Research Institute, Manasbal SKUAST-K, Jammu and Kashmir, India

Tehseen Khan
Department of Biochemistry, Government Medical College (GMC), Karan Nagar, Srinagar–190010, Jammu and Kashmir, India

Naziya Khurshid
Parasitology Laboratory, Department of Zoology/Microbiology Laboratory, Center of Research for Development, University of Kashmir, Hazratbal, Srinagar–190006, Jammu and Kashmir, India

Moline Severino Lemos
Department of Cellular Biology, Institute of Biological Science, Federal University of Minas Gerais, Belo Horizonte, Brazil

Sabhiya Majid
Department of Biochemistry, Government Medical College (GMC), Karan Nagar, Srinagar–190010, Jammu and Kashmir, India

Muniza Manzoor
Wildlife Laboratory, Department of Zoology/Cytogenetics and Molecular Biology Laboratory,
Center of Research for Development, University of Kashmir, Hazratbal, Srinagar–190006,
Jammu and Kashmir, India

Mubashir Hussain Masoodi
Department of Pharmaceutical Sciences, Faculty of Applied Sciences, University of Kashmir,
Srinagar–110006, Jammu and Kashmir, India

Shehzada Munawar Mehdi
Rapid Soil Fertility Research Institute, Lahore, Punjab, Pakistan

Hassan Mehmood
Department of Soil Science, The Islamia University of Bahawalpur, Pakistan

Manzoor Ur Rahman Mir
Division of Veterinary Biochemistry, Faculty of Veterinary Science and Animal Husbandry,
SKUAST-Kashmir, Alustang, Shuhama, Srinagar–190006, Jammu and Kashmir, India

Zahoor Ahmad Mir
Fish Biology and Limnology Research Lab., Department of Zoology, University of Kashmir, Hazratbal,
Srinagar–190006, Jammu and Kashmir, India

Ghulam Murtaza
Institute of Soil and Environmental Sciences, University of Agriculture, Faisalabad, Pakistan

Muhammad Naveed
Soil and Water Testing Laboratory for Research, Lahore, Pakistan

Abdul Ghaffar Niazi
Rapid Soil Fertility Research Institute, Lahore, Punjab, Pakistan

Ashok K. Pandit
Center of Research for Development (CORD), University of Kashmir, Srinagar–190006,
Jammu and Kashmir, India

Jayachithra Ramakrishna Pillai
RAK College of Pharmaceutical Sciences, RAK Medical and Health Sciences University,
Ras Al Khaimah–11172, United Arab Emirates

Wajhul Qamar
Department of Pharmacology and Toxicology, Central Lab, College of Pharmacy, King Saud University,
Riyadh–11451, Saudi Arabia

Faizan Rafi
Institute of Soil and Environmental Sciences, University of Agriculture, Faisalabad, Pakistan

Iflah Rafiq
Department of Botany, University of Kashmir, Srinagar–190006, Jammu and Kashmir, India

Muneeb U. Rahman
Department of Biochemistry, Government Medical College, Srinagar, Jammu & Kashmir, India;
Department of Clinical Pharmacy, College of Pharmacy, King Saud University, Kingdom of Saudi Arabia

Irfan Rashid
Department of Botany, University of Kashmir, Srinagar–190006, Jammu and Kashmir, India

Masrat Rashid
Food and Drug Control Organization, Government of Jammu and Kashmir, India

Shahzada Mudasir Rashid
Assistant Professor (Scientist), Division of Veterinary Biochemistry, Faculty of Veterinary Science and Animal Husbandry, SKUAST-Kashmir, Alustang, Shuhama, Srinagar–190006, Jammu and Kashmir, India, Mobile: 9622464045, E-mails: mudasir@skuastkashmir.ac.in, drsmrashid786@gmail.com

Tabasum Rashid
Department of Biochemistry, Government Medical College (GMC), Karan Nagar, Srinagar–190010, Jammu and Kashmir, India

Saiema Rasool
Department of School Education, Govt. of Jammu & Kashmir, Srinagar, 190001 India

Shabhat Rasool
Department of Biochemistry, Government Medical College (GMC), Karan Nagar, Srinagar–190010, Jammu and Kashmir, India

Muneeb U. Rehman
Department of Biochemistry, Government Medical College (GMC), Karan Nagar, Srinagar–190010, Jammu and Kashmir, India
Department of Clinical Pharmacy, College of Pharmacy, King Saud University, Riyadh–11451, Saudi Arabia, E-mail: muneebjh@gmail.com

Zafar Ahmad Reshi
Department of Botany, University of Kashmir, Srinagar–190006, Jammu and Kashmir, India

Umair Riaz
Soil and Water Testing Laboratory for Research, Bahawalpur–63100, Pakistan, E-mail: umairbwp3@gmail.com

Fozia Shah
Division of Veterinary Physiology, FVSc & AH, SKUAST-K, Shuhama, Jammu and Kashmir, India

G. Mustafa Shah
Department of Zoology, University of Kashmir, Srinagar–190006, Jammu and Kashmir, India

Javaid Ahmad Shah
Center of Research for Development (CORD), University of Kashmir, Srinagar–190006, Jammu and Kashmir, India; Government Degree College (Boys), Pulwama–192301, Jammu and Kashmir, India, E-mail: javaidshah31@gmail.com

Sheeba Shakeel
Department of Pharmaceutical Sciences, Faculty of Applied Sciences, University of Kashmir, Srinagar–190006, Jammu and Kashmir, India

Irshad Ahmad Sofi
Department of Botany, University of Kashmir, Srinagar–190006, Jammu and Kashmir, India

Fernanda Maria Policarpo Tonelli
Department of Cellular Biology, Institute of Biological Science, Federal University of Minas Gerais, Belo Horizonte, Brazil, E-mail: tonellibioquimica@gmail.com

Flávia Cristina Policarpo Tonelli
Department of Biochemistry, Federal University of São João del Rei, Divinópolis, Brazil

Adil Farooq Wali
RAK College of Pharmaceutical Sciences, RAK Medical and Health Sciences University, Ras Al Khaimah–11172, United Arab Emirates

Hilal Ahmad Wani
Department of Biochemistry, Govt. Degree College (Sumbal), Sumbal, Bandipora,
Jammu and Kashmir, India

Javaid Wani
Department of Biochemistry, Government Medical College (GMC), Karan Nagar, Srinagar–190010,
Jammu and Kashmir, India

Ali Mohd Yatoo
Centre of Research for Development (CORD), University of Kashmir, Srinagar–190006, Hazratbal,
Jammu and Kashmir, India
Department of Environmental Science, University of Kashmir, Srinagar–190006, Hazratbal,
Jammu and Kashmir, India

Zarka Zaheen
Center of Research for Development/Department of Environmental Science, University of Kashmir,
Hazratbal, Srinagar–190006, Jammu and Kashmir, India

Abbreviations

ACh	acetylcholine
AChE	acetylcholinesterase
ACOPS	Advisory Committee on Protection of the Sea
ADP	adenosine di-phosphate
AHR	anti-hydroxyl radical
Al	aluminum
ALT	alanine aminotransferase
AMF	arbuscular mycorrhizal fungi
AMP	adenosine mono-phosphate
AOPs	advanced oxidation processes
As	arsenic
AST	aspartate aminotransferase
ATSDR	Agency for Toxic Substances and Disease Registry
BAFs	bioaccumulation factors
BBB	blood-brain barrier
BCFs	bioconcentration factors
BMFs	bio magnifications
BOD	biological oxygen demand
BSAFs	biota/sediment accumulation factor
$CaCO_3$	calcium carbonate
CAT	catalase
CBA	3-chlorobenzoate
CBP	4-chlorobenzophenone
Cd	cadmium
$CdCl_2$	cadmium chloride
CNS	central nervous system
CPF	chlorpyrifos
Cr	chromium
Cu	copper
DALYs	disability-adjusted life years
DDT	dichlorodiphenyltrichloroethane
DETP	diethyl phosphate
DMN	dimethyl nitrosamine

DNA	deoxyribonucleic acid
DOM	dissolved organic matter
DTI	drug toxicity index
ECDC	European Center for Disease Prevention and Control
EDTA	ethylene diamine tetraacetic acid
EMS	environmental mutagen society
EMS	ethyl methane sulphonate
EPA	Environmental Protection Agency
FDA	Food and Drug Administration
Fe	iron
FeRP	iron reducers
FFG	functional feeding groups
GAC	granular active carbon
GC-ECD	gas chromatography with electron capture detector
GFAP	glial fibrillary acidic protein
GMB	genetically modified bacteria
GMOs	genetically modified organism
GMPs	genetically modified plants
GPX	glutathione peroxidase
GSH	glutathione
GSI	gonado-somatic index
GSSG	glutathione disulfide
GTN	glycerol trinitrate
H_2O_2	hydrogen peroxide
HCH	hexachlorocyclohexane
Hct	hematocrit
Hg	mercury
Hg_2Cl_2	mercurous chloride
$HgC_4H_6O_4$	mercuric acetate
$HgCl_2$	mercuric chloride
HgS	mercuric sulfide
HPT	hypothalamic pituitary-thyroid
IAEMS	International Association of Environmental Mutagen Societies
IBI	index of biotic integrity
IP3	inositol triphosphate
ISO	International Organization for Standardization
LPO	lipid peroxidation
MCH	mean corpuscular hemoglobin
MCHC	mean corpuscular hemoglobin concentration

ME	middle east
MLA	maximum allowable limits
MMS	methanesulfonate
MN	micronucleus
MNT	micronucleus test
MT	metallothioneins
NH_4^+	ammonium
Ni	nickel
NIH	National Institute of Health
NK	natural killer
NO_3^-	nitrate
NOAA	National Ocean and Atmospheric Administration
NTE	neuropathy target esterase
NTP	National Toxicology Program
O_2	oxygen
OCPs	organochlorine pesticides
OECD	Organization for Economic Cooperation and Development
OH	hydroxyl radical
PAC	powder activated carbon
PAH	polyaromatic hydrocarbons
Pb	lead
PCBs	polychlorinated biphenyls
PCR	polymerase chain reaction
PKC	protein kinase C
POPs	persistent organic pollutants
PTI	pollution tolerance index
PTWI	provisional tolerable weekly intake
RNA	ribonucleic acid
ROS	reactive oxygen species
SMS	spent mushroom substrate
SO_2	sulfur dioxide
SO_4^{-2}	sulfate
SOB	sulfur-oxidizing bacteria
SOD	superoxide dismutase
SRB	sulfur-reducing bacteria
TBA	thiobarbituric acid
TBARS	TBA reactive substances
TCE	trichloroethylene
TCP	3,5,6-trichloro-2 pyridinol

TNT	2,4,6-trinitrotoluene
TSH	thyroid-stimulating hormone
U	uranium
WHO	World Health Organization
Zn	zinc

Preface

The importance of water for the sustenance of life cannot be overemphasized. Whether it is in use of running water in our homes, rearing cattle and growing crops in our farms, or the increased uses in industry, water remains immeasurable. With the onset of the industrial era and development activities, when the so-called standard of life was on the rise, the quality of the environment was being degraded. Human activities, including industrialization and agricultural practices, contributed in no small measure to the degradation and pollution of the environment, which adversely has an effect on water bodies (rivers and ocean) that is a necessity for life.

According to the United Nations Educational Scientific and Cultural Organization (UNESCO), water covers 72% of the earth's surface; however, fresh water only accounts for 0.5% of all the water resources. Despite the vast quantity of water on earth, the portion that can be directly used is very small.

In the 21st century, water is becoming a precious scarce resource. Problems with water resources are not only a problem of resources; they are also a major strategic issue related to the sustainable development of a country's economy, society, and long-term stability. With the global population explosion and rapid economic development, the demand for industrial water, agricultural water, and domestic water has increased rapidly, and the shortage of freshwater is becoming more and more serious. At the same time, the deterioration of water quality is exacerbating the shortage of water resources. Indeed, quality-induced water shortages have recently been a matter of great concern at local, regional, and global levels, especially in developing countries. Therefore, water resources protection and rational utilization, and the efficient restoration of polluted water, have become majors focuses of the world.

Since water pollution has direct consequences on humans' well-being, effective education regarding it in the formal education system is essential for a better understanding so as to develop the right attitude towards water. In this backdrop, the book titled, *Aquatic Pollution: Mechanism and Stress to Aquatic Environs* is an effort towards documenting the problems of pollution, its impacts, and remediation mechanisms for different types of pollutants.

Chapter 1 discusses water pollution as a threat to the environment and human health, taking the cognizance of various microbial diseases. This chapter emphasizes the need for proper treatment and disposal of wastes that are diverse in behavior and toxicity before releasing them into aquatic systems. Besides, it also emphasized the importance of hygiene education as well as the development of strategies for the prevention of water borne diseases.

Chapter 2 deals with the challenges and concerns of the toxicity of heavy metals in freshwater fishes. It discusses in detail the mechanism of action and impacts of these heavy metals (Pb, Hg, Cd, Al, Cr, Cu, Zn, Ni) on the reproductive, nervous, respiratory, and gastrointestinal system of the fishes.

Chapter 3 discusses the environmental distribution, degradation, and transformation, pharmacokinetics, and toxicity of chlorpyrifos (CPF). It also elaborates its effect on the behavior of different biological systems. Chapter 4 details the mechanistic approach of the neurotoxicity of heavy metals in fishes.

Chapter 5 discusses the pesticides having different levels of toxicity, their mechanism of mutagenicity, their genotoxic effects, and their relation with the fluctuations of the various anti-oxidative enzymes like superoxide, glutathione (GSH), dismutase, and catalases (CAT).

Chapter 6 details the pesticides in the aquatic systems and their environmental persistence. Chapter 7 summarizes the responses of benthic macroinvertebrates towards pollution and describes various bioassessment approaches, including diversity index, saprobic index, FBI index, BMWP score, ASPT score, NSFWQI score, taxa richness, and EPT ratio.

Chapter 8 deals with bioindicators, their advantages, classification, and the use of various organisms as bio-indicators in the environment. Chapter 9 explains in detail the role zooplanktons play as bioindicators in freshwater ecosystems. Chapter 10 provides an overview of different types of aquatic pollutants, their consequences, and mitigation of these pollutants through mycoremediation. Chapter 11 focuses on the use of recombinant DNA technology to genetically modify organisms to make them suitable to perform bioremediation of contaminated waters. Chapter 12 laid emphasis on the technique of bio-monitoring or biological monitoring as an important and essential means through which the quality of water can be assessed, in which morphological, biochemical, and physiological alterations occur in indicator and are being related to particular environmental pollution.

The present book is an important reference source that highlights the issues of aquatic pollution as well as the technologies, techniques, and strategies that can be used to manage this pollution. This book is a valuable reference for academicians, researchers, students, professionals, and policy-makers who can benefit from the innovative content of the book. Suggestions for the improvement of the book are always welcome.

—**Gowhar Hamid Dar, PhD**
Khalid Rehman Hakeem, PhD
Mohammad Aneesul Memood, PhD
Humaira Qadri, PhD

CHAPTER 1

Water Pollution: Diseases and Health Impacts

SADAF ALI,[1] SHIEKH AMIR,[2] SHAFAT ALI,[1,3] MUNEEB U. REHMAN,[1,4] SABHIYA MAJID,[1] and ALI MOHD YATOO[3,5]

[1]*Department of Biochemistry, Government Medical College (GMC), Karan Nagar, Srinagar–190010, Jammu and Kashmir, India*

[2]*Department of Forensic Medicine & Toxicology, Government Medical College (GMC), Karan Nagar, Srinagar–190010, Jammu and Kashmir, India*

[3]*Centre of Research for Development (CORD), University of Kashmir, Srinagar–190006, Hazratbal, Jammu and Kashmir, India*

[4]*Department of Clinical Pharmacy, College of Pharmacy, King Saud University, Riyadh–11451, Saudi Arabia" E-mail: muneebjh@gmail.com*

[5]*Department of Environmental Science, University of Kashmir, Srinagar–190006, Hazratbal, Jammu and Kashmir, India*

ABSTRACT

Environmental pollution is a serious problem nowadays, especially due to growing industrialization and urbanization. The use of water bodies for waste disposal has led to devastating effects both on human and aquatic life. The contamination of drinking water has an equal impact on the health of children as well as adults. Aquatic fauna also gets affected due to deprivation of oxygen supply to these organisms as a result of water pollution. Various microbial diseases like cholera, hepatitis (usually subtype E), amoebiasis, gastroenteritis, shigellosis, and various other viral and parasitic infections spread through contaminated drinking water or food. The contamination of water bodies by different types of heavy metals cannot be under estimated. In Japan, contamination of water bodies by cadmium has led to cadmium nephropathy and Itai-Itai disease. Similarly, Minimata disease is a severe form of neuropathy induced by contamination of water bodies due to mercury and its various compounds. Lead poisoning also influences the health of people, and one of the sources being contamination in the water bodies. Strategies should be developed for the prevention of waterborne diseases, and awareness programs are the need of the day to prevent such illnesses in the future. Nature is divine, and therefore, it is important to prevent pollution in any form and nurture both aquatic life and human health and, as a whole, our ecosystem.

1.1   INTRODUCTION

> *"Water, Water, everywhere, nor any drop to drink."*
> *—Samuel Taylor*

Water is an essential vital element to the human body. Water makes life possible on this earth, making water and life go hand in hand. Inside the human body, water acts as both a solvent, as well as a buffer. Water provides sustenance to all living organisms, including plants, animals, pathogens, as well as microorganisms. A human body is made up of around 70% water. Some amount of water is lost through metabolism, respiration, bowel movements, and urine. Humans use water for drinking and various other purposes (Bibi et al., 2016). Even being vital elements, most of the people on earth don't consume water in its purest form. Poor quality of drinking water drastically has an effect on human health. According to a recent report, a minimum of 2 billion humans worldwide utilize water for drinking purposes from sources infected with feces (WHO, 2018). Decreasing the quality or quantity of water poses a serious threat to the human body and health. Water is polluted due to the addition of harmful substances that alter water quality (Alrumman, 2016) and cause deleterious effects on the environment and human health (Briggs, 2003). Human activities contribute to environmental pollution directly or indirectly. Automobile exhausts contain various harmful gases like nitrogen, sulfur dioxide, carbon dioxide, and carbon monoxide, in addition to black soot, which pollutes the atmosphere. Domestic wastes from household activities, agricultural wastes also pollute the atmosphere. The use of pesticides and fertilizers contribute to water pollution. Waste materials from tanneries also contain many unsafe chemicals and release a foul smell. When water is contaminated with bacteria, protozoans, or viruses, it leads to the development of waterborne diseases. The majority of infectious diseases originate from polluted water. World Health Organization (WHO) reported that about 80% of diseases arise from water. The quality of drinking water lacks WHO standards in different nations of the world (Khan et al., 2013). Poor water quality causes about 3.1% of deaths (Pawari and Gawande, 2015). Various diseases such as polio, diarrhea, dysentery, and cholera are transmitted via poor hygiene and infected drinking water (WHO, 2018). Diseases like vomiting, kidney problems, gastroenteritis, and skin disorders also spread via utilizing infected water (Juneja and Chauhdary, 2013).

1.2   WATER POLLUTION INCIDENCE

World Health Organization (WHO) and UNICEF reported that during the year 2000, about 2.1 billion human populace has no access to an adequate way of hygiene, and approximately 1.1 billion lack a better water supply. Increased mobility of people and traveling to various countries has led to the global menace of waterborne diseases. National surveillance systems are required to prevent a wide array of diseases due to water pollution. Waterborne diseases occur due to pollution of water bodies, either via pathogenic protozoa, bacteria, viruses, or by other means like chemical wastes. The use of this contaminated water for household purposes such as cooking, drinking, bathing, and washing leads to direct transmission of pathogens to humans.

1.3   MEANS OF WATER POLLUTION

Organic as well as inorganic carbon affects the eutrophication of lakes (Goldman, 1972), which ultimately affects the chemical composition of water bodies and rivers (Crowder, 1991). Pollution is also caused by spillage into the water bodies (Sharma, 1999). Toxic organic contaminants of farms and agricultural wastages in the water stream were also reported by Thanas et al. (2001). The presence of organic flora and microorganisms in oceanic sediments has been reported by Volterra et al. (1985). Some microorganisms are helpful in the removal of nutrients from the water bodies (Tam and Wang, 1989). Underground water bodies contain various bacteria (Anderson and Stentrom, 1987). The number of Coliforms in the river has been studied in Jordan by Hades et al. (2000). Heavy metals have always contributed to the pollution of water bodies. These are present in a variety of industrial effluents. They are absorbed by hydrophytes. These metals also precipitate in the sediments (Gonzalez et al., 2000). Sinha et al. (1993) conducted a study on the uptake of magnesium and chromium by the aquatic plant such as Hydrilla and was supported by Say and Witton (1983). The main sources of water pollution are outlined in Figure 1.1.

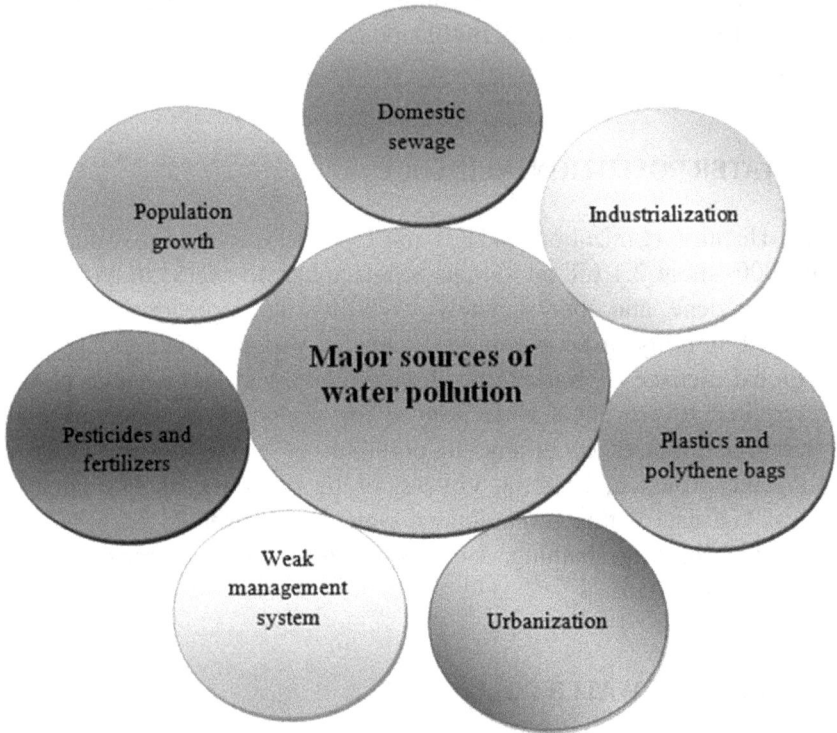

FIGURE 1.1   Major sources of water pollution.

1.4   DISEASES DUE TO WATER POLLUTION (WATERBORNE DISEASES)

Waterborne diseases include a range of ailments or infections commonly caused by the consumption of infected water. Diseases associated with water may be categorized as water-based, water-vector, waterborne, and water-washed based on their mode of transmission. Waterborne pathogenic organisms are a key public health concern that has emerged in all global nations. Worldwide epidemiology of cholera was studied with a novel dimension after the appearance of vibrio cholera O139. The outburst of chlorine-resistant cryptosporidium leads to a reassessment of the sufficiency of present-day water-quality policy. A range of microscopic organisms like hepatitis viruses, *Campylobacter jejuni*, Mycobacteria, *Giardia lamblia*, Legionella, and *Pseudomonas aeruginosa* have been known as causative agents for waterborne infections (Bagchi, 1993; Bert, 1998). The common waterborne diseases are described in Table 1.1.

1.5   COMMON DISEASES CAUSED BY USE OF CONTAMINATED DRINKING WATER

1.5.1   GASTROENTERITIS

This ailment is characterized by the inflammation of the gastrointestinal tract (Chlossberg, 2015). Gastroenteritis is caused by the intake of polluted foodstuffs or drinking water or close contact with an infected person (Ciccarelli, 2013). The disease is manifested in the form of diarrhea, vomiting, and abdominal pain (Singh and Amandeep, 2010). Symptoms also include fever, fatigue, dehydration, and water and salt loss (Ciccarelli, 2013; Ferris, 2015). The symptoms can last up to two weeks (Chlosberg, 2015). Gastroenteritis is also called stomach or gastric flu, but it is unlikely that of influenza infection (Shors, 2013). Gastroenteritis is commonly caused by viruses (Richard, 2006). However, this infection can be caused by bacteria, parasites, and fungus, also (Ciccarelli, 2013). *Campylobacter jejuni* infection is the most common cause of adult diarrhea (Marshall, 2011).

Children are more prone to develop rotavirus diarrhea, and the disease can be severe (Tate et al., 2012). Rotavirus vaccine is recommended for the prevention of infection in children (Cicarelli, 2013). Treatment includes rehydration in the form of fluids and ORS in mild to moderate cases and intravenous fluids in severe infection. Zinc supplementations and maintenance of electrolyte balance is the key in treatment. Antibiotics are usually not recommended unless the disease is severe and accompanied by fever and bloody stools (Singh, 2010). Sometimes symptoms include both diarrhea and vomiting (Eckardt, 2011). Abdominal discomfort and cramps may also present (Singh, 2010).

In viral infections, usually, the symptoms may be resolved in a period of one week. Occasionally, viral infections are accompanied by headache, fever, generalized weakness, and arthralgias. Bloody stools are more of a bacterial infection and less likely that of viral infection (Eckardt, 2011). Some of the bacterial diseases cause acute pain in the abdomen pain and discomfort that may continue for some weeks (Galanis, 2007).

TABLE 1.1 Common Waterborne Diseases

Disease	Causative Agent	Means of Transmission	Symptoms
Gastroenteritis	Viruses	Polluted foodstuffs or drinking water or close contact with an infected person	Diarrhea, vomiting, and abdominal pain, fever, fatigue, dehydration.
Typhoid fever	*Salmonella typhii*	Contaminated food or water	Generalized weakness, constipation, pain in the abdomen, headache, vomiting, and skin rashes with rose-pink spots.
Shigellosis	*Shigella*	Ingestion of the organism with infected food or water	Nausea, diarrhea, flatulence, abdominal cramps, and vomiting.
Cryptosporidiosis	*Cryptosporidium*	Drinking infected water	Watery diarrhea with or without unsolved cough
Escherichia coli infection	*Escherichia coli*	Packed dairy products, juices, contaminated drinking water	Sepsis, hemolytic uremic syndrome, bowel necrosis, mastitis, gram-negative pneumonia, and peritonitis.
Campylobacteriosis	*Campylobacter jejuni*	Sexual contact, consumption of unpasteurized milk, infected food, water, and poorly cooked chicken	Inflammatory condition and bloody diarrhea
Giardiasis	*Giardia lamblia*	Consumption of infected foodstuffs or drinking water	Abdominal pain leading to diarrhea and weight loss
Amoebiasis	Entamoeba		Abdominal cramps and mild diarrhea
Cholera	*Vibrio cholera*	Ingestion of a toxic strain through food or drinks	Watery diarrhea, cramps, abdominal discomfort, and vomiting.
Hepatitis A	Hepatovirus A	Contamination of food or water	Fever, fatigue, diarrhea, nausea, and gastric upset
Adenoviral disease	Adenovirus	Contaminates drinking water, etc.	Tonsillitis, conjunctivitis, ear infections, meningitis, and encephalitis, and also to bronchopneumonia

1.5.2 TYPHOID FEVER

It is also named enteric fever and results from infection by *Salmonella typhii* that grow up in blood and intestines. Infection is caused by the consumption of contaminated food or water (Wain, 2015). Symptoms of the infection vary and may be mild, moderate, or severe that generally begins from about a week to a month following exposure (Anna, 2014). The disease is often manifested by a slow onset of high-grade fever that lasts over some days (Wain, 2015). Patients develop generalized weakness, constipation, pain in the abdomen, headache, and sometimes it is also associated with vomiting and skin rashes with rose-pink spots. The symptoms of untreated typhoid may persist for weeks or even months. Some infected persons may act as carriers of typhoid bacterium showing no symptoms of the disease but can transmit the infection to other people (WHO, 2008).

The factors that increase the risk of infection include improper sanitation, inadequate hygiene, and traveling to developing countries. Enteric fever is diagnosed via blood culture in the first week followed by the Widal test in the second week and/or identifying bacterial DNA in bone marrow, blood or stool in consequent weeks (Crump, 2010; Magill, 2013). However, the most precise approach is testing bone marrow (Crump, 2010). Typhoid vaccination may prevent almost 40–90% of cases during the two initial years (Milligan, 2018). Vaccination is optional for high-risk persons and those traveling to places with high typhoid prevalence. In addition to vaccination, preventative measures, including the supply of pure drinking water, excellent hygiene, and proper hand wash can prevent the spreading of the infection. The infected people have been treated with the application of several antibiotics, including fluoroquinolones, third-generation cephalosporins, and azithromycin. The disease is most widespread in developing countries like India. Children are at higher risk for disease development (Wain, 2015). The emergence of antibiotic-resistant bacteria leads to difficulty in treating the disease properly (Chatham et al., 2019). In the year 2015, around 149,000 people died throughout the globe as a result of this disease, and the figure is comparatively less than the 181,000 deaths in the year 1990 (approximately 0.3% globally) (Abubakar et al., 2015; Wang et al., 2016). The untreated disease can be deadly up to 20%; nevertheless, the proper treatment can decline its incidence to 1%–4% (Wain, 2015).

1.5.3   SHIGELLOSIS

This disease is caused by a rod-shaped, gram-negative, aerobic, non-motile, facultative, and nonsporic bacteria known as *Shigella,* that shows genetic similarity with *Escherichia coli* (Yabuuchi, 2002). There are different species of Shigella like *Shigella bodyii, S. Sonnei, S. dysenteriae,* and *S. flexneri.* Shigella infection is one of the principal causes of diarrhea in the world (Bowen, 2016). Symptoms occur in the form of nausea, diarrhea, flatulence, abdominal cramps, and vomiting. Infection is caused by the ingestion of the organism. Fewer than 100 bacterial cells cause infection. This infection results in diarrhea and dysentery as the bacteria invades the epithelial lining of the colon (Mims et al., 2004). Some of the Shigella strains produce enterotoxins like Shiga toxin, which shows similarity with verotoxin of entero-hemorrhagic *E. coli* and leads to the hemolytic uremic syndrome. Shigellosis is one of the common disease-causing microorganisms that causes moderate to severe diarrhea among children in South Asia and Africa (Kotloff et al., 2013).

1.5.4   CRYPTOSPORIDIOSIS (CRYPTO)

The disease is caused by the protozoan parasite Cryptosporidium that infects the small intestines distally, causing watery diarrhea with or without unsolved cough (Sponseller, 2014). In immune-suppressed persons, the symptoms may be acute and lethal. The parasite enters the body via the fecal-oral route, mostly by drinking infected water. Current evidences suggest that cryptosporidiosis may spread through fomites present in the secretion of the respiratory system (Hawkins, 1987). Cryptosporidiosis can occur as an acute, asymptomatic or acute infection wherein symptoms recur subsequently to a short duration of recuperation for up to a month and as a chronic infection (period more than two weeks) wherein the symptoms are persistent and acute (Hawkins, 1987; Sponseller, 2015).

Symptoms of the disease are generally quite severe and emerge 5–10 days subsequent to infection. The symptoms usually continue for a maximum period of 14 days in the case of immune-competent persons (Hawkins, 1987; Sponseller, 2015). After the resolution of diarrhea, the symptoms may recur later in some days or weeks as a result of re-infection with the organism. The chance of getting re-infected is quite higher in immune-compromised individuals and little in persons within those with good resistance. In the case of immune-competent persons, cryptosporidiosis mainly localizes in the distal part of the small intestines and occasionally the respiratory passage (Cabada, 2015). However, in such individuals, the disease may spread to other organs like the pancreas, liver, upper digestive tract, gall bladder, and urinary bladder. The biliary and pancreatic infection may lead to pancreatitis, acalculous cholecystitis, papillary stenosis, or sclerosing cholangitis (Hawkins, 1987; Rigs, 2002; Cabbada, 2016).

1.5.5 ESCHERICHIA COLI INFECTION

Infection is caused by some specific strains of *E. coli* that spreads mainly through packed dairy products, juices, contaminated drinking water. This bacterium is a facultative anaerobic gram-negative that takes about a duration of seven days to develop the symptoms of the infection. Some virulent strains can cause sepsis, hemolytic uremic syndrome, bowel necrosis, mastitis, gram-negative pneumonias, and peritonitis. Usually, the incubation period of the bacterium is 3–4 days; however, it may vary from 1–10 days. Treatment protocol remains rehydration therapy and antibiotics (Tenaillon, 2010).

1.5.6 CAMPYLOBACTERIOSIS

This infectious disease is caused by the bacterium Campylobacter. The disease is spread by consuming food infected by the bacterium. This bacterium is gram-negative, motile, rod-shaped, curved, and non-sporic. The most common agent of the infection is *C. jejuni,* which is a comma and spiral shaped bacterium usually found in farm animals, birds, and swine, wherein it doesn't cause disease. However, there are other species of Campylobacter, such as *C. lari, C. upsaliensi,* and *C. coli* that can also cause the disease. It leads to inflammatory conditions and bloody diarrhea. This may manifest like dysentery syndrome, and the person experiences abdominal cramps, fever, and abdominal pain (Singleton, 1999).

The common transmission routes used by the bacterium for spreading the disease include sexual contact, consumption of unpasteurized milk, infected food, and water and poorly cooked chicken, person-to-person, and fecal-oral route. The infectious dose is a minimum of 1,000 bacteria and can be up to 10,000. However, human infection is possible by even 500 hundred bacteria. Species of Campylobacter are sensitive to gastric HCl and cannot grow in our body in such a less acidic environment. Campylobacter infection is often acquired during traveling, and hence also called travelers' diarrhea. The infection is usually resolved spontaneously, and treatment includes intake of oral liquids and electrolyte replacement.

The prodromal symptoms are fever, headache, and myalgia, dysentery, abdominal cramps, and high-grade fever. Diarrhea is inflammatory or invasive diarrhea, which is also known as bloody diarrhea or dysentery. If the disease is left untreated, complications like toxic megacolon, dehydration, and sepsis may result, especially in young children and immune-compromised people. Diarrhea and dehydration cause a reduction of blood volume and may lead to sudden death in otherwise healthy young individuals. A few individuals (1–2 in 1 lakh patients) develop Guillain-Barre syndrome wherein the nerves of the patient are affected and resulting in ascending paralysis; however, in rare cases, the paralysis is permanent. The development of such a condition is only associated with the infection caused by *C. upsaliensis* and *C. jejuni* (Murray, 2002). Campylobacteriosis leads to injuries in gut tissue in any part, such as jejunum, ileum, and even colon. This can complicate Guillain Barre syndrome and ascending paralysis and can complicate in the form of respiratory breakdown. Several *C. jejuni* strains produce entero-toxins, leading to watery diarrhea (Saenz, 2000).

Infection with campylobacter leads to exudative enteritis and can be identified via gram staining of a stool sample and usually by stool culture. The presence of fecal leukocytes indicates that diarrhea is inflammatory in nature (Wilson, 2008). Currently, different techniques, such as PCR and antigen testing, are utilized to detect whether campylobacter is present (Murray, 2002).

1.5.7 GIARDIASIS

A parasitic disease caused by the parasitic organism *Giardia lamblia*. The incubation period of the disease ranges from 1–4 weeks, and without treatment, it may last up to 6 weeks. This disease is also commonly known as beaver fever. Less than 10% of the patients are asymptomatic. The symptoms appear in the form of abdominal pain leading to diarrhea and weight loss. The route of infection is oro-fecal and consumption of infected foodstuffs or drinking water. For establishing the disease diagnosis, a routine stool examination is carried out. Children are more vulnerable to develop this disease. Once diagnosed, the disease can be treated with drugs that are readily available in the market. Maintenance of cleanliness and proper hygiene is helpful in preventing giardiasis. The differential diagnosis is like irritable bowel syndrome.

1.5.8 AMOEBIASIS

This infectious disease is also well-known as amoebic dysentery and is caused by every group member of Entamoeba (Farar et al., 2013). Amoebiasis is also believed to be caused by *Entamoeba histolytica* (Berger and Marr, 2006). Nearly 90% of the patients with this disease are asymptomatic, but may possibly become serious (Haque et al., 2003). Infection, if left untreated, can, at times, persist for years. Usually, a few days to a few weeks are sufficient for the development of symptoms. Symptoms include abdominal cramps and mild diarrhea that may progress to bloody dysentery (Farar et al., 2013). Cysts of entamoeba may continue to exist for up to 30 days in soil and approximately 45 minutes underneath fingernails. The organism invades the epithelial lining of the intestines and causes bloody diarrhea. Inside the circulation, amoeba frequently results in liver infections causing amoebic liver abscesses. However, liver abscesses may be produced with no previous history of diarrhea (Farar et al., 2013).

The disease may lead to complications such as colon inflammation that eventually causes ulceration as well as perforation or death of tissue and can further complicate peritonitis. Infected individuals develop anemia as a result of blood loss. Diagnosis is often made by microscopic examination of a stool sample. An increase in white blood cells is suggestive of an acute infection state. The most accurate test for the diagnosis of the disease is locating particular antibodies in the blood. However, these may still remain positive for sometime after treatment. Similar symptoms may be produced in bacterial colitis. The outburst of amoebiasis can be prevented via improving sanitation and encouraging hygienic living conditions. Different antibiotics, in combination with other medicines, are required for effectively treating the disease at all stages. Infected individuals that show no symptoms of the disease do not need to be given any medication, however, they may transmit the infection to others, and for that reason, such persons might be given treatments (Haque et al., 2003). Acute amoebic colitis is associated with a higher death rate. Diagnosis is possible with colonoscopy. Trophozoites can be recognized at the edge of ulcer or in tissue by means of immune histochemical staining with specific anti-*Entamoeba histolytica* antibodies. In the case of asymptomatic individuals, diagnosis is made by the detection of cysts in the stool. Different techniques such as sedimentation or flotation, along with staining procedures, facilitate the visualization of isolated cysts under a microscope. At least three samples of stool should be examined. In the case of symptomatic infection, trophozoites are often observed in fresh fecal samples. The infection shows positive serological tests for the antibodies present. Individuals with the disease progressed to liver abscesses show quite a high antibody titer. The development of kits that identify the occurrence of amoeba proteins and DNA in the fecal samples of infected persons is a recent advancement in diagnostics. However, the cost factor is an obstacle to these tests. PCR (polymerase chain reaction) is assumed as a gold standard in the testing methods.

1.5.9   CHOLERA

It is caused by *Vibrio cholera* due to the ingestion of a toxic strain by food or drinks. Mostly the bacteria survive up to three weeks from the excretion. It is mainly found in seafood. Cholera leads to watery diarrhea, cramps, abdominal discomfort, and vomiting. This may cause electrolyte imbalance, and therefore proper intervention is the necessary step. Oral and intravenous rehydration is the main protocol in treatment, along with antibiotics. It is a waterborne infection and may be checked by the maintenance of sanitation and proper hygiene (Finkelstein, 1996).

1.5.10   HEPATITIS A VIRUS

This infection is caused by the viruses originated from fecal contaminated food and water and usually enter the human body via oral route. The majority of the cases show signs and symptoms in around 48 days, but upon infection itself, the patient experiences fever, fatigue, diarrhea, nausea, and gastric upset (Cullen and Lemon, 2019).

1.5.11   ADENO VIRUS

This virus is a member of family adenoviral, non-enveloped medium sized virus. This virus can be transmitted by adenoviral that contaminates drinking water and also by swimming pools with deficient chlorine levels being one of the ways. Mostly causes infections of the upper respiratory tract like tonsillitis, conjunctivitis, ear infections. In children, it can result in viral meningitis and encephalitis, and also to bronchopneumonia. From mild respiratory infections, it can lead to severe life-threatening multi-organ failure due to weakened immunity. Treatment is mainly symptomatic. It resides in the host for 5 to 9 days before showing the signs and symptoms (Bennett et al., 2014, 2016).

1.6 IMPACT OF WATER POLLUTION ON THE HEALTH OF HUMAN BEINGS

Water pollution is common nowadays due to industrialization and urbanization. Hospital waste and even household wastes contaminate the water bodies that lead to various waterborne illnesses. Heavy metals such as lead, zinc, cadmium, copper, arsenic, and mercury that are present in wastewater from industries have an effect on aquatic flora and fauna as well as human beings. Arsenic has led to the poisoning of groundwaters and has been reported from areas of Orissa, West Bengal, Bihar, and some parts of western UP. Arsenic accumulation in the human body due to drinking of contaminated water leads to chronic arsenic poisoning. Arsenic concentration increases in blood and gets deposited in nails and hair that leads to skin lesions and, ultimately, skin cancers. Mercury compounds in wastewaters cause diseases in humans, such as Minimata, and fish such as dropsy. This can lead to a lack of limb sensation, tingling of tongue, and lips. Other symptoms like blurred vision, deafness, and mental derangement have also been reported. Heavy metal cadmium is carcinogenic and causes lung and liver cancers. Itai-Itai and ouch-ouch diseases have been reported due to increased cadmium levels in water and have led to cadmium nephropathy.

The determination of the effects of food and waterborne diseases on public health is quite challenging. In order to estimate the burden of such infections, their rate, and mortality is a mandatory step (EFSA and ECDC, 2014). Global Burden of Disease study developed disability-adjusted life years (DALYs), which is a précis evaluation of public health (Murray et al., 2012) match these provisions. These are helpful for decision making in evidence-based health policy. In 2006, the European Center for Disease Prevention and Control (ECDC) specially made a pilot assessment of seven-selected infectious diseases for exploration of the possible burden of infectious diseases and to study their methodology (Van Lier, 2007; Jakab, 2007). A reliable methodology to estimate the burden of these infectious diseases was expressed in DALYs. Pathogen and incidence-based methodology were developed (Mangen et al., 2013). Recently, the burden outcomes of selected FWDs were presented at the 2015 European Public Health Conference (Colzani, 2015). Using the study of BCoDE (2015) as a reference point, a comparison of these methods with other procedural frameworks was carried out; and it was believed that options for the burden of disease studies might be possible (Cassini, 2016).

1.7 EFFECTS OF WATER POLLUTION ON THE ENVIRONMENT

Some poisonous substances that are commonly found in wastewater from industries undergo biomagnifications in the food chains of aquatic ecosystems. This fact is well supported for DDT and mercury. Higher DDT levels create a disturbance to calcium metabolism in birds, ultimately leading to premature breaking of eggshell due to thinning. This may result in a decline of the bird population—thermal wastewater results in the death of temperature-sensitive life forms. However, primarily it causes damage to aquatic plants and animals. Pesticides present in water are taken up by aquatic dwellers and become incorporated into the food chain, and go up into higher trophic levels of the food chain. The concentration of pesticides increases at higher trophic levels. Biodegradation of untreated waste in sewage consumes a huge amount of oxygen, which makes water bodies deficient in oxygen resulting in the indirect killing of aquatic organisms, including fish.

1.7.1 ALGAL BLOOM

Harmful algal blooms result due to the rapid production of cyanobacteria, dinoflagellates, and diatoms in aquatic ecosystems. These form a part of natural processes and pose a grave threat to the health of human beings, the sustainability of the environment, and aquatic dwellers because of toxin formation and biomass accumulation. Blooms of autotrophic algae and some heterotrophic plants are increasingly at an alarming rate in the coastal waters around the world and are named harmful algal blooms. Eutrophication is commonly assumed to be the primary cause of all blooms.

1.8 WATER POLLUTION AND HEAVY METAL POISONING

1.8.1 MERCURY POISONING AND MINIMATA DISEASE

Human methyl mercury poisoning was, for the first time, reported and published early in 1940 (Hunter et al., 1940). However, it was in the

1960s that the mercuric methyl poisoning became well-known when a fundamental association was recognized between intake of polluted seafood and an unexplained disease that led to neurological symptoms and death in a few thousand residents of the Minamata Bay in Kumamoto region of Kyushu in Japan (Semionov, 2018). Mercury exists in organic as well as inorganic form. The organic form consists of alkylated forms, and the inorganic form includes an elemental form of mercury salts. All forms have distinct poisonous properties, and poisoning with each form produces a typical neuropathological and clinical picture. Toxic effects of inorganic mercury on nerves have been known long back in history. A few of the classic symptoms of inorganic mercury intoxication among laborers of the felting industry due to chronic exposure to the vapors of elemental mercury were illustrated by Mad Hatter. They developed health issues like slurred speech, blurred vision, and irritability in a varying manner. The other symptoms that were seen include excitability decreased sleep, predisposition to cry, nervousness, and social withdrawal. Changes in personality, loss of memory, incapability to focus, and infrequent hallucinations were also seen in these people (Semionov, 2018). Danbury shakes (fine tremors) in limbs of such patients were also seen.

Severe methyl mercury poisoning results in typical permanent neuro-pathological changes that are often manifested clinically as ataxia, visual field constriction, and sensory disorders. Radiological images show them typically as selective atrophy of calcarine, postcentral gyri, and cerebellum. The extent of neuropathological changes is usually proportional to the quantity of mercury and intensity of exposure (Semionov, 2018).

1.8.2 LEAD POISONING

Lead is an extremely lethal metal, and its extensive use has led to worldwide contamination of the environment and health disorders (Jaishankar, 2014). It discolors on exposure to air and results in the formation of a complex mixture of compounds based on the specified conditions. Storage batteries, gasoline, lead bullets, toys, paints, and plumbing pipes are the chief sources of lead (Thurmer et al., 2002). Some amount is consumed by plants; a little goes to the soil, and some flow into water bodies. Therefore, in general, populace lead exposure of human beings occurs via food consumption or drinking water (Goyer, 1990).

1.8.3   CADMIUM NEPHROPATHY AND ITAI-ITAI DISEASE

Itai-Itai disease is characterized by severe pain and was mainly due to water pollution of the river Jinzu in Japan during World War II. It was caused by cadmium poisoning. Cadmium is an important environmental toxic mediator that mostly targets the kidney (Chan, 1981). Cadmium mainly gets concentrated in the renal cortex. Excessive deposition of cadmium in the renal cortex leads to distinct ultrastructural, morphological, and pathological alterations in proximal tubules. Cadmium neuropathy results in various functional alterations that include polyuria, proteinuria, hypercalciuria, aminoaciduria, enzymatic, and glycosuria. It also increases the concentration of urinary cadmium and uric acid (Gonick, 2008).

Practically, proteinuria may have two constituents: low and high molecular weight proteinuria. The former is of tubular origin, e.g., 32 microglobulin, and the latter is of glomerular origin, e.g., transferring, albumin, IgG excretion. A broad array of lethal signs are caused by cadmium involving the pancreas, heart, liver, blood vessels, kidneys, digestive tract, bones, and testes.

1.9   PREVENTION OF AQUATIC POLLUTION AND WATERBORNE DISEASES

Water pollution poses an extreme menace to aquatic dwellers and humans as well, and an increase in the human population leads to climatic changes (Palmate et al., 2017). Increased industrialization that leads to urbanization has constantly stressed the aquatic environment and reduced the availability of pure water. The protection of the aquatic environment is indispensable for sustainable development, and all people must make efforts for the conservation of water resources. More effective treatments for wastewater can save aquatic systems.

However, sound environmental policies, along with regular public awareness (Figure 1.2) about the grave effects of aquatic pollution, will surely help in saving the aquatic environment to a large extent (Inyinbor et al., 2018). The prevention of waterborne infections is possible via controlling pathogenic organisms in drinking water. The disease-causing organisms can be well effectively controlled by a multiple barrier approach. Water being the base of survival of all life forms, including humans, its sustainable availability and conservation cannot be underestimated (Inyinbor et al., 2018).

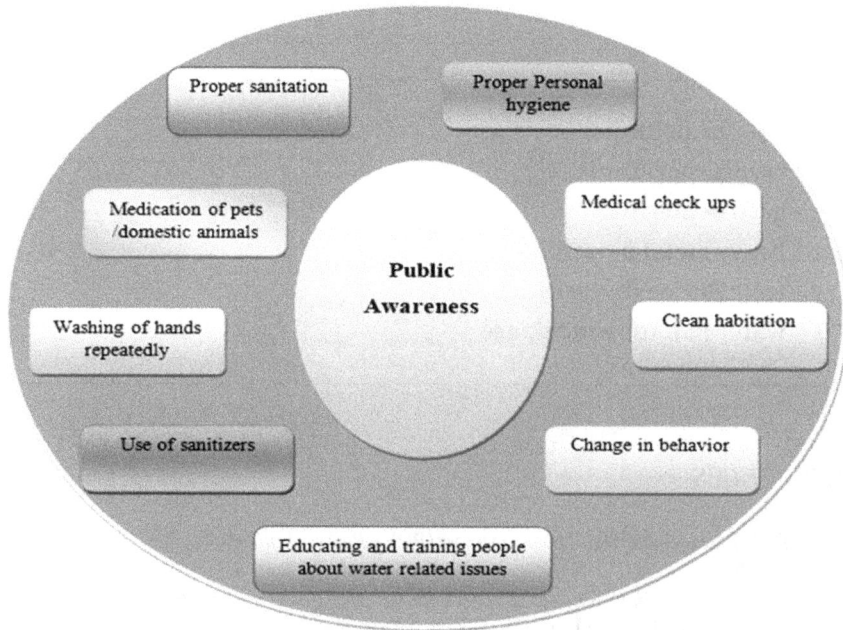

FIGURE 1.2   Public awareness for the prevention of waterborne diseases.

1.10   CONCLUSION

Water pollution has posed a global threat to the environment and human health and worried the global nations. Billions of human beings worldwide utilize drinking from infected sources. Domestic wastes from household activities, agricultural wastes also pollute the water. The use of pesticides and fertilizers contribute to water pollution. Waste materials from industries also contain many unsafe chemicals that add to water pollution. Pathogenic organisms such as bacteria, protozoans, or viruses infect water and lead to a wide range of waterborne diseases. Thus, there is a need for the proper treatment of wastes before releasing them into aquatic systems. The wastes should also be disposed of properly. Educating people about personal hygiene and proper sanitation is essential to check waterborne diseases. Besides, the pollutants that show diversity in behavior, toxicity, and approaches of remediation must be studied in detail, and their exposure pathways and impact on public health should be assessed.

KEYWORDS

- aquatic pollution
- cryptosporidiosis
- gastroenteritis
- heavy metal poisoning
- waterborne diseases
- water pollution incidence

REFERENCES

(2008). Typhoid vaccines: WHO position paper (PDF). *Releve. Epidemiologique Hebdomadaire, 83*(6), 49–59.

(2015). *"Cryptosporidium: Sources of Infection and Risk Factors."* United States Centers for Disease Control and Prevention.

(2015). Global burden of disease study C: Global, regional, and national incidence, prevalence, and years lived with disability for 301 acute and chronic diseases and injuries in 188 countries, 1990–2013: A systematic analysis for the global burden of disease study 2013. *Lancet, 386,* 743–800.

Alan, J. M., (2013). *Hunter's Tropical Medicine and Emerging Infectious Diseases* (9th edn., pp. 568–572). London: Saunders/Elsevier.

Altmann, P., Cunningham, J., Dhanesha, U., Ballard, M., Thompson, J., & Marsh, F., (1999). Disturbance of cerebral function in people exposed to drinking water contaminated with aluminum sulphate: Retrospective study of the camel ford water incident. *British Med. J., 319,* 807–812.

Anderson, Y., & Stentrom, T. A., (1987). Waterborne outbreak in Sweden: Causes and etiology. *Wat. Sci. Tech., 19,* 375–380.

Andersson, N., (1993). Disaster epidemiology: Lessons from Bhopal. In: Murray, J., (ed.), *Major Chemical Disasters: Medical Aspects of Management* (Vol. 42, pp. 653–656). London, UK: Royal Society of Medicine Services Ltd. Anon. Public health consequences of a flood disaster; Iowa.

Anil, K. D., (2017). researches in water pollution: A review. *International Research Journal of Natural and Applied Sciences, 4*(1).

Anna, E. N., (2014). Three infectious diseases related to travel. *CDC Health Information for International Travel 2014: The Yellow Book.* ISBN: 9780199948499.

Bagchi, K., Echeverria, P., Arthur, J. D., Sethabutr, O., Serichantalergs, O., & Hoge, C. W., (1993). Epidemic diarrhea caused by *Vibrio* cholerae non-O1 that produced heat-stable toxin among Khmers in a camp in Thailand. *J. Clin. Microbiol., 31*, 1315–1317.

Bartram, J. R. G., (1999). *Monitoring Bathing Waters: A Practical Guide to the Design and Implementation of Assessments and Monitoring Programs.* Published on behalf of the WHO E& FN Spon, UK.

Batz, M. B., Hoffmann, S., & Morris, J. G., (2011). *Ranking the Risks: The 10 Pathogen-Food Combinations with the Greatest Burden on Public Health.* Gainesville, FL: Emerging Pathogens Institute, University of Florida.

Baxter, P., (1990). Review of major chemical incidents and their medical management. In: Murray, V., (ed.), *Major Chemical Disasters-Medical Aspects of Treatment* (pp. 615–620). London: Royal Society of Medicine.

Bennet, G., (1970). Controlled survey of effects on health of local community disaster. *Br. Med. J., 22*, 4.

Bentley, R., & Meganathan, R., (1982). Biosynthesis of vitamin K (menaquinone) in bacteria. *Microbiological Reviews, 46*(3), 241–280.

Bert, F., Maubec, E., Bruneau, B., Berry, P., & Lambert-Zechovsky, N., (1998). Multi-resistant *Pseudomonas aeruginosa* outbreak associated with contaminated tap water in a neurosurgery intensive care unit. *J. Hosp. Infect., 39*, 53–62.

Bijkerk, P., Van, L. A., McDonald, S., Kardamanidis, K., Fanoy, E. B., Wallinga, J., et al., (2014). *State of Infectious Diseases in the Netherlands, 2013.* Bilthoven, The Netherlands: National Institute for Public Health and the Environment, RIVM.

Bowen, A., (2016). Infectious diseases related to travel. *The Yellow Book: Health Information for International Travel CDC.* ISBN: 978-0-19-937915-6.

Cabada, M. M., White, A. C., Venugopalan, P., Sureshbabu, J., & Bronze, M. S., (2016). *Cryptosporidiosis Clinical Presentation.* Medscape. WebMD.

Chlossberg, D., (2015). *Clinical Infectious Disease* (2nd edn., p. 334). ISBN:978-1-107-03891-2.

Ciccarelli, S., Stolfi, I., & Caramia, G., (2013). Management strategies in the treatment of neonatal and pediatric gastroenteritis. *Infection and Drug Resistance, 6*, 133–161.

Colzani, E., (2015). *Results from the 2015 Burden of Communicable Diseases in Europe (BCoDE) study.* European public health conference; Milan.

Crowder, A., (1991). Acidification, metals, and macrophytes. *Environmental Pollution, 71*(2–4), 171–203.

Crump, J. A., & Mintz, E. D., (2010). Global trends in typhoid and paratyphoid fever. *Clinical Infectious Diseases, 50*(2), 241–246.

Cryptosporidiosis, (2009). Centers for Disease Control and Prevention.

Day, D. W., Basil, C. M., Jeremy, R. J., Geraint, W., & Ashley, B. P., (2003). *Morson and Dawson's Gastrointestinal Pathology.* John Wiley & Sons, Inc. ISBN: 978-0-632-04204-3.

Eckardt, A. J., & Baumgart, D. C., (2011). Viral gastroenteritis in adults. *Recent Patents on Anti-Infective Drug Discovery, 6*(1), 54–63.

EFSA (European Food Safety Authority) and ECDC (European Center for Disease Prevention and Control), (2015). The European Union summary report on trends and sources of zoonoses, zoonotic agents, and food-borne outbreaks in 2014. *EFSA J., 12*, 191.

Fàbrega, A., Sánchez-Céspedes, J., Soto, S., & Vila, J., (2008). Quinolone resistance in the food chain. *Int. J. Antimicrob. Agents, 31*(4), 307–15.

Farrar, J., Hotez, P., Junghanss, T., Kang, G., Lalloo, D., & White, N. J., (2013). *Manson's Tropical Diseases* (pp. 664–671). Elsevier Health Sciences.

Galanis, E., (2007). Campylobacter and bacterial gastroenteritis. *Canadian Medical Association Journal, 177*(6), 5701.

Ghosh, S., Padalia, J., & Moonah, S., (2019). Tissue destruction caused by entamoeba histolytica parasite: Cell death, inflammation, invasion, and the gut microbiome. *Current Clinical Microbiology Reports, 6*(1), 51–57.

Gkogka, E., Reij, M. W., Havelaar, A. H., Zwietering, M. H., & Gorris, L. G., (2011). Risk-based estimate of effect of foodborne diseases on public health, Greece. *Emerg. Infect. Dis., 17*, 1581–1590.

Goldman, J. C., (1972). The effect of inorganic carbon on eutrophication. In: Brown, R. L., & Tunzi, M. G., (eds.), *Proceedings of a Seminar on Eutrophication and Biostimulation* (pp. 3–53). California Department of Water Resources, San Francisco.

Gonick, H. C., (2008). Nephrotoxicity of cadmium and lead. *Indian J. Med. Res., 128*, 335–352.

Gonzalez, A. E., Rodriguez, M. T., Sanchez, J. C. J., Espinosa, A. J. F., & Dela, R. F. J., B., (2000). Assessment of metals in sediments in a tributary of Guadalquivir River (Spain). Heavy metal partitioning and relation between the water and sediment system. *Water, Air, Soil Pollut., 121*(1–4), 11–30.

Haagsma, J. A., Polinder, S., Cassini, A., Colzani, E., & Havelaar, A. H., (2014). Review of disability weight studies: Comparison of methodological choices and values. *Popul. Health Metr., 12*, 20.

Hades, O., Shteinman, B., & Pinkas, R., (2000). Distribution of fecal coliforms in the Jordan River mouth originating from anthropogenic activities in the Watershed. *Water Sci. Technol., 42*(1/2), 129–134.

Haque, R., Huston, C. D., Hughes, M., Houpt, E., & Petri, W. A., (2003). *Amebiasis. The New England Journal of Medicine, 348*(16), 1565–1573.

Harada, M., (1995). Minamata disease: Methylmercury poisoning in Japan caused by environmental pollution. *Critical Reviews in Toxicology, 25*, 1–24.

Havelaar, A. H., Haagsma, J. A., Mangen, M. J., Kemmeren, J. M., Verhoef, L. P., Vijgen, S. M., et al., (2012). Disease burden of foodborne pathogens in the Netherlands-2009. *Int. J. Food Microbiol., 156*, 231–238.

Hawkins, S., Thomas, R., & Teasdale, C., (1987). Acute pancreatitis: A new finding in cryptosporidium enteritis. *Br. Med. J., (Clin. Res. Ed.), 294*(6570), 483–484.

Helms, R., (2006). *Textbook of Therapeutics: Drug and Disease Management* (8th edn., p. 2003). Philadelphia [u.a.]: Lippincott Williams & Wilkins.). ISBN: 978-0-7817-5734-8.

Herman, S. D., Sebestina, A. D., Geraldine, M., & Thuppil, V., (2011). Diagnosis, evaluation, and treatment of lead poisoning in general population. *Ind. J. Clin. Biochem., 26*(2), 197–201.

Hudault, S., Guignot, J., & Servin, A. L., (2001). *Escherichia coli* strains colonizing the gastrointestinal tract protect germfree mice against *Salmonella typhimurium* infection. *Gut., 49*(1), 47–55.

Hunter, D., & Bomford, R. R., (1940). Poisoning by methyl mercury compounds. *Quarterly Journal of Medicine, 33*, 193–213.

Jaishankar, M., et al., (2014). *Interdiscip. Toxicol., 7*(2), 60–72.

Jakab, Z., (2007). Why burdens of disease study? *Euro Surveill., 12*, E1, E2.

Kotloff, K. L., Nataro, J. P., Blackwelder, W. C., et al., (2013). Burden and aetiology of diarrheal disease in infants and young children in developing countries (the Global Enteric Multicenter Study, GEMS), a prospective, case-control study. *The Lancet, 382*(9888), 209–222.

Kretzschmar, M., Mangen, M. J., Pinheiro, P., Jahn, B., Fevre, E. M., Longhi, S., et al., (2012). New methodology for estimating the burden of infectious diseases in Europe. *Plos Med., 9*, e1001205.

Lozano, R., Naghavi, M., Foreman, K., Lim, S., Shibuya, K., Aboyans, V., Abraham, J., et al., (2012). Global and regional mortality from 235 causes of death for 20 age groups in 1990 and 2010: A systematic analysis for the global burden of disease study 2010. *Lancet, 380*(9859), 2095–2128.

Man, S. M., (2011). The clinical importance of emerging campylobacter species. *Nature Reviews Gastroenterology and Hepatology, 8*(12), 669–685.

Mangen, M. J. J., Plass, D., & Kretzschmar, M. E., (2014). *Estimating the Current and Future Burden of Communicable Diseases in the European Union and EEA/EFTA.* [Section 2.4.3] Bilthoven, The Netherlands: National Institute for Public Health and the Environment.

Mangen, M. J., Plass, D., Havelaar, A. H., Gibbons, C. L., Cassini, A., Muhlberger, N., et al., (2013). The pathogen- and incidence-based DALY approach: An appropriate [corrected] methodology for estimating the burden of infectious diseases. *Plos One, 8*, e79740.

Markell, E. K., (1986). The 1933 Chicago outbreak of amebiasis. *Western Journal of Medicine, 144*(6), 750.

Marshall, J. A., & Bruggink, L. D., (2011). The dynamics of norovirus outbreak epidemics: Recent insights. *International Journal of Environmental Research and Public Health, 8*(4), 11419.

McDermott, P., Bodeis, S., English, L., White, D., Walker, R., Zhao, S., Simjee, S., & Wagner, D., (2002). Ciprofloxacin resistance in *Campylobacter jejuni* evolves rapidly in chickens treated with fluoroquinolones. *J. Infect. Dis., 185*(6), 837–840.

Milligan, R., Paul, M., Richardson, M., & Neuberger, A., (2018). Vaccines for preventing typhoid fever. *The Cochrane Database of Systematic Reviews, 5*, CD001261.

Mims, C., Dockrell, H., Goering, R., Roitt, I., Wakelin, D., & Zuckerman, M., (2004). *Medical Microbiology* (3rd edn., p. 287). Mosby. ISBN: 978-0-7234-3259-3.

Mondal, D., Petri, Jr. W A., Sack, R. B., et al., (2006). *Entamoeba histolytica*-associated diarrheal illness is negatively associated with the growth of preschool children: Evidence from a prospective study. *Trans. R. Soc. Trop Med. Hyg., 100*(11), 1032–1038.

Monisha, J., Tenzin, T., Naresh, A., Blessy, B. M., & Krishnamurthy, N. B., (2014). Toxicity, mechanism, and health effects of some heavy metals. *Interdisciplinary Toxicology.*

Murray, C. J. L., (2016). Global, regional, and national life expectancy, all-cause mortality, and cause-specific mortality for 249 causes of death, 1980–2015: A systematic analysis for the global burden of disease study-2015. *Lancet, 388*(10053), 1459–1544.

Murray, P. R., et al., (2002). *Centers for Disease Control and Prevention Medical Microbiology.* Mosby St. Louis.

Nespola, B., Betz, V., Brunet, J., Gagnard, J. C., Krummel, Y., Hansmann, Y., Hannedouche, T., et al., (2015). First case of amebic liver abscess 22 years after the first occurrence. *Parasite, 22, 20; EJM, 348*(16), 1565–1573.

Nikolas, C., Papanikolaou, E. G., Hatzidaki, S. B., George, N. T., & Aristidis, M. T., (2005). *Med. Sci. Monit., 11*(10), RA.

O'Carroll, R. E., Masterton, G., Dougall, N., & Ebmeier, K. P., (1995). The neuropsychiatric sequelae of mercury poisoning. The mad hatter's disease revisited. *British Journal of Psychiatry, 167*, 95–98.

Oostvogels, A. J., De, W. G. A., Jahn, B., Cassini, A., Colzani, E., De, W. C., et al., (2015). Use of DALYs in economic analyses on interventions for infectious diseases: A systematic review. *Epidemiol. Infect., 143*, 1791–1802.

Ravi, K., & Amol, D., (2018). Cadmium, the real link to chronic kidney disease causation in hotspots of rural Pondicherry and coastal Cuddalore areas. *International Journal of Medical Science and Public Health, 8*, 255–258.

Recavarren-Arce, S., Velarde, C., Gotuzzo, E., & Cabrera, J., (1999). Amoeba angeitic lesions of the central nervous system in *Balamuthia mandrilaris* amoebiasis. *Hum. Pathol., 30*(3), 269–273.

Reid, G., Howard, J., & Gan, B. S., (2001). Can bacterial interference prevent infection? *Trends in Microbiology, 9*(9), 424–428.

Riggs, M. W., (2002). Recent advances in cryptosporidiosis: The immune response. *Microbes and Infection, 4*(10), 1067–1080.

Saenz, Y., Zarazaga, M., Lantero, M., Gastanares, M. J., Baquero, F., & Torres, C., (2000). Antibiotic resistance in campylobacter strains isolated from animals, foods, and humans in Spain in 1997–1998. *Antimicrob. Agents Chemother., 44*(2), 267–271.

Say, P. J., & Witton, B. A., (1983). Accumulation of heavy metals by aquatic mosses. I. *Fontinalis antipyretica. Hedw. Hydro Biol., 100*, 245–260.

Semionov, A., (2018). Minamata disease: Review. *World Journal of Neuro Science, 8*, 178–184.

Sharma, G. R., (1999). Oil pollution at sea and its control. *Emp. News, XXIV*(24), 1, 2; *Pollut., 71*, 171–203.

Sharma, K. D., Lal, N., & Pathak, R. D., (1981). Water quality of sewage drains entering Yamuna at Agra. *Indian J. Environ. Hlth., 23*(2), 118–122.

Shirley, D. A., Moonah, S., & Meza, I., (2016). Fulminant amebic colitis after corticosteroid therapy: A systematic review. *PLOS Neglected Tropical Diseases, 10*(7), e0004879.

Shirley, D. T., Farr, L., Watanabe, K., & Moonah, S., (2018). A Review of the global burden, new diagnostics, and current therapeutics for amebiasis." *Open Forum Infectious Diseases, 5*(7), 161.

Shors, T., (2013). *The Microbial Challenge: A Public Health Perspective* (3rd edn., p. 457). Burlington, MA: Jones & Bartlett Learning. ISBN: 978-1-4496-7333-8.

Singh, A., (2010). Pediatric emergency medicine practice acute gastroenteritis: An update. *Pediatric Emergency Medicine Practice, 7*, 7.

Singleton, P., (1999). *Bacteria in Biology, Biotechnology and Medicine* (5th edn., pp. 444–454). Wiley.

Sinha, S., Rai, U. N., & Tripathi, R. D., (1993). Chromium and manganese uptake by *Hydrilla verticillata* (I.f.) Royle: Amelioration of chromium toxicity by manganese. *J. Env. Sci. Health., A28*, 1545–1552.

Sponseller, J. K., Griffiths, J. K., & Tzipori, S., (2014). The evolution of respiratory Cryptosporidiosis: Evidence for transmission by inhalation. *Clin. Microbiol. Rev., 27*(3), 575–586.

Takkinen, J., Kramarz, P., Cassini, E. C., & Kretzschmar, M. E., (2016). Impact of food and waterborne diseases on European population health. *Current Opinion in Food Science, 12*, 21–29.

Tate, J. E., Burton, A. H., Boschi-Pinto, C., Steele, A. D., Duque, J., & Parashar, U. D., (2012). Estimate of worldwide rotavirus-associated mortality in children younger than 5 years before the introduction of universal rotavirus vaccination programs: A systematic review and meta-analysis. *The Lancet Infectious Diseases, 12*(2), 136–141.

Tenaillon, O., Skurnik, D., Picard, B., & Denamur, E., (2010). The population genetics of commensal *Escherichia coli*: Nature reviews. *Microbiology, 8*(3), 207–217.

Ternhag, A., Asikainen, T., Giesecke, J., & Ekdahl, K., (2007). A meta-analysis on the effects of antibiotic treatment on duration of symptoms caused by infection with campylobacter species. *Clin. Infect. Dis., 44*(5), 696–700.

Thanas, K. V., Hurst, M. P., & Sheahan, D., (2001). Toxicity characterization of organic contaminants in storm waters from an agricultural headwater stream in south East England. *Wat. Res., 35*(10), 2411–2416.

Timothy, G. S., (2001). *Minamata: Pollution and the Struggle for Democracy in Postwar Japan* (p. 385). Harvard University Asia Center, Cambridge.

Typhoid Fever, (2013). *cdc.gov*. Archived from the original on 2 April 2015.

Typhoid Fever, (2015). *cdc.gov*. Archived from the original on 6 June 2016.

Van, L. E. A., Havelaar, A. H., & Nanda, A., (2007). The burden of infectious diseases in Europe: A pilot study. *Euro Surveill.*

Visvesvara, G. S., Moura, H., & Schuster, F. L., (2007). Pathogenic and opportunistic free-living amoebae: *Acanthamoeba* spp., *Balamuthia mandrill* Aris, *Naegleria fowleri*, and *Sappinia diploidea*. *FEMS Immunol. Med. Microbiol., 50*(1), 1–26.

Vogt, R. L., & Dippold, L., (2005). *Escherichia coli* O157: H7 outbreak associated with consumption of ground beef. *Public Health Reports, 120*(2), 174–178.

Volterra, L., Tosti, E., Verma, A., & Izzo, G., (1985). Microbiological pollution of marine sediments in the southern stretch of the Gulf of Naples. *J. Water, Air and Soil Pollution, 26*, 175–184.

Wain, J., Hendriksen, R. S., Mikoleit, M. L., Keddy, K. H., & Ochiai, R. L., (2015). Typhoid fever. *Lancet, 385*(9973), 1136–1145.

Wai-Yee, C., & Owen, M. R., (1979). Cadmium nephropathy. *Annals of Clinical and Laboratory Science, 11*, 3329–3336.

Wilson, D. J., Gabriel, E., Leatherbarrow, A. J. H., Cheesbrough, J., Gee, S., Bolton, E., Fox, A., et al., (2008). Tracing the source of campylobacteriosis. *Plos Genet., 4*(9), e1000203.

Yabuuchi, E., (2002). *Bacillus dysentericus* (sic) 1897 was the first taxonomic rather than bacillus dysenteriae 1898. *International Journal of Systematic and Evolutionary Microbiology, 52*(Pt 3), 1041.

CHAPTER 2

Toxicity of Heavy Metals in Freshwater Fishes: Challenges and Concerns

AIMEN FIRDOUS,[1] JAYACHITHRA RAMAKRISHNA PILLAI,[2]
MUNEEB U. REHMAN,[3,4] SHAHZADA MUDASIR RASHID,[5]
SAIEMA RASOOL,[6] SABHIYA MAJID,[3] TABASUM RASHID,[3]
ADIL FAROOQ,[2] and MUBASHIR HUSSAIN MASOODI[7]

[1]Department of Processing Technology, Kerala University of Fisheries and Ocean Studies (KUFOS), Panangad, Kerala–682506, India

[2]RAK College of Pharmaceutical Sciences, RAK Medical and Health Sciences University, Ras Al Khaimah–11172, United Arab Emirates

[3]Department of Biochemistry, Government Medical College (GMC), Karan Nagar, Srinagar–190010, Jammu and Kashmir, India

[4]Department of Clinical Pharmacy, College of Pharmacy, King Saud University, Riyadh–145111, Kingdom of Saudi Arabia,
E-mail: muneebjh@gmail.com

[5]Division of Veterinary Biochemistry, Faculty of Veterinary Science and Animal Husbandry, SKUAST-Kashmir, Shuhama, Srinagar–190006, Jammu and Kashmir, India

[6]Forest Biotech Lab, Department of Forest Management, Faculty of Forestry, University Putra Malaysia, Serdang, Selangor–43400, Malaysia

[7]Department of Pharmaceutical Sciences, Faculty of Applied Sciences, University of Kashmir, Srinagar–110006, Jammu and Kashmir, India

ABSTRACT

Heavy metal is a naturally occurring earth crust element that exists in nature and is well known for its polluting and carcinogenic effect in water bodies and soil. Limitless anthropogenic activities, industrial runoff, municipality discharge, untreated effluent of chemical and automobile industries, and overutilization of inorganic fertilizers in agricultural land have made alarming heavy metal toxicity in the aquatic environment. Through a cyclic process, heavy metal transfers from originating source to aquatic animal and then to the human being through the food web. Being the highest trophic level, fish accumulates the highest concentration of toxic constituents in their body, which eventually comes to a human being after bioaccumulation and bioconcentration. Among all the alarming heavy metals, Pb, Hg, Cd are those three elements that are considered as the most potent hazardous elements by Environment Protection Agency (EPA). Other trace elements, i.e., Zn, Cr, Ni, Al, are important micronutrients for our body, but they turn to be carcinogenic at a higher dose. Heavy metal pollution is a global problem that has reached its maximum level now. This is the high time to combat heavy metal pollution to keep the aquatic environment safe and protect living beings.

2.1 INTRODUCTION

Presently water pollution has reached an alarming stage due to speedy urbanization and industrialization, which are leading to the over-discharge of pollutants like heavy metals, e-waste, radio nucleotides, hazardous chemicals, and organic-inorganic substances to the environment. The growing population on earth and their anthropogenic activities are increasing the pressure on the planet's water resources and the toxic level of chemicals; heavy metals are threatening the life on earth tremendously over the decades. Survey says that every day, over 2 million tons of sewage, industrial disposes, and agricultural wastes are released into the water bodies, which is almost an equal ratio of the entire human resources on the earth (UN WWAP, 2003). As per World Health Organization (WHO), 15% of child death each year is because of waterborne diseases, whereas, in India alone, the main reason behind a large number of ill health and death is due to waterborne diseases like diarrhea (WHO, 2002). Freshwater species show an extinction rate, which is five times more than the terrestrial species extinction rate (Ricciardi and Rasmussen, 1999). Among all the sources of aquatic contamination, heavy metal possesses the maximum percentage of chemical toxic waste in water bodies (Sönmez et al., 2012a). Heavy metals are those naturally occurring elements with a specific gravity of more than 5 gcm^{-3} (Alissa and Ferns, 2011), and these earth crust elements cannot be despoiled easily from aquatic systems by natural methods, hence cause tenacious environmental pollution.

Heavy metals like Lead (Pb), Mercury (Hg), Cadmium (Cd), Chromium (Cr), Arsenic (As), Copper (Cu), Zinc (Zn), and Nickel (Ni) are the most hazardous elements causing chemical pollution in water bodies due to their ability to bioconcentrate and not being able to be eliminated from the water body by metabolic activities. Uncontrolled anthropogenic activities, over the use of agricultural fertilizers, especially phosphorus and nitrogen-containing fertilizers, pesticides, over-irrigation, industrial runoff, wastewater treatment plant, coal combustion, tannery effluents, metal mining, smelting, leaching from landfills, geological weathering of the earth crust, shipping, and harbor activities (Fahr et al., 2013; Elhussien and Adwok, 2018) are the prime causes for the aquatic pollution. They remain floating in the water column, and eventually, with the course of the time, it gets settled down to the bottom of the aquatic system. Gradually, fish, shellfish, and other aquatic organisms take up this metal through the food web of different niche, which gets transported through blood from the alimentary tract or ingested organ, gill to tissues (Adeyemo et al., 2010; Fazio et al., 2014) and causes health hazards upon consumption by a human being (Chandra, 1999). The cyclic chain of heavy metal contamination invades from industry, agricultural land, or automobile units to water and via phytoplankton-zooplankton and fish; it reaches up to the human body (Kadar et al., 2001). Heavy metals mostly have the tendency to get accumulated in fish tissue depending on the nature of species, fishing zone, habitat, and environmental condition, bioavailability, the trophic position of the fish, water concentration (Zuluaga et al., 2015; Qiu, 2015; Gray, 2002). Among the wide range of concerned heavy metals, Pb, Cd, and Hg are included in the European Union regulations for lethal metals list (Canli et al., 2001). Even though some of the heavy metals such as Zn, Fe, Co, Cu are necessary for enzymatic activity and other metabolic mechanisms in our body at a lower dose but become deadly when they exceed the permissible limit. On the other hand, Pb, Hg, Cd, As have no known role in living organisms and are noxious even at too low concentrations (Bryan, 1976) as they inappropriately interact with the biological system and intracellular structure of a marine organism and human being even at a lower dose (Mehouel et al., 2019). Environmental Protection Agency" (EPA) and Agency for Toxic Substances and Disease Registry (ATSDR) have declared top 20 Unsafe heavy metals' as 'Priority List' in 2001 – As, Pb, and Hg possess 1st, 2nd, and 3rd position, respectively, and Cd is at 7th position. Hence, As, Cd, Pb, and Hg are considered most poisonous to humans, animals, and the environment even at a lesser dose.

2.2   HEAVY METAL TOXICITY AND FISH ORGAN SYSTEM

Aquatic organisms can accumulate heavy metal in their body thousands of times more than the environmental concentration (Gremyachikh et al., 2006; Moiseenko and Zakislenye, 2003). Fish uptakes trace metal generally through gills, gastrointestinal tract, and skin and are carried to organs by blood circulation with the help of carrier protein present in blood path by bonding to proteins, which are having an affinity towards various metals in tissues (Sönmez et al., 2016). As fish occupies the highest position in the aquatic system, hence, the heavy metal concentration is likely to be more in fish tissue than other aquatic organisms (APHA, 1981), and the rate of the bioaccumulation in fish can be measured by the rate of absorbance and excretion by fish (Afshan et al., 2014). Fish can easily pass major dietary sources of some heavy metals such as Pb, Cd, Hg, As to the human body upon consuming due to their readily uptake mechanism of these particular trace elements through their tissue. Trace elements accumulation order in gill and liver of fish is more for lead and followed by cadmium, nickel, and chromium. On the other hand, kidney and flesh seem to have more affinity towards the lead, but the kidney is prone towards chromium contamination followed by cadmium and nickel, whereas flesh is vulnerable to cadmium contamination followed by chromium and nickel (Afshan et al., 2014). Heavy metal of higher concentration is believed to affect adversely to the kidney and other organs of fish as they form metal-compounds by reacting with enzymes, nucleic acid, and other cellular proteins (Akahori et al., 1999). Ever-increasing heavy metal contamination is mostly directed to lotic aqua systems where the industrial effluents are mostly discharged and cause an ecological imbalance in the aquatic system (Yousafzai et al., 2012). These metals have the tendency of diffusing radially, and fish gets affected the most as they are the apex part of the aquatic trophic chain (Rauf et al., 2009) (Table 2.1). Heavy metal ingestion by fish affects its physiological activities, growth rate, biological performances, and reproduction pattern (Afshan et al., 2014; Basha and Rani, 2003), and consuming this heavy metal contaminated fish causes serious health hazards to a human being (Kamaruzzaman et al., 2011). The toxic reaction mainly takes place by osmotic disturbances and changing the pattern of enzyme synthesis (Jezierska et al., 2009).

2.2.1   *HEAVY METAL TOXICITY AND REPRODUCTIVE ORGANS*

Heavy metals adversely affect the embryonic development of fish as an elevated level of different heavy metals is often noticed in gonad and

TABLE 2.1 Heavy Metal, Sources, and Clinical Toxic Effects

Metal	Manufacturing Industries	Primary Sources	Target Organ	Clinical Effects
Lead	Paints, battery	Industrial dust and fumes and polluted food	Nervous system, hemato-poietic system, renal	Encephalopathy, peripheral neuropathy, central nervous disorders, anemia.
Mercury	Chlor-alkali, scientific instruments, chemicals	Industrial dust and fumes and polluted water and food	Nervous system, renal	Proteinuria
Cadmium	phosphate fertilizer, electronics, pigments, and paints	Industrial dust and fumes and polluted water and food	Renal, skeletal pulmonary	Proteinuria, Glucosuria, Osteomalacia, Aminoaciduria, Emphysema
Aluminum	Metal hardening, paints, and textile	Water treatment plant	Nervous system	Central and peripheral neuropathies
Chromium	Metal plating, tanning, rubber, and photography	Industrial dust and fumes and polluted food	Pulmonary	Perforation of the nasal septum, respiratory cancer
Copper	Rayon and electrical	Agricultural pesticide, fungicide	Liver, kidney	Anemia
Zinc	Galvanizing, plating iron and steel	Galvanized ironworks, alloy industry	Heart, respiratory tract	Cardiovascular disorder
Nickel	Electroplating, iron steel	Industrial dust, aerosols	Pulmonary skin	Cancer, dramatis

testes of the spawners (Ellenberger et al., 1994; Miller et al., 1992). A high concentration of Zn, Cu, Pb, and Cd gets deposited on the ovary, which results in contamination in the reproductive system and abnormal pattern of maturation (Allen, 1995). Methyl mercury hinders the motility time of spermatozoa and causes unsuccessful fertilization (Khan and Weis, 1987). Sometimes just after the fertilization process, when the egg starts swelling by absorbing water through perivitelline space where colloidal suspension of protein exists, heavy metal enters into the eggshell and changes chorion structure and permeability (Jezierska et al., 2009; Peterson and Martin-Robichaud, 1982). Again heavy metal reacts with the hatching gland, which produces chorionase enzyme for disintegrating the eggshell and disrupts the function of this gland during the hatching process (Kapur and Yadab, 1982). Another consequence of heavy metal is a malformation of embryos such as craniofacial anomalies, curved vertebrae, and the distorted yolk sac at the blastula and organogenesis stage (Jezierska et al., 2009). Mercury causes the vacuolated and enlarged spinal cord and damaged cardiovascular system in the embryo (Devlin and Mottel, 1992). Overall, exposure to heavy metal causes a huge rate of embryonic mortality (Ługowska, 2005).

2.2.2   HEAVY METAL TOXICITY AND NERVOUS SYSTEM

Increasing the discharge of hazardous metal substances in the aquatic system affects the nervous system of higher vertebrates in the aquatic system, including fish. Fish depends on the sense organs, both internal and external sense organs, for sensing food, locating predators, orientation, and other migratory behavior as sensory structures help them to receive stimuli that are then conducted to the central nervous system (CNS) (Baatrup, 1991). Heavy metal toxicity is affecting the normal function of the chemosensory activity of the nervous system by masking and hindering the biochemical signals as this vulnerable part does not have any external protective barrier (Sutterlin, 1974). Pb, Hg, and Cr are three main heavy metal component which plays a major role in neurological disorder in vertebrates, including fish. Acetylcholine (ACh) is one of the most important neurotransmitters, which helps to keep the muscle active, gets diminished due to continuous hypoxia, including increases in metallothionein isoforms followed by physiological stress, immunotoxicity, and ultimately cell damage (Green and Planchart, 2018). Cd exposure increases the chance of neuron damage and gets accumulated on the mucus membrane results in interference with oxygen diffusion (Green and Planchart, 2018). Various studies observed that Cd is highly capable of damaging sensory macula and neuromast by continuous depositing of olfactory bulbs (Faucher et al., 2006, 2008). The neurotoxicity of Pb is well known. It can alter the GABA gene expression (Wirbisky et al., 2014), followed by decreased axon length (Zhu et al., 2016). A recent study brought the fact into light that Pb is one such heavy metal which is efficient enough to counteract the function of superoxide dismutase (SOD) by removing Cu and Zn ions from its catalytic zone (Zheng et al., 2014); hence, Pb appears to be one of the potentially hazardous substances, which can firmly alter the epigenomic structure and disturbs swimming pattern, hyperventilation, and also muscle spasms (Holcombe et al., 1976; Giusi et al., 2008; Green and Planchart, 2018). Nevertheless, continuous Hg accumulation delays hatching rate, increase mortality, reduces the level of neurotransmitters (serotonin and dopamine) (Maximino et al., 2011). Mercury exists in the form of methylmercury, which causes behavioral and visual deficits (Carvan et al., 2017). Several studies showed a higher dose of Hg could alter the protein associated with gap junction and oxidative phosphorylation dysfunction (Amlund et al., 2015; Rasinger et al., 2017). These effects of heavy metal can persist throughout three consecutive generations, and hence it causes an alarming situation for heavy metal toxicity in the aquatic system.

2.2.3   HEAVY METAL TOXICITY AND RESPIRATORY ORGAN

Metal uptake by fish is entirely different from the terrestrial animal heavy metal uptake pathway. As gill is their main respiratory organ and always submerged in water, it serves the main pathway for heavy metal entry in fish body. As water can hold less quantity of dissolved oxygen as compared to air, hence, to gain the same amount of oxygen what terrestrial organism is getting out of 1-liter air, fish circulates twenty times more water to get the same quantity of oxygen. While moving this large amount of water through, gill is continuously getting exposed to aquatic toxic heavy metals that keep on passing through the gill (Olsson et al., 1998). Heavy metal adversely affects the ion transporting chloride cell on the superficial gill area, and subsequently, the change in gill affects ion fluxing markedly across the gill (Goss et al., 1992; Perry et al., 1993). Cd, Co, Zn all these heavy metals can traverse gill epithelium cell through chloride channel and eventually mixed with blood and causes acute toxicity in fish body (Verbost et at., 1989; Hogstrand et al., 1994).

2.2.4   *HEAVY METAL TOXICITY AND GASTRO-INTESTINAL TRACT*

The gastrointestinal tract is another major route of heavy metal ingestion in the fish body, which is again subjected to the water hardness and alkalinity the uptake of heavy metal through the gastrointestinal tract differs. Methyl mercury affects this intestinal route adversely due to its high lipophilic nature (Olsson et al., 1998). It was found that the major portion of Cd that is uptaken through the intestinal path is not transferred to the body, but the obvious sign of heavy metal exposure to the epithelial cell is highly mucus secretion, which further traps heavy metal (Sorensen, 1991). 95% of ingested Cd gets adhere to the gut of fish and on mucosal cells and is uptaken by membrane flexibility processes (Handy, 1996). Nevertheless, mercury in the form of methyl mercury stores in muscle tissue and biomagnifies and diffuses across the cells due to its high lipophilicity (Boudou and Ribeyre, 1985).

2.3   BIOACCUMULATION OF HEAVY METALS IN FISH

Bioaccumulation, bioconcentration, and biomagnification are three inevitable terms in the field of heavy metal toxicity in the aquatic system, especially in the fish body. When any kind of substances, including chemicals, organic-inorganic compound, heavy metal, or any toxic constituents gets absorbed in an organism and absorbance rate is higher than catabolism (excretion) rate, it gets piled up in the body and causes a carcinogenic effect, which terms as Bioaccumulation. On the other hand, if the accumulation of the toxic constituents is taken only from water bodies, then it is termed as bioconcentration (Alexander, 1999). When the concentration of toxic substances gets amplified in the tissue or cell or organ with the trophic level of the food chain, it is called biomagnification. There are two limits of significance in order to monitor the bioaccumulation or biomagnification status to keep the consumers in a safer site, i.e., maximum allowable limits (MLA) and provisional tolerable weekly intake (PTWI) (Tressou et al., 2004) (Table 2.2). Different heavy metals show a different level of affinity to get accumulated in different organs of the fish body, such as Pb, Cd mainly in gill, likewise, Zn, Cu, Ni in the liver and kidney. Accumulation pattern of essential trace elements depends on their role in metabolism, such as, Zn and Cu have a pivotal role in a hepatic cell as it binds with protein and forms metallothioneins (MT), which helps in enzymatic and other metabolic activities (Gorur et al., 2012). Likely, Fe is related to physiological activities of hemoglobin synthesis and blood cell formation; hence it is having a tendency of accumulating in hepatic cells (Gorur et al., 2012). Benthic fish shows more bioaccumulation than surface-dwelling species or column dwelling species as benthic species are in direct contact with the heavy metal toxic sediments and due to the preying upon zoobenthic predators (Yi and Zhang, 2012).

TABLE 2.2 Heavy Metal and its Maximum Daily Limit

Heavy Metals	Maximum Daily Dose or Exposure (ug/day)	
	Parenteral	Oral/Topical/Dermal/Mucosal
Lead	1	10
Mercury	1.5	15
Arsenic	1.5	15
Cadmium	0.5	5
Chromium	0.1	1
Copper	250	2500
Zinc	30	300
Nickel	25	250

Source*:* Elemental Impurities (2010). Pharmacopeial Forum *36*(1).

2.4 EFFECTS AND CORRESPONDING MECHANISM OF ACTION OF HEAVY METALS

2.4.1 LEAD (Pb)

Lead is one of the highest-rated hazardous heavy metals among all due to its severe chronic toxicity. Sources of lead are mainly metal plating industries, battery acid-producing industries, automobiles, additives in gasoline production, exhausts of chimneys of different chemical factories, ores production, fertilizers, and pesticides, etc., (Sharma and Dubey, 2005). Pb elevates dissolved organic matter (DOM) percentage by remaining suspended in the environment in organic form, and its toxicity increased with the number of methyl or ethyl group addition with it (Sánchez-Marín, 2007; Munoz-Olivas et al., 2001). Pb mainly accumulates in the kidney, liver, spleen, digestive tract, and gill (Jezierska and Witeska, 2006). Lead has an inhibitory effect on gonadal growth, like cadmium and chromium contamination, lead shows the same effect of hepatocyte vacuolization, hepatic lesions, ulcer, nuclear pyknosis, and increase of sinusoidal spaces in lead-exposed fish (Delistraty, and Stone, 2007). Acute Pb toxicity affects gill epithelial tissues and causes respiratory disorders. The chronic stage causes the curling of lamellar cells and disintegration of the epithelium, and it destroys thiol-containing antioxidants and enzymes and affects CNS (El-Badawl, 2005). It also hinders impulse conductivity by prohibiting mono aminooxidase and ACh esterase, which directly affect the cellular function and impair embryonic development and juvenile growth (Dave and Xiu, 1991). Agency for Toxic Substance and Disease Registry (2007) revealed that Pb chronic toxicity in humans frequently causes depression, irritation, lack of interest/enthusiasm, attentional deficiency, epigastric pain, digestion problem, convulsions even it may cause coma and eventually death if Pb dose reaches to the maximum. In children, its effect is too severe as encephalopathy is common in Pb contamination with lethargic nature and kids show mental dullness, and sometimes it causes anorexic condition; to some extent, prolonged exposure to Pb may reduce the cognitive function. In some cases, it was reported that children become aggressive, psychic, and mental disorders. It is found that lead blocks the function of certain growth and hem synthesizing enzymes such as Na+/K+-ATPase enzyme and d-aminolevulinic acid dehydratase and affects lipid peroxidation (LPO), hence causes alteration of blood parameters in the body (Afshan et al., 2014). It also changes the alanine aminotransferase (ALT) and aspartate aminotransferase (AST) concentrations in tissues, heart, liver, to some extent kidney and other organs (Çoğun and Şahin, 2012) and affects the amino acid metabolism. Due to the health hazards of Pb intake, Codex Alimentarius sets a maximum permissible limit of Pb in fish of 0.3 mg kg^{-1} of wet weight (CODEX Alimentarius, WHO, 2014). The present PTWI recommendation in humans of 25 µg Pbkg^{-1} of the wet weight determined by The Joint FAO/WHO Expert Committee on Food Additives (JECFA) has been taken back after assessing it in 2011 and then stated that it is not possible to determine a new PTWI that may be considered health protector (FAO, 2014). Lead causes toxicity through the ionic mechanism and by following oxidative stress in living cells by creating a gap between the production of enough anti-oxidative agents to counteract and detoxify the free radicals in the body, hence, generation of reactive oxygen species (ROS) and antioxidant defense mechanism gets imbalanced. In the presence of Pb, ROS level elevates, and the antioxidant level drops down, glutathione (GSH) remains in two forms, i.e., reduced condition (90% of total GSH content) and oxidized condition (10% of total GSH content) (Jaishankar, 2014). Reduced GSH gives away the reducing component (H^{++} and e^{-}) from its thiol groups of cysteine to ROS to keep it balanced and stable and forms glutathione disulfide (GSSG) and gradually under the oxidative stress, GSSG concentration becomes higher than its reduced counterpart (Mathew et al., 2011). Basically, Pb can replace the bivalent ion of Ca, Mg, Fe, and monovalent ion of Na, which makes the biological process disturbed, and this ionic mechanism of Pb adversely affect apoptosis, intra-inter cell signaling, the release of neurotransmitter, even Pb can replace Ca at the picomolar level and alters the *PKC* (Protein Kinase C) which plays an important role in brain recognition function (Flora et al., 2012).

2.4.2 MERCURY (Hg)

Hg poisoning is well known for Minamata disease, which is a kind of neuro-toxicity similar to *'Hunter Russell syndrome.'* The organic form of mercury compounds is considered to be more toxic as it gets bioaccumulated in fish tissue easily and gets transferred to higher trophic level, especially to the consumer's body, whereas, inorganic form of mercury is considered to be less toxic compared to the organic counterpart as it accumulates in the lower level in fish and has better excretion rate in the human body; hence it doesn't rise to the level where it could be carcinogenic (Adeyemo et al., 1998; Boening et al., 2000). Inorganic mercury such as mercuric chloride ($HgCl_2$), mercurous chloride (Hg_2Cl_2), mercuric acetate ($HgC_4H_6O_4$), and mercuric sulfide (HgS) are less toxic, but in nature, they can get methyl-ated and transform into the organic compound, i.e., methylmercury (MeHg) which is again considered to be highly carcinogenic (Escobar et al., 1995). Mercury poisoning can significantly affect fetal growth. It is not an essential element in the human body, hence causes acute and chronic toxicity when accumulates even if in small quantities. Mercury has an adverse effect on human physiology, especially monomethyl and dimethylmercury, which are more hazardous. The major sources of Hg in water bodies are organic fertil-izers, industrial runoff, especially, chemical, battery, thermometer industries, and pharmaceutical dealing with dentistry (amalgam). It was reported that monomethyl mercury could bioaccumulate over a million-fold than the level at origin (Pandey and Madhuri, 2014). It creates neurotoxicological disorders in human along with it has an adverse effect on gingivitis, tremors, congenital, and fetal death related carcinogenic effect origin (Pandey and Madhuri, 2014). In the environment, organometallic Hg often undergoes methylation in the presence of anaerobic bacteria, named as sulfur-reducing bacteria (SRB) and iron reducers (FeRP) and forms the highest toxic form of mercury called Methyl mercury (MeHg+) (Avramescu et al., 2011; Pack et al., 2014) which is the main component (70–100%) of mercury toxicity in fish (Amlund et al., 2007; Nøstbakken et al., 2015). Methylation depends on water temperature, a rise in water body temperature causes more methylation of inorganic Hg, and the concentration of MeHg+ in the fish body also will elevate (Pack et al., 2014; Booth and Zeller, 2005). The target organ in fish is the liver as fish intake Hg through the food chain, which then transfers to the liver, binds, stores, and distributes this mercurial substance through circulation as it can easily cross the blood barrier (Sorensen, 1991). The peculiar character of Hg is irreversible damage to the nervous system, including ataxia, convulsion, cognitive changes in the body (Eccles and Annau, 1982) on long-term retention. Furthermore, it has a negative impact on reproductive biology as it reduces egg production, affects the viability of spermatozoa and mortality to the hatched larvae and fry stage (Raldúa et al., 2007). However, mercury toxicity affects gill, kidney (Necrosis and fibrosis of renal tubular lumen), and blood parameters (decrease in hemoglobin (Hb), hematocrit (Ht), mean). Mercury disrupts the cell membrane and hinders intracellular Ca homeostasis and can freely bind with available thiols (Patrick, 2002). It damages the protein structure, mainly 3rd and 4th-degree structure (tertiary and quaternary), and attaches with selenohydryl and sulfhydryl groups and again, these groups react with methyl mercury and damages cell structure, affects transcription and translation, which leads to the destruction of ribosomes and endoplasmic reticulum and damage of natural killer (NK) cells (Jaishankar, 2014). Free sulfhydryl groups provoke the mobility of metal within ligands (Bernhoft, 2011).

2.4.3   CADMIUM (Cd)

Among all the heavy metals, Cadmium is well known for its higher toxic nature even at a very lower dose and is considered to be a potent element for public health hazards. Sources of Cadmium in the aquatic system are mainly industrial discharge, agricultural runoff, smelting of other metallic substances, and burning of fossil fuel (Sastry and Gupta, 1979; Karadede et al., 2004; Nriagu and Pacyna, 1988). Cd is an inevitable byproduct of Zn refining and PVC pipe making industries as Cd is a good stabilizer and pigment widely used in PVC. It is also being used hugely in marine and aerospace applications as an anti-corrosion agent. In the environment, Cd exists mainly in two forms viz. $CdCl_2$ and $CdCl^+$ (Simpson, 1981). Almost 75% of aquatic sediment available Cd gets accumulated mostly in the kidney, liver, and gill of freshwater fish, which causes necrotic liver disease by damaging the hyaline, liver cell vacuolation, cellular inflammation, and cramming of blood vessels (Thophon et al., 2004). Several studies found that Cd causes acute and chronic toxicity in the fish body, including the impaired reproductive system, kidney dysfunction, tumor, ulcer, etc., (Sridhara et al., 2008). Cadmium is a non-essential component for the human body; therefore, once a human being takes cadmium through the food chain, it gets biomagnified and causes chronic toxicity in bodies (Amiard et al., 1987). The maximum daily intake limit of Cd for a human being is 0.5–1 Ug. Cd can easily penetrate barriers of cell and tissue and gets bind with ligand like protein and transfers from blood to kidney and binds with protein again and hinders purifying mechanisms in the kidney followed by excretion of sugar and other essential protein from the body (Afshan et al., 2014). Other symptoms of Cd contamination in the human body are diarrhea, stomach upset, failure in fertility, myocardial disease, pulmonary obstructive disease, reproduction failure, and to some extent Cd damage nervous system if ingesting in higher doses (Afshan et al., 2014). It causes osteomalacia, osteoporosis in human beings and other animals (Pandey and Madhuri, 2014) and prohibits Ca ion uptake in the gill of fish. It hinders trace elements such as Zn, Cu distribution in the fish body causes abnormal metabolism activity in fish (Verbost et al., 1987; Li et al., 1995). Fish (Dicentrarchuslabrax) exposed to Cd showed epithelial swelling of the mitochondrial and endoplasmic reticulum (ER) and chloride cell (Giari et al., 2007). Furthermore, Cd exposure affects hematocrit value, erythrocytes count, and hemoglobin concentration level and ultimately leads to anemia (Gill and Epple, 1993). The study found that Cd exposed goldfish showed significant hyperactivity of plasma enzymes, especially glutamic acid, oxaloacetic acid – transaminase (GOT), and glutamic acid – pyruvic acid transaminase (GPT) (Žikić et al., 2001). Even Cd can disrupt carbohydrate metabolism and leads to hyperglycemia, and also adversely affects on the endocrine system (Thomas et al., 1982; Vetillard and Bailhache, 2005). Long time exposure to Cd, even if at a lower dose, can cause DNA damage and oxidative stress in fish (Jia et al., 2011). Even at xenotoxic level, Cd was found to form micronuclei in erythrocytes of *Anguilla anguilla* (Sanchez-Galan et al., 2001). Cd can intensify its concentration up to three thousand times when it binds with cysteine-enriched protein and creates hepatotoxicity in the kidney and invades to renal tissues leads to nephrotoxicity (Jaishankar, 2014). Other than cysteine, Cd has an affinity towards glutamate, histidine, and aspartate ligands (Castagnetto et al., 2002). Furthermore, Cd can replace zinc ion in metallothionein as both are having the same oxidation state and prevents it from free radical scavenging within the cellular system (Jaishankar, 2014).

2.4.4 ALUMINUM (Al)

Aluminum is considered to be more toxic in its ionic soluble forms, which mainly comes from water treatment plants as it uses Al as a flocculating agent in water treatment (Walton et al., 2009; Silva et al., 2007). Al toxicity depends on water parameters, mostly on water pH (Svobodová, 1993), as it can get dissolved at pH 6 or below and can be hazardous for fish. Fish get exposed once Al gets dissolved in water and creates an ionic-osmotic imbalance, and due to the coagulation of mucous, it causes fusion in gill lamellae and filaments, which leads to a severe respiratory problem in fish (Abdel-Latif, 2008). Al ingestion causes adverse effects on the cardiovascular system and hematology of fish as it has a tendency to increase total erythrocyte counts; mean corpuscular hemoglobin concentration (MCHC), hematocrit (Hct); and mean corpuscular hemoglobin (MCH) (Alwan et al., 2009). Al ion can invade a cell, and gradually, it comes to the gastrointestinal tract, and even it was found that it comes in contact with the plasma membrane, which then starts a toxic reaction in the system itself (Kochian et al., 2005). Aluminum ion can replace magnesium ion and ferric ion and causes disturbance in cellular growth and damages neurons, and leads to degeneration of neurons in the basal ganglia portion of the brain (Jaishankar et al., 2014).

2.4.5 CHROMIUM (Cr)

Ever growing urbanization and poor dumping practice of wastes by a human being are the main source of chromium toxicity in the aquatic system. In addition to that, a huge amount of chromates, dichromates are released by painting, electroplating, mining, tanneries, textile, battery making, photographic, and metal plating industry as an anti-corrosion coating, dyeing agents. As per the Environ Protection Act (EPA), the permissible range of chromium in water should be within a range of 50–100 ug Cr/L. Fish acquires heavy metal through the food web or by absorbing through body and gill from surroundings (Said et al., 1992). Fish can accumulate this heavy metal in their gill in higher concentration compared to body parts and causes hyperplasia of the primary lamellar epithelium (Afshan et al., 2014; Ackermann, 2008). However, gill and kidney are considered to be critical organs, but reproductive and excretory organs to get adversely affected when exposed to Cr contamination. It causes hypertrophy and vacuolation of spermatocytes in the testis and deposition of interstitial tissues in ovaries. Again it damages renal tubules, including reducing tubular lumen diameter, hypertrophy of epithelial cells, and contraction of glomeruli (Mishra and Mohanty, 2009). Next to the gill, the kidney is the most targeted organ for Cr to get accumulated and causes liver necrosis, congestion of blood vessels, increased meta-macrophage centers, and fatty liver (Ackermann, 2008). Chromium causes renal lesions, bronchial damages and disrupts the cardio-vascular system; resulting in anemia, lymphocytosis, eosinophilia (Afshan et al., 2014). As earlier discussed, about the bioaccumulation rate in the fish body is high because of its highest position at the trophic level in an aquatic system. Hence, it gets magnified when transfers to the human body upon feeding the contaminated fish. In surface water, there are two oxidative stages of chromium that exists, i.e., chromium tetravalent (Cr^{3+}) and hexavalent (Cr^{6+}). Among these, Cr^{6+} is considered to be highly carcinogenic due to its cell permeability, which is harmful to a human being; it causes nose irritations and nose bleeding, skin irritations. The weakening of the immune system, damage of liver and kidney, respiratory problem, ulcer, to some extent it causes lung cancer and leads to death (Karadede et al., 2014; WHO, 1990; Eisler, 2000). Hexavalent chromium can deplete the level of glycogen, lipid, and protein in the fish body and causes disruption in the metabolic process (Nath and Kumar, 1987; Saxena and Tripathi, 2007). Cr (VI) invades the anions which are having a similar structural assembly and possess isoelectric nature such as SO_4^{2-} and HPO_4^{2-} and by phagocytic process gets absorbed. Though chromium can be reduced to penta- and tetravalent states, but trivalent and hexavalent are more stable (Jaishankar, 2014). Cr (VI) is considered to be cancer-causing as it reacts with thiols and ascorbates and produces ROS (superoxidase, hydrogen peroxide) and leads to nucleic acid and protein damage (Stohs and Bagchi, 1995). International Agency for the Research on Cancer has declared Cr (VI) as Group 1 human oncogenic compound (Zhang, 2011).

2.4.6 COPPER (Cu)

Unlike Pb, Cd, and Hg, copper is a crucial micronutrient in our body as it is responsible for cellular metabolism, and it is a key constituent of the metabolic enzyme (Monteiro et al., 2009b). However, Cu is an essential component in the body, but in high doses, it can be carcinogenic as it causes anemia, damage of hepatic and renal functions, and gastrointestinal irritations may occur. During Wilson's disease, it affects greatly. It comes to drinking water from Copper water supply pipes and additives designed to control the algal growth, mollusk, and other insect killers in the fish pond to control the phytoplankton and filamentous algae in the form of $CuSO_4$ (Michael, 1986). Cu shows an affinity towards the liver, kidney, gill, chemoreceptor, and hematopoietic tissue of fishes (Sorensen, 1991). Cu exposure mainly affects the liver, gill, and kidney of the fish body. Nuclear pyknosis, necrosis, hepatocytic vacuoles, and more sinusoidal space are the sign of Cu exposed liver damage (Figueiredo-Fernandes et al., 2007). Cu exposed Nile tilapia showed edema in gills and vacuolation in the liver, structural alteration, and physiological changes are mostly observed in Cu contaminated toxicity in fish (Pieterse, 2004). Waterborne Cu exposure can dystrophies chloride cells (Arellano et al., 1999). Excess Cu affects enzymatic mechanisms as Na+/K+-ATPase enzyme activity is sensitive towards Cu accumulation (Arellano et al., 1999) and hinders the function of catalase (CAT) enzyme in hepatic cell and muscle (Radi and Matkovics, 1988). Cu toxicity hampers cardiovascular physiology too by altering blood parameters; it elevates Hct and hemoglobin levels, followed by erythrocyte swelling (Cyriac et al., 1989). As per the reproductive system is concerned, the egg production rate reduces, and abnormalities of newly hatched larvae become common once fish is affected with Cu toxicity at a higher dose (Sorensen, 1991). Other hazardous symptoms of Cu toxicity include, reduced heart rate, the mortality of fingerlings, deformities of hatched larvae with kyphosis, lordosis, and C-shaped larvae, abnormal behavior, weight loss, weak immune responses, etc. (Eyckmans et al., 2011).

2.4.7 ZINC (Zn)

Though Zn is a vital micronutrient for the human body as it plays a key role in the nucleic acid synthesis, enzymatic functions, neurotransmitter, and cell signaling (Sfakianakis et al., 2015; Celik and Oehlenschläger, 2004; Hogstrand, 2011) but excess accumulation causes toxicity. Fish has a different pattern of accumulating Zn in their body; Zn level decreases with the increase of fish body length (Toth and Brown, 1987). The main source of Zn to the environment is geological weathering of rocks, dye castings metals, plastic, galvanized-iron works, and alloy industries (Jennings and Rainbow, 1979). Zn causes reduced growth in fish, structural deformities, and affect gill, adversely leading to hypoxia by changing the state of pH equilibrium and ionic concentration and eventually death (Olaifa et al., 1998). Zn has a negative impact on the cardiovascular system, including heart physiology and hematology. Even it affects the hatchability of fish; bring changes in behavior as they show erratic movement pattern, gulping air, dormant condition, and eventual death (Kori and Ubogu, 2008). Accumulation of Zn sometimes gets influenced by ecological factors such as alkalinity, pH, DO, and hardness (Blackmore and Wang, 2003). Zn is such a kind of element which is required for the human body in a little dose. Absence of Zn, cause still-birth and other birth-related problems, growth impediments, and weakening immune system in human being (Afshan et al., 2014). An excess amount of Zn causes pancreatic lesions, slow wound healing, skin rashes, nausea, stomach pain (Yildirim et al., 2009). Zn gets accumulated mostly in the gill portion and inhibits Ca^{2+} uptake, and eventually causes hypocalcemia and death (Niyogi, 2006). Chloride cell gets detached from epithelium tissues and widens because of the disintegration of epithelial cells from the basal lamina, making the gaseous exchange more difficult (Sorensen, 1991). Like other toxic metal, Zn too adversely affects reproductory function, survivability, and hatching rate (Hogstrand, 2011). Next to gill, the kidney is the second affected organ in fish by Zn toxicity along with cardiovascular disturbance, growth retardation, spawning inhibition, etc., (Sorensen, 1991). The function of serum transaminases in some freshwater fishes increases with Zn contamination along with the proliferation of gill (Nemcsók et al., 1981). Zn exposure damages ovarian tissue (degeneration and hyperemia) (Cariño and Cruz, 1990) and liver tissue (Dyk et al., 2007).

2.4.8 NICKEL (Ni)

Nickel is an omnipresent constituent which occurs naturally in environs, in soil, water, and biosphere. The source of Ni to the aquatic body is mainly Ni alloy making industries, trash incinerators, and oil-coal burning power plants (Al-Attar, 2007). Ni reacts very easily with different ligands in the environment and becomes more viable than other heavy metals (Palaniappan and Karthikeyan, 2009). The Ni has an important role in RBC production; hence it is an essential component in our body, but it becomes slightly poisonous in excess quantity (Magyarosy, 2002). Its chronic exposure can cause weight reduction, cardiac and hepatic mutilation, respiratory disorder, and itchy skin (ATSDR, 2004) as it penetrates the epidermis layer and reacts with the protein (Nielson, 1977). Unlike other heavy metals, Ni accumulation in the body varies with time, and it gets affected by the physicochemical properties of water (Svobodo, 1993). The prominent sign of Ni contamination in fish is over slimy gill, hyperplasia, shortening, and decolorization of lamellae, and tissue hypertrophy (Yang et al., 2007; Athikesavan et al., 2006; Al-Attar, 2007). Ni exposure affects serum glucose, cholesterol, aminotransferase as their level increases, and on the other hand, serum NaCl level drops and causes osmolality and its effects on blood parameters (reduced mean corpuscular volume, leucocytes, hematocrit, and hemoglobin level) (Al-Ghanim, 2011).

2.5 CONCLUSION

Heavy metals such as Pb, Hg, Cr, Cd, Al, Zn, Ni, and Al all are highly lethal to a human being, fish, and other aquatic animals. Heavy metal interaction with the different organs of the fish body and human health and bioavailability are all factors dependent. It causes damage in every way, starting from larval mortality, spawning impairment, gonadal damage to the respiratory, gastrointestinal, and nervous system. Hence, it has a dangerous outcome when it comes to a human being because each higher trophic level gets more and more carcinogenicity due to bioaccumulation and biomagnification. Since enormous dispose of modern urbanized e-waste, automobile, industrial wastes, and agricultural wastes are ever increasing; hence the hazardous effect of heavy metal toxicity in the aquatic environment has reached its highest alarming peak. Several studies have found most of the heavy metal prone regions have crossed the set limit of acceptability of heavy metal intake in food. It is of high importance and necessity that national and international organizations to take up precaution against these ever-increasing heavy metal threats, strict regulation has to be implemented to prevent this aforementioned heavy metal pollution.

ACKNOWLEDGMENT

The special thanks are due to Research Center, College of Pharmacy, King Saud University, Riyadh, and Deanship of Scientific Research, King Saud University, Kingdom of Saudi Arabia.

KEYWORDS

- **aquatic environment**
- **bio-accumulation**
- **fish toxicity**
- **gastrointestinal tract**
- **heavy metal toxicity**
- **mercury**

REFERENCES

Abdel-Latif, H. A., (2008). The influence of calcium and sodium on aluminum toxicity in Nile tilapia (*Oreochromis niloticus*). *Aust. J. Basic. Appl. Sci., 2*, 747–757.

Ackermann, C., (2008). *Quantitative and Qualitative Histological Assessment of Selected Organs of Oreochromis mossambicus After Acute Exposure to Cadmium, Chromium and Nickel* (Vol. 49, pp. 139–143). MSc dissertation, University of Johannesburg, South Africa.

Adeyemo, O. K., Adedeji, O. B., & Offor, C. C., (2010). Blood lead level as biomarker of environmental lead pollution in feral and cultured African catfish (*Clarias gariepinus*). Nigerian. *Vet. J., 31*, 139–147.

Afshan, S., Ali, S., Ameen, U. S., Farid, M., Bharwana, S. A., Hannan, F., & Ahmad, R., (2014). Effect of different heavy metal pollution on fish. *Res. J. Chem. Environ. Sci., 2*(1), 74–79.

Agency for Toxic Substance and Disease Registry, (2007). *Toxicological Profile for Lead.* Atlanta: U.S. Department of Health and Humans Services, Public Health Service, Centers for Diseases Control.

Akahori, A., Gabryelak, T., & Jozwiak, Z., (1999). Zinc-induced damage to carp (*Cyprinus carpio* L.) erythrocytes *in vitro. Biochem. Mol. Biol. J. Int., 47*, 89–98.

Al-Attar, A. M., (2007). The influences of nickel exposure on selected physiological parameters and gill structure in the teleost fish, *Oreochromis niloticus. J. Biol. Sci., 7*, 77–85.

Alexander, (1999). Bioaccumulation, bioconcentration, biomagnification. *Environ. Geol., 43*, 44.

Al-Ghanim, K. A., (2011). Impact of nickel (Ni) on hematological parameters and behavioral changes in *Cyprinus carpio* (common carp). *Afr. J. Biotechnol., 10*, 13860–13866.

Alissa, E. M., & Ferns, G. A., (2011). Heavy metal poisoning and cardiovascular disease. *J. Toxicol.*, 2011.

Almeida, J., Novelli, E., Dal, M., Silva, P., & Alves, R. Jr., (2001). Environmental cadmium exposure and metabolic responses of the Nile tilapia, *Oreochromis niloticus. Environ. Pollut., 114*(2), 169–175.

Alwan, S. F., Hadi, A. A., & Shokr, A. E., (2009). Alterations in hematological parameters of fresh water fish, *Tilapia zillii*, exposed to aluminum. *J. Sci. Applica., 3*, 12–19.

Amiard, J., Amiard-Triquet, C., Berthet, B., & Metayer, C., (1987). Comparative study of the patterns of bioaccumulation of essential (Cu, Zn) and non-essential (Cd, Pb) trace metals in various estuarine and coastal organisms. *J. Exp. Marin. Biol. Ecol., 106*, 73–89.

Amlund, H., Lundebye, A. K., & Berntssen, M. H. G., (2007). Accumulation and elimination of methylmercury in Atlantic cod (*Gadus morhua L.*) following dietary exposure. *Aquat. Toxicol., 83*, 323–330.

APHA, (1981). *Standards Methods for the Examination of Water and Wastewater* (14th edn.), Washington, D.C., USA.

Arellano, J. M., Storch, V., & Sarasquete, C., (1999). Histological changes and copper accumulation in liver and gills of the *Senegales sole, Solea senegalensis. Eco. Toxicol. Environ. Saf., 44*, 62–72.

Athikesavan, S., Vincent, S., Ambrose, T., & Velmurugan, B., (2006). Nickel induced histopathological changes in the different tissues of freshwater fish, *Hypophthalmichthys molitrix* (Valenciennes). *J. Environ. Iol., 27*, 391–395.

ATSDR (Agency for Toxic Substances and Disease Registry), (2004) *Agency for Toxic Substances and Disease Registry, Division of Toxicology*. Clifton Road, NE, Atlanta, GA.

Avramescu, M. L., Yumvihoze, E., Hintelmann, H., Ridal, J., Fortin, D., & Lean, D. R. S., (2011). Biogeochemical factors influencing net mercury methylation in contaminated freshwater sediments from the St. Lawrence River in Cornwall, Ontario, Canada. *Sci. Total. Environ., 409*, 968–978.

Basha, P. S., & Rani, A. U., (n.d.). Cadmium-induced antioxidant defence mechanism in freshwater teleost *Oreochromis mossambicus* (Tilapia). *Ecotoxicol. Environ. Saf., 56*(2), 218–221.

Blackmore, G., & Wang, W. X., (2003). Comparison of metal accumulation in mussels at different local and global scales. *Environ. Toxicol. Chem., 22*, 388–395.

Boening, D. W., (2000). Ecological effects, transport, and fate of mercury: A general review. *Chemosphere, 40*, 1335–1351.

Booth, S., & Zeller, D., (2005). Mercury, food webs, and marine mammals: Implications of diet and climate change for human health. *Environ. Health Perspect., 113*, 521–526.

Boudou, A., & Ribeyre, F., (1985). Experimental study of trophic contamination of *Salmo gairdneri* by two mercury compounds—$HgCl_2$ and $CH_3 HgCl$—analysis at the organism and organ levels. *Water, Air, and Soil Pollution, 26*(2), 137–148.

Bryan, G. W., (1976). Some aspects of heavy metal tolerance in aquatic organisms. In: Lockwood, A. P. M., (ed.), *Effects of Pollutants on Aquatic Organisms* (pp. 7–34). Cambridge University Press., UK.

Canli, M., & Atli, G., (2003). The relationships between heavy metal (Cd, Cr, Cu, Fe, Pb, Zn) levels and the size of six Mediterranean fish species. *Environ. Pollut., 121*(1), 129–136.

Cariño, V. S., & Cruz, N. C., (1990). Effects of low levels of zinc on the ovarian development of *Tilapia nilotica* Linneaus. *Sci. Diliman., 3*, 34–40.

Carvan, M. J., (2017). Mercury-induced epigenetic transgenerational inheritance of abnormal neurobehavior is correlated with sperm epimutations in zebrafish. *Plos One., 12*, e0176155.

Castagnetto, J. M., Hennessy, S. W., Roberts, V. A., Getzoff, E. D., Tainer, J. A., & Pique, M. E., (2002). MDB: The metalloprotein database and browser at the Scripps Research Institute. *Nucleic Acids Res., 30*(1), 379–382.

Celik, U., & Oehlenschläger, J., (2004). Determination of zinc and copper in fish samples collected from Northeast Atlantic by DPSAV. *Food Chem., 87*, 343–347.

Ceron, J. J., Sancho, E., Ferrando, M. D., Gutierrez, C., & Andreu, E., (1997). Changes in carbohydrate metabolism in the eel Anguilla, during short-term exposure to diazinon. *Environ. Toxicol. Chem., 60*(1–4), 201–210.

Chandra, K., (1999). Status of heavy metals and pesticides in aquatic food chain in context of Indian rivers. In: *Summer School on Ecology, Fisheries, and Fish Stock Association in Indian Rivers* (pp. 92–98). CIFRI, Barrackpore.

Clarkson, T. W., Nordberg, G. F., & Sager, P. R., (1985). Reproductive and developmental toxicity of metals. *Scand. J. Work Environ. Health, 11*, 145–154.

Codex Alimentarius Commission, (2012). *Joint FAO/WHO Food Standards Program Codex Committee on Contaminants in Foods Sixth Session [Internet].* Available from: ftp://ftp.fao.org/codex/meetings/cccf/cccf6/cf06_INFe.pdf (accessed on 12 November 2020).

Çoğun, H. Y., & Şahin, M., (2012). Nil tilapia (*Oreochromis niloticus* Linnaeus, 1758)'da Kurşun Toksisitesinin Azaltılmasında Zeolitin Etkisi. *Kafkas Univ Vet. Fak. Derg., 18*(1), 135–140.

Connell, D. W., (1989). Biomagnification by aquatic organisms-a proposal. *Chemosphere, 19*, 1573–1584.

Connell, D. W., (1990). Environmental routes leading to the bioaccumulation of lipophilic chemicals. In: Connell, D. W., (ed.), *Bioaccumulation of Xenobiotic Compounds* (pp. 60–73). CRC Press, Boca Raton, Florida.

Cyriac, P. J., Antony, A., & Nambisan, P. N. K., (1989). Hemoglobin and hematocrit values in the fish *Oreochromis mossambicus* (Peters) after short-term exposure to copper and mercury. *Bull. Environ. Contam. Toxicol., 43*, 315–320.

Dave, G., & Xiu, R., (1991). Toxicity of mercury, copper, nickel, lead, and cobalt to embryos and larvae of zebrafish, *Brachydanio rerio. Arch Environ. Contam. Toxicol., 21*, 126–134.

Delistraty, D., & Stone, A., (2007). Dioxins, metals, and fish toxicity in ash residue from space heaters burning used motor oil. *Chemosphere, 68*(5), 907–914.

Dyk, J. C. V., Pieterse, G. M., & Van, V. J. H. J., (2007). Histological changes in the liver of *Oreochromis mossambicus* (Cichlidae) after exposure to cadmium and zinc. *Ecotoxicol. Environ. Saf., 66*, 432–440.

Eccles, C. U., & Annau, Z. P., (1982). Prenatal methyl mercury exposure: I. Alternations in neonatal activity. *Neurobehav. Toxicol. Teratol., 4*, 371–376.

Eisler, R., (2000). *Handbook of Chemical Risk Assessment: Health Hazards to Humans, Plants, and Animals.* Boca Raton: CRC Press, Lewis Publishers.

El-Badawi, A. A., (2005). Effect of lead toxicity on some physiological aspects of Nile tilapia fish, *Oreochromis niloticus. Inter. Conf. Vet. Res. Div.* NRC, Cairo, Egypt.

Elemental Impurities, (2010). *Pharmacopeial Forum, 36*(1). https://www.usp.org/chemical-medicines/elemental-impurities-updates (accessed on 25 February 2021).

Elhussien, M. E., & Adwok, B. A., (2018). Determination of heavy metals in fish and water of White Nile during watery diarrhea outbreak from June to July, 2017, Gezira Aba-Sudan. *Sci. J. Anal. Chem., 6*(1), 1–6.

Escobar, O., Sandoval, M., Vargas, A., & Hempe, J. M., (1995). Role of metallothionein and cysteine-rich intestinal protein in the regulation of zinc absorption by diabetic rats. *Pediatr Res., 37*(3), 321.

Eyckmans, M., Celis, N., Horemans, N., Blust, R., & De, B. G., (2011). Exposure to waterborne copper reveals differences in oxidative stress response in three freshwater fish species. *Aquat. Toxicol., 103*, 112–120.

Fahr, M., Laplaze, L., Bendaou, N., Hocher, V., El Mzibri, M., Bogusz, D., & Smouni, A., (2013). Effect of lead on root growth. *Front. Plant Sci.*, *4*, 175.

FAO/WHO Expert Committe on Food Additives. Lead [Internet], (2011). Available from: http://apps.who.int/food-additives-contaminants-jecfa-database/chemical.aspx?chemID=3511 (accessed on 26 October 2020).

Faucher, K., Fichet, D., Miramand, P., & Lagardère, J. P., (2006). Impact of acute cadmium exposure on the trunk lateral line neuromasts and consequences on the "C-start" response behavior of the sea bass (*Dicentrarchus labrax* L.; Teleostei, Moronidae). *Aquat. Toxicol.*, *76*, 278–294.

Faucher, K., Fichet, D., Miramand, P., & Lagardère, J. P., (2008). Impact of chronic cadmium exposure at environmental dose on escape behavior in sea bass (*Dicentrarchus labrax* L.; Teleostei, Moronidae). *Environ. Pollut.*, *151*, 148–157.

Fazio, F., Piccione, G., Tribulato, K., Ferrantelli, V., Giangrosso, G., Arfuso, F., & Faggio, C., (2014). Bioaccumulation of heavy metals in blood and tissue of striped mullet in two Italian Lakes. *J. Aquat. Anim. Health.*, *26*(4), 278–284.

Figueiredo-Fernandes, A., Ferreira-Cardoso, J. V., Garcia-Santos, S., Monteiro, S. M., Carrola, J., & Matos, P., (2007). Histopathological changes in liver a gill epithelium of Nile tilapia, *Oreochromis niloticus*, exposed to waterborne copper. *Pesq. Vet. Bras.*, *27*, 103–109.

Flora, S. J. S., Mittal, M., & Mehta, A., (2008). Heavy metal induced oxidative stress and it's possible reversal by chelation therapy. *Indian J. Med. Res.*, *128*, 501–523.

Forstner, U., & Wittmann, G. T. W., (1983). *Metal Pollution in the Aquatic Environment* (pp. 30–61). Springer-Verlag, Berlin.

Giari, L., Manera, M., Simoni, E., & Dezfuli, B. S., (2007). Cellular alterations in different organs of European sea bass *Dicentrarchus labrax* (L.) exposed to cadmium. *Chemosphere*, *67*, 1171–1181.

Gill, T. S., & Epple, A., (1993). Stress-rela ted changes in the hematological profile of the American eel (*Anguilla rostrata*). *Ecotoxicol. Environ. Saf.*, *25*, 227–235.

Giusi, G., Alò, R., Crudo, M., Facciolo, R. M., & Canonaco, M., (2008). Specific cerebral heat shock proteins and histamine receptor cross-talking mechanisms promote distinct lead-dependent neurotoxic responses in teleosts. *Toxicol. Appl. Pharmacol.*, *227*, 248–256.

Gonzalez-Munoz, M. J., Rodríguez, M. A., Luque, S., & Álvarez, J. R., (2006). Recovery of heavy metals from metal industry waste waters by chemical precipitation and nanofiltration. *Desalination*, *200*(1–3), 742–744.

Gorur, F. K., Keser, R., Akcay, N., & Dizman, S., (2012). Radioactivity and heavy metal concentrations of some commercial fish species consumed in the black sea region of Turkey. *Chemosphere*, *87*, 356e61.

Goss, G. G., Perry, S. F., Wood, E. M., & Laurent, P., (1992). Mechanisms of ion and acid-base regulation at the gills of freshwater fish. *J. Exp. Zool.*, *263*, 143–159.

Govind, P., & Madhuri, S., (2014). Heavy metals causing toxicity in animals and fishes. *Res. J. Animal Vet. Fishery Sci.*, *2*(2), 17–23.

Gray, J. S., (2002). Biomagnification in marine systems: The perspective of an ecologist. *Mar. Pollut. Bull.*, *45*, 46–52.

Green, A. J., & Planchart, A., (2018). The neurological toxicity of heavy metals: A fish perspective. *Comparative Biochemistry and Physiology Part C: Toxicol. Pharmacol.*, *208*, 12–19.

Gremyachikh, V. A., Grebenyuk, L. P., Komov, V. T., & Stepanova, I. K., (2006). Accumulation of mercury and its teratogenic effect upon larvae of *Chironomus riparius* meigen (Diptera: Chironomidae). *Biologia Vnutrennih Vod.*, *1*, 99–107.

Handy, R. D., (1996). Dietary exposure to toxic metals in fish. In: *Seminar Series-Society for Experimental Biology* (Vol. 57, pp. 29–60). Cambridge University Press.

Hempel, M., Chau, Y. K., Dutka, B. J., McInnis, R., Kwan, K. K., & Liu, D., (1995). Toxicity of organomercury compounds: Bioassay results as a basis for risk assessment. *Analyst, 120,* 721–724.

Hogstrand, C., (2011). *Zinc.* Academic Press, New York, USA.

Hogstrand, C., Wilson, R. W., Polgar, D., & Wood, C. M., (1994). Effects of zinc on the branchial calcium uptake in freshwater rainbow trout during adaptation to waterborne zinc. *J. Exp. Biol., 186,* 55–73.

Holcombe, G. W., Benoit, D. A., Leonard, E. N., & McKim, J. M., (1976). Long-term effects of lead exposure on three generations of brook trout (*Salvelinus fontinalis*). *J. Fish. Res. Board Can., 33,* 1731–1741.

Jaishankar, M., Tseten, T., Anbalagan, N., Mathew, B. B., & Beeregowda, K. N., (2014). Toxicity, mechanism, and health effects of some heavy metals. *Interdisciplinary Toxicol., 7*(2), 60–72.

Jennings, J. R., & Rainbow, P. S., (1979). Studies on the uptake of cadmium by the crab *Carcinus maenas* in the laboratory. I. Accumulation from seawater and a food source. *Mar. Biol., 50*(2), 131–139.

Jezierska, B., & Witeska, M., (2006). *The Metal Uptake and Accumulation in Fish Living in Polluted Waters.* NATO Science Series, Netherlands: Springer.

Jia, X., Zhang, H., & Liu, X., (2011). Low levels of cadmium exposure induce DNA damage and oxidative stress in the liver of Oujiang colored common carp *Cyprinus carpio* var. color. *Fish Physiol. Biochem., 37,* 97–103.

Kadar, I., Koncz, J., & Fekete, S., (2001). Experimental study of Cd, Hg, Mo, Pb and Se movement in soil-plant-animal systems. *Krmiva, 43*(3), 185–190.

Kamaruzzaman, B. Y., Rina, Z., John, B. A., & Jalal, K. C., (2011). A. Heavy metal accumulation in commercially important fishes of southwest Malaysian coast. *Int. Res. J. Environ., 5*(6), 595–602.

Karadede, H. L., Oymak, S. A., & ÃoenlÃ, E., (2004). Heavy metals in mullet, Liza Abu and catfish, *Silurus triostegus*, from the Atat Ãrk Dam Lake (Euphrates), Turkey. *Environ. Inter., 30,* 183–188.

Kinne, O., (1984). *Marine Ecology, Ocean Management* (Vol. 5, pp. 618–627). John Wiley.

Kochian, L. V., Pineros, M. A., & Hoekenga, O. A., (2005). The physiology, genetics, and molecular biology of plant aluminum resistance and toxicity. *Plant and Soil, 274,* 175–195.

Kori, S., & Ubogu, O. E., (2008). Sub-lethal hematological effects of zinc on the freshwater fish, hetero *Clarias* sp. (Osteichthyes: Clariidae), *Afr. J. Biotech., 7*(12), 2068–2073.

Li, D., Katakura, M., & Sugawara, N., (1995). Improvement of acute cadmium toxicity by pretreatment with copper salt. *Bull. Environ. Contam. Toxicol., 54,* 878–883.

Magyarosy, A., Laidlaw, R. D., Kilaas, R., Echer, C., & Clark, D. S., (2002). Nickel accumulation and nickel oxalate precipitation by *Aspergillus niger. Appl. Microbiol. Biotechnol., 59,* 382–388.

Mahurpawar, M., (2015). Effects of heavy metals on human health. *Int. J. Res. Granthaalayah, 1*(7).

Mathew, B. B., Tiwari, A., & Jatawa, S. K., (2011). Free radicals and antioxidants: A review. *J. Pharm. Res., 4*(12), 4340–4343.

Maximino, C., (2011). Possible role of serotoninergic system in the neurobehavioral impairment induced by acute methylmercury exposure in zebrafish (Danio rerio). *Neurotoxicol. Teratol., 33,* 727–734.

Mehouel, F., Bouayad, L., Hammoudi, A. H., Ayadi, O., & Regad, F., (2019). Evaluation of the heavy metals (mercury, lead, and cadmium) contamination of sardine (*Sardina pilchardus*) and swordfish (*Xiphias gladius*) fished in three Algerian coasts. *Vet. World*, *12*(1), 7.

Michael, P., (1986). *Ecological Methods for Field and Laboratory Investigations*. TATA McGraw-Hill Publishing Company Ltd., New Delhi.

Mishra, A. K., & Mohanty, B., (2009). Chronic exposure to sublethal hexavalent chromium affects organ histopathology and serum cortisol profile of a teleost, *Channa punctatus* (Bloch). *Sci. Total Environ.*, *407*, 5031–5038.

Moiseenko, T. I., (2003). *Mehanismy, Ecologicheskiye Posledstviya (Water Acidification, Factors, Mechanisms, Ecological Consequences)* (p. 276). Moscow: Nauka. Zakislenye Wod, Factory.

Monteiro, S. M., Dos, S. N. M. S., Calejo, M., Fontainhas-Fernandes, A., & Sousa, M., (2009b). Copper toxicity in gills of the teleost fish, *Oreochromis niloticus*: Effects in apoptosis induction and cell proliferation. *Aquat. Toxicol.*, *94*, 219–228.

Morel, F. M. M., Kraepiel, A. M. L., & Amyot, M., (1998). The chemical cycle and bioaccumulation of mercury. *Annu. Rev. Ecol. Syst.*, *29*, 543–566.

Munoz-Olivas, R., & Camara, C., (2001). Speciation related to human health. In: Ebdon, L., Pitts, L., Cornelis, R., Crews, H., Donard, O. F. X., & Quevauviller, P., (eds.), *Trace Element Speciation for Environment, Food and Health* (pp. 331–345). By Royal Society of Chemistry, Cambridge.

Nath, K., & Kumar, N., (1987). Effect of hexavalent chromium on the carbohydrate metabolism of a freshwater tropical teleost *Colisa fasciatus*. *Bull. Inst. Zool. Acad. Sin. (Taipei)*, *26*, 245–248.

Nemcsók, J., Benedeczky, I., Boross, L., Asztalos, B., & Orban, L., (1981). Subcellular localization of transaminase enzymes in fishes and their significance in the detection of water pollution. *Acta. Biol. Szeged.*, *27*, 9–15.

Nielson, F. H., (1987). *Nickel Toxicity*. John Wiley and Sons Inc, New York.

Niyogi, S., & Wood, C. M., (2006). Interaction between dietary calcium supplementation and chronic waterborne zinc exposure in juvenile rainbow trout, *Oncorhynchus mykiss*. *Comp. Biochem. Physiol.*, *143*, 94–102.

Nøstbakken, O. J., Hove, H. T., Duinker, A., Lundebye, A. K., & Berntssen, M. H. G., (2015). Contaminant levels in Norwegian farmed Atlantic salmon (*Salmo salar*) in the 13-year period from 1999 to 2011. *Environ. Int.*, *74*, 274–280.

Nriagu, J. O., & Pacyna, J. M., (1988). Quantitative assessment of worldwide contamination of air, water, and soils by trace metals. *Nature, 333*, 134–139.

Olaifa, F., Olaifa, A., & Onwude, T., (1998). Lethal and sub-lethal affects of copper to the African catfish (*Clarias gariepinus*) Juveniles. *Afri. J. Biomed. Res.*, *7*, 45–53.

Olsson, P. E., Kling, P., & Hogstrand, C., (1998). Mechanisms of heavy metal accumulation and toxicity in fish. In: *Metal Metabolism in Aquatic Environments* (pp. 321–350). Springer, Boston, MA.

Pack, E. C., Kim, C. H., Lee, S. H., Lim, C. H., & Sung, D. G., (2014). Effects of environmental temperature change on mercury absorption in aquatic organisms with respect to climate warming. *J. Toxicol. Environ. Health A.*, *77*, 1477–1490.

Palaniappan, P. L. R. M., & Karthikeyan, S., (2009). Bioaccumulation and depuration of chromium in the selected organs and whole-body tissues of freshwater fish *Cirrhinus mrigala* individually and in binary solutions with nickel. *J. Environ. Sci.*, *21*, 229–236.

Palmer, C. D., & Puls, R. W., (1994). *Natural Attenuation of Hexavalent Chromium in Ground Water and Soil.* USEPA Ground Water Issue EPA/540/5-94/505, U.S. Government Printing Office.

Pandey, G., & Madhuri, S., (2014). Heavy metals causing toxicity in animals and fishes. *Res. J. Animal, Vet. Fishery Sci., 2*(2), 17–23.

Patrick, L., (2002). Mercury toxicity and antioxidants: Part 1: Role of glutathione and alpha-lipoic acid in the treatment of mercury toxicity. *Altern Med Rev., 7*(6), 456–471.

Perry, S. F., Goss, G. G., & Fenwick, L. C., (1993). Interrelationships between gill chloride cell morphology and calcium uptake in freshwater teleosts. *Fish Physiol. Biochem., 10*, 327–337.

Pieterse, G. M., (2004). *Histopathological Changes in the Testis of Oreochromis mossambicus (Cichlidae) as a Biomarker of Heavy Metal Pollution.* PhD Thesis, Rand Afrikaans University, South Africa.

Qiu, Y. W., (2015). *Bioaccumulation of Heavy Metals Both in Wild and Mariculture Food Chains in Daya Bay, South China* (Vol. 163, pp. 7–14.). Estuar Coast Shelf. Sci. Elsevier Ltd.

Radi, A. A. R., & Matkovics, B., (1988). Effects of metal ions on the antioxidant enzyme activities, protein contents, and lipid peroxidation of carp tissues. *Comp. Biochem. Physiol. C., 90*, 69–72.

Raldúa, D., Diez, S., Bayona, J. M., & Barceló, D., (2007). Mercury levels and liver pathology in feral fish living in the vicinity of mercury cell chlor-alkali factory. *Chemosphere, 66*, 1217–1225.

Rand, G. M., Wells, P. G., & McCarthy, L. S., (1995). Introduction to aquatic ecology. In: Rand, G. M., (ed.), *Fundamentals of Aquatic Toxicology* (pp. 3–53). Taylor and Francis, London.

Rasinger, J., Lundebye, A. K., Penglase, S., Ellingsen, S., & Amlund, H., (2017). Methyl mercury induced neurotoxicity and the influence of selenium in the brains of adult zebrafish (*Danio rerio*). *Int. J. Mol. Sci., 18*, 725.

Rauf, A., Javed, M., & Ubaidullah, M., (2009). Heavy metal levels in three major carps (catlacatla, labeorohita and cirrhinamrigala) from the river Ravi, Pakistan. *Pak. Vet. J., 29*(1), 24–26.

Ricciardi, A., & Rasmussen, J. B., (1999). Extinction rates of North American freshwater fauna. *Conserv. Boil., 13*(5), 1220–1222.

Said, S., Aribarg, A., Virutamsen, P., Chutivongse, S., Koetsawang, S., Meherjee, P., Kumar, T., Cuadros, A., Shearman, R., & Conway, A., (1992). The influence of varicocele on parameters 0f fertility in a large group of men presenting to infertility clinics. *Fert. Steril., 57*, 1289–1293.

Sanchez-Galan, S., Linde, A. R., Ayllon, F., & Garcia-Vazquez, E., (2001). Induction of micronuclei in eel (*Anguilla anguilla*) by heavy metals. *Ecotox Environ Saf.*

Sánchez-Marín, P., Lorenzo, J. I., Blust, R., & Beiras, R., (2007). Humic acids increase dissolved lead bioavailability for marine invertebrates. *Environ Sci. Technol., 41*, 5679–5684.

Saxena, D., & Tripathi, M., (2007). Hexavalent chromium induces biochemical alterations in air-breathing fish, *Channa punctatus. J. Ecophysiol. Occup. Health., 7*, 171–175.

Sfakianakis, D. G., Renieri, E., Kentouri, M., & Tsatsakis, A. M., (2015). Effect of heavy metals on fish larvae deformities: A review. *Enviro. Res., 137*, 246–255.

Sharma, P., & Dubey, R. S., (2005). Lead toxicity in plants. *Braz. J. Plant Physiol., 17*(1), 35–52.

Silva, V. S., Nunes, M. A., Cordeiro, J. M., Calejo, A. I., & Santos, S., (2007). Comparative effects of aluminum and ouabain on synaptosomal choline uptake, acetylcholine release and (Na$^+$/K$^+$) ATPase. *Toxicol., 236*, 158–177.

Simpson, W. R., (1981). A critical review of cadmium in the marine environment. *Prog. Oceanogr., 10*, 1–70.

Sönmez, A. Y., Hisar, O., & Yanık, T., (2012). Karasu Irmağında Ağır Metal Kirliliğinin Tespitive Su Kalitesine Göre Sınıflandırılması. Atatürk Üniv. *Ziraat Fak. Derg., 43*(1), 69–77.

Sorensen, E. M. B., (1991). *Metal Poisoning in Fish: Environmental and Life Sciences Associates.* Boca Raton: CRC Press Inc.

Stohs, S. J., & Bagchi, D., (1995). Oxidative mechanisms in the toxicity of metal ions. *Free Radic. Biol. Med., 18*(2), 321–336.

Svobodová, Z., (1993). *Water Quality and Fish Health* (p. 67). FAO, Rome, EIFAC technical paper No., 54.

Thomas, P., Bally, M., & Neff, J. M., (1982). Ascorbic acid status of mullet, *Mugil cephalus* Linn., exposed to cadmium. *J. Fish Biol., 20*, 183–196.

Thophon, S., Pokethitiyook, P., Chalermwat, K., Upatham, E. S., & Sahaphong, S., (2004). Ultrastructural alterations in the liver and kidney of white sea bass, *Lates calcarifer*, in acute and sub chronic cadmium exposure. *Environ Toxicol., 19*, 11–19.

Toth, J. F., & Brown, R. B., (1997). Racial and gender meanings of why people participate in recreational fishing. *Leisure Sci., 19*, 129–146.

Tressou, J., Crépet, A., Bertail, P., Feinberg, M. H., & Leblanc, J. C., (2004). Probabilistic exposure assessment to food chemicals based on extreme value theory. Application to heavy metals from fish and sea products. *Food Chem. Toxicol., 42*, 1349–1358.

Verbost, P. M., Flik, G., Lock, R. A. C., & Wendelaar, B. S. E., (1987). Cadmium inhibition of Ca^{2+} uptake in rainbow trout gills. *Am. J. Physiol. Regul. Integ. Comp. Physiol., 253*, 216–221.

Verbost, P. M., Van, R. J., & Flik, G., (1989). The movement of cadmium through freshwater branchial epithelium and its interference with calcium transport. *J. Exp. Bioi., 145*, 185–197.

Vetillard, A., & Bailhache, T., (2005). Cadmium: An endocrine disrupter that affects gene expression in the liver and brain of juvenile rainbow trout. *Biol. Repr., 72*, 119–126.

Walton, R. C., McCrohan, C. R., Livens, F. R., & White, K. N., (2009). Tissue accumulation of aluminum is not a predictor of toxicity in the freshwater snail, *Lymnaea stagnalis. Environ Pollut., 157*, 2142–2146.

WHO (World Health Organization), (1990). *Chromium, Nickel, and Welding.* International Agency for Research on Cancer. IARC Monographs on the Evaluation of Carcinogenic Risks to Humans, France.

Wikipedia, (2013). *Metal Toxicity.* Available at: www.google.com (accessed on 26 October 2020).

Wirbisky, S. E., Weber, G. J., Lee, J. W., Cannon, J. R., & Freeman, J. L., (2014). Novel dose-dependent alterations in excitatory GABA during embryonic development associated with lead (Pb) neurotoxicity. *Toxicol Lett., 229*(1), 1–8.

Yang, R., Yao, T., Xu, B., Jiang, G., & Xin, X., (2007). Accumulation features of organochlorine pesticides and heavy metals in fish from high mountain lakes and Lhasa River in the Tibetan plateau. *Environ Int., 33*, 151–156.

Yi, Y. J., & Zhang, S. H., (2012). Heavy metal (Cd, Cr, Cu, Hg, Pb, Zn) concentrations in seven fish species in relation to fish size and location along the Yangtze River. *Environ. Sci. Pollut. Res., 19*(9), 3989–3996.

Yildirim, Y., Gonulalan, Z., Narin, I., & Soylak, M., (2009). Evaluation of trace heavy metal levels of some fish species sold at retail in Kayseri, Turkey. *Environ. Monitor. Asses., 149*, 223–228.

Yousafzai, A. M., Siraj, M., Ahmad, H., & Chivers, D. P., (2012). Bioaccumulation of heavy metals in common carp: Implications for human health. *Pak. J. Zool., 44*(2), 489–494.

Zeng, G. M., Tang, W. W., Gong, J. L., Liang, J., Xu, P., Zhang, C., & Huang, B. B., (2014). Impact of humic/fulvic acid on the removal of heavy metals from aqueous solutions using nanomaterials: A review. *Sci. Total Environ., 468*, 1014–1027.

Zhang, H., Liu, Y., Liu, R., Liu, C., & Chen, Y., (2014). Molecular mechanism of lead-induced superoxide dismutase inactivation in zebrafish livers. *J. Phys. Chem. B., 118*, 14820–14826.

Žikić, R. V., Štajn, A. Š., Pavlović, S. Z., Ognjanović, B. I., & Saićić, Z. S., (2001). Activities of superoxide dismutase and catalase in erythrocytes and plasma transaminases of goldfish (*Carassius auratus gibelio* Bloch.) exposed to cadmium. *Physiol. Res., 50*, 105–111.

Zuluaga, R. J., Gallego, R. S. E., & Ramírez, B. C. M., (2015). Content of Hg, Cd, Pb and as in fish species: A review. *Vitae, 22*(2), 148–149.

CHAPTER 3

Chlorpyrifos Toxicity in Fishes: A Perspective

RUKHSANA AKHTAR,[1] ADIL FAROOQ WALI,[2] SHEEBA SHAKEEL,[3] SAIEMA RASOOL,[4] MUNEEB U. REHMAN,[5,6] YUSRA AL DHAHER,[7] HILAL AHMAD WANI,[8] SABHIYA MAJID,[5] TEHSEEN KHAN,[5] HENNA AMIN,[3] and WAJHUL QAMAR[8]

[1]Department of Clinical Biochemistry, Govt. Degree College (Baramulla), Khawaja Bagh, Baramulla, Jammu and Kashmir, India

[2]RAK College of Pharmaceutical Sciences, RAK Medical and Health Sciences University, Ras Al Khaimah–11172, United Arab Emirates

[3]Department of Pharmaceutical Sciences, Faculty of Applied Sciences, University of Kashmir, Srinagar–190006, Jammu and Kashmir, India

[4]Department of School Education, Govt. of Jammu & Kashmir, Srinagar, 190001 India

[5]Department of Biochemistry, Government Medical College (GMC), Karan Nagar, Srinagar–190010, Jammu and Kashmir, India

[6]Department of Clinical Pharmacy, College of Pharmacy, King Saud University, Riyadh–11451, Saudi Arabia, E-mail: muneebjh@gmail.com

[7]Department of Biology, College of Science, UAE University, P.O. Box–17551, United Arab Emirates University, Al-Ain–15551, United Arab Emirate

[8]Department of Biochemistry, Govt. Degree College (Sumbal), Sumbal, Bandipora, Jammu and Kashmir, India

[9]Department of Pharmacology and Toxicology, Central Lab, College of Pharmacy, King Saud University, Riyadh–11451, Saudi Arabia

ABSTRACT

The annual increase in the human population in the world increases by 1.1%. Currently, every one person out of seven remains hungry. To reduce this hunger and meet the demand for increased food production, pesticides are being used by agriculturists who are losing 20–40% of food production to pests, weeds, and diseases every year. These pesticides enable farmers to increase the quality and quantity of food. One of such pesticides regularly used throughout the globe is chlorpyrifos (CPF). CPF belongs to the class of organophosphates, chemically known as O, O-diethyl-O-3,5,6-trichlor-2-pyridyl is a large spectrum pesticide used commercially in the control of subterranean termites and foliar insects affecting agricultural crops. Following its application to crops, it finds its way to water surfaces where it adversely affects the aquatic biota. It enters into the body of organisms and causes inhibition of acetylcholinesterase (AChE), oxidative stress, and hormonal imbalance and thus affects growth, reproduction, and neural behavior of aquatic organisms, when present in a sub-lethal concentration in water. In the present chapter, environmental distribution, degradation and transformation, pharmacokinetics, and toxicity of CPF is discussed. Moreover, its effect on the behavior of different biological systems is also assessed.

3.1   INTRODUCTION

One of the main problems of this century is water pollution. The natural quali-ties of water are altered in a drastic manner due to the addition of pollutants (Voltz et al., 2005). In aquatic ecosystems, an increasing trend of chemical pollution used in industry and agriculture is observed. Pesticides released into the environment tend to contaminate the environment through various processes. Organochlorines, organophosphates, and carbamates are three of the important categories of pesticides that cause serious problems (Dyk and Pletschke, 2011). For community wellbeing, agricultural pesticides are of pronounced significance, but because of their widespread usage, they have an undesirable influence on the ecosystem (Ogueji et al., 2007). The marine biota is challenged with the danger of biodiversity loss due to the discharge of agri-cultural pesticides into water resources (Rahman et al., 2002). Seventy percent of agricultural chemicals enter the water bodies and affect the aquatic organ-isms (Bhatnagar et al., 1992). In fish, physiological and behavioral changes may occur on frequent exposures to high concentrations of insecticides (Jindal and Kaur, 2014). Pollutants get piled up in the food chain and cause the loss of the aquatic organisms. One of the flagship species of the aquatic system is fish, so fish is generally used to assess the status of aquatic ecosystems (Farkas et al., 2002). The presence of agricultural chemicals in the water in sub-lethal concentration reduces the metabolism, growth, and calorific value of fish

(Moore and Waring, 2002). Peroxides have a very important role to play in pesticide toxicity. They have a dominant part in reactive oxygen species (ROS) generation therefore leading to oxidative stress (Elia et al., 2002; Banerjee et al., 2001; Sayeed et al., 2003; Das and Mukherjee, 2000).

One of the most regularly used insecticides throughout the globe is chlorpyrifos (CPF). Belonging to the class of organophosphates, CPF is a widely used insecticide that controls subterranean termites (Venkateswara et al., 2005), insects that adversely affect the crop yield (Rusyniak and Nanagas, 2004), and household pests all over the world (Yen et al., 2011). It is usually branded as Dursban in non-agriculture, while as in and in agricultural commodities, it is identified as Lorsban. In India, it is the second major retailed OP agrochemical. It is more lethal than organochlorine compounds and is typically poisonous to fish (Tilak et al., 2001; Firdous, 2012). The chemical formula of the compound is $C_9H_{11}CL_3NO_3PS$ (shown in Figure 3.1) CPF is a colorless to white crystalline solid, having a mild mercaptan (thiol) odor, that resembles the odor of sulfur compounds found in rotten eggs, onions, garlic, and skunks (Pragati et al., 2018). It has a density of 1.398 g/c m³, a vapor pressure of 1.87×10^{-5} mmHg at 25°C and an octanol-water partition coefficient (K_{ow}) of 4.70. The chemical shows a molecular weight of 350.6 g/mol, and Solubility in water is 1.4 mg/L at 25°C (Axel et al., 2018).

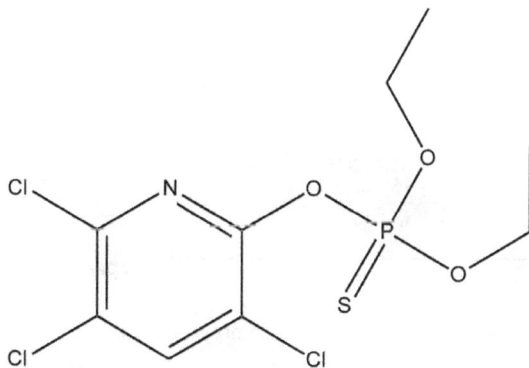

FIGURE 3.1 Chemical structure of chlorpyrifos.

3.2 DISTRIBUTION IN ENVIRONMENT

Following the application of CPF to crops, it rapidly combines to soil particles and enters the deep down in the soil. The CPF chemicals make their entry in water bodies through different processes like soil erosion, leaching, and runoff

(Moore et al., 2001; Fogg et al., 2004; Reichenberger et al., 2007). Therefore, the insecticides are under frequent monitor in groundwater and surface water (Kammerbauer et al., 1998; Zhang et al., 2012; Masia et al., 2015). Although CPF undergoes degradation quickly in the system, but its residual levels can remain for a considerable period of time. The exposure of CPF to humans occurs from agricultural applications through many pathways viz. dermal, oral, or inhalation (Dwayne et al., 2014). The larger concentrations of agricultural chemicals like CPF in the aquatic environment may disrupt normal behavior, reduce the food consumption, and affect the normal functioning of the organisms; however, it cannot cause the death of the organism (Susan et al., 2010). The behavior of fishes under conditions of stress provides a platform for generating adequate information for aquaculturists (Kristiansen et al., 2004). In order to evaluate the quality of water, the monitoring methods and quantifying behavioral response are a potential tool (Kane et al., 2004) Major pathways of entry of CPF into water is depicted in Figure 3.2.

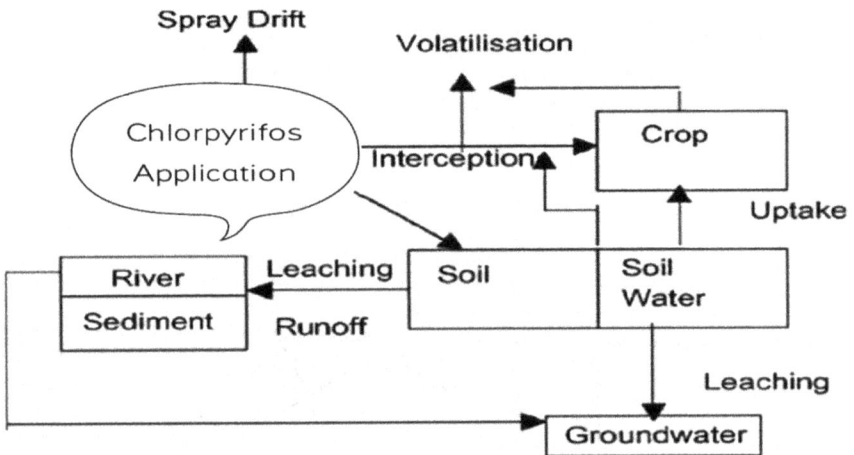

FIGURE 3.2   Different pathways of entry of chlorpyrifos into water.

3.3   STABILITY AND ENVIRONMENTAL FATE

CPF shows the vapor pressure of 1.9×10^{-5} mm Hg at 2°C, which is suggestive of its rapid volatilization in the air. In less alkaline soils, it has been seen that an average soil half-life of CPF varies from one to two months (Arcury et al., 2007). It gains access to surrounding waters through air drift or surface runoff and finds its place in one of the most detected chemicals in water bodies (Ensminger et al., 2011; Philips et al., 2007). In aquatic systems, degradation of CPF occurs from aqueous hydrolysis and aqueous photolysis, where its half-life varies from 29.6 d from aqueous photolysis to 72 d from abiotic degradation (Xingang et al., 2018). It has also been found that CPF gets piled up in a variety of aquatic animals, especially fish, and affects them adversely (Varo et al., 2002).

3.4   ENVIRONMENTAL DEGRADATION AND TRANSFORMATION OF CPF

CPF is exposed to many degradation pathways in nature. In the presence of OH radicals, CPF and its metabolites undergo photodegradation in the environment. Along with microbial degradation, the processes like hydrolytic cleavage, dechlorination, and oxidative degradation of CPF are also carried out through photodegradation. However, in indoor environments, the main CPF degrading factors like sunlight, water, or soil microorganisms are absent, and as a result, CPF can persist for several months in indoor environments (Rezk et al., 2018).

CPF undergoes degradation both by abiotic and biotic transformation processes in aquatic and terrestrial systems. The CPF gets piled up in soil microorganisms due to frequent application and thus affects some nonspecific ecologically important organisms like bees, wasps, beetles, and aquatic organisms as well (Singh et al., 2011).

3.4.1   MICROBIAL DEGRADATION

It has been reported that the parent molecule of CPF undergoes a microbial transformation into other by-products such as TCP (3,5,6-trichloro-2 pyridinol), DETP (diethyl phosphate), CPF oxon, trichloro-2-methoxy-pyridine. Different bacteria have been found to transform CPF into the major transformed metabolite of Trichloro pyridinol (Singh et al., 2011). In one of the studies of bioremediation of CPF through the association of a bacterium *Pennisetum pedicellatum* with rhizosphere bacterium *Stenotrophomonas maltophilia* MHF ENV20, it has been found that during initial days of incubation, different concentrations of CPF gets transformed into TCP and DETP. These two metabolites were later used as a mineral source of carbon and nitrogen (Dubey and Fulekar, 2012). In another study, CPF was subjected to biodegradation by a bacterial consortium consisting of *Pseudomonas stutzeri* (NII 1119), *Pseudomonas putida* (NII 117), *Klebsiella species* (NII 1118), and *Pseudomonas aeruginosa* (NII 1120). This bacterial suspension was capable of degrading the insecticide at a higher concentration of about 450–500 mg per liter into metabolites viz. CPF oxon and DETP CPF oxon and DETP. After 3 weeks, it was seen that under alkaline conditions, bacterial suspension transformed the compound into TCP (Sasikala et al., 2012).

In a similar experiment, The CPF at a lower concentration of 50 mg per liter was subjected to degradation in two different media by four different bacterial strains viz. *Pseudomonas aeruginosa, Klebsiella* species, *B. cereus,* and *S. marcescens.* In liquid medium, it was found that CPF was degraded by 84, 84, 81, and 80% in 20 days, and 37–92% degradation occurred in 30 days. TCP was formed as a major metabolite in the initial days of incubation (Chishti et al., 2013). In one of the studies, different gut strains of bacteria viz. *L. lactis, Lactobacillus fermentum, Lactobacillus plantarum, Enterococcus faecalis,* and *E. coli* were tested for their competence in CPF degradation. Out of these five strains, few bacterial strains like *Lactobacillus lactis, E. coli,* and *Lactobacillus fermentum* showed optimal efficacy in degrading CPF. The maximum (70%) of CPF degradation was achieved by the isolate *L. fermentum* into TCP (Figure 3.3). Only 16% of CPF were degraded into CPF oxon and DETP by *E. coli* (Drevenkar et al., 1993). In a mineral medium, the bacterial strain *C. fimi* degraded CPF concentration of 50 mg per liter into TCP.

3.4.2 DEGRADATION BY FUNGI

Biodegradation of CPF by fungi is rarely reported. In one of the studies, CPF of varying concentrations were subjected to transformation by *Verticillium* sp. CPF with three different concentrations from 1 to 100 mg/l in the mineral media were used. The incubation in mineral media for 7 days leads to the change in concentration of CPF from 0.139 to 11.014 mg (dl/l). In the initial days of incubation, the prominent degraded product found was TCP, which vanished within the time period of 9 days (Haohua et al., 2012). Different strains of fungi viz. *A. species, Penicillium species,* and *E. species* were checked for their efficiency to degrade CPF and its transformed metabolites. After 1 week of incubation, 60 to 90% of degradation of insecticide and its metabolites were observed. One of the strains, namely *Acremonium* sp. GFRC-1 utilized the CPF with high concentration as the nitrogen and carbon source and degraded the compound within 3 weeks with the formation of 3,5,6 TCP (Xu et al., 2007). Another species, namely Cladosporium Hu-01 was tested for CPF degradation. Analysis by gas chromatography-mass spectroscopy

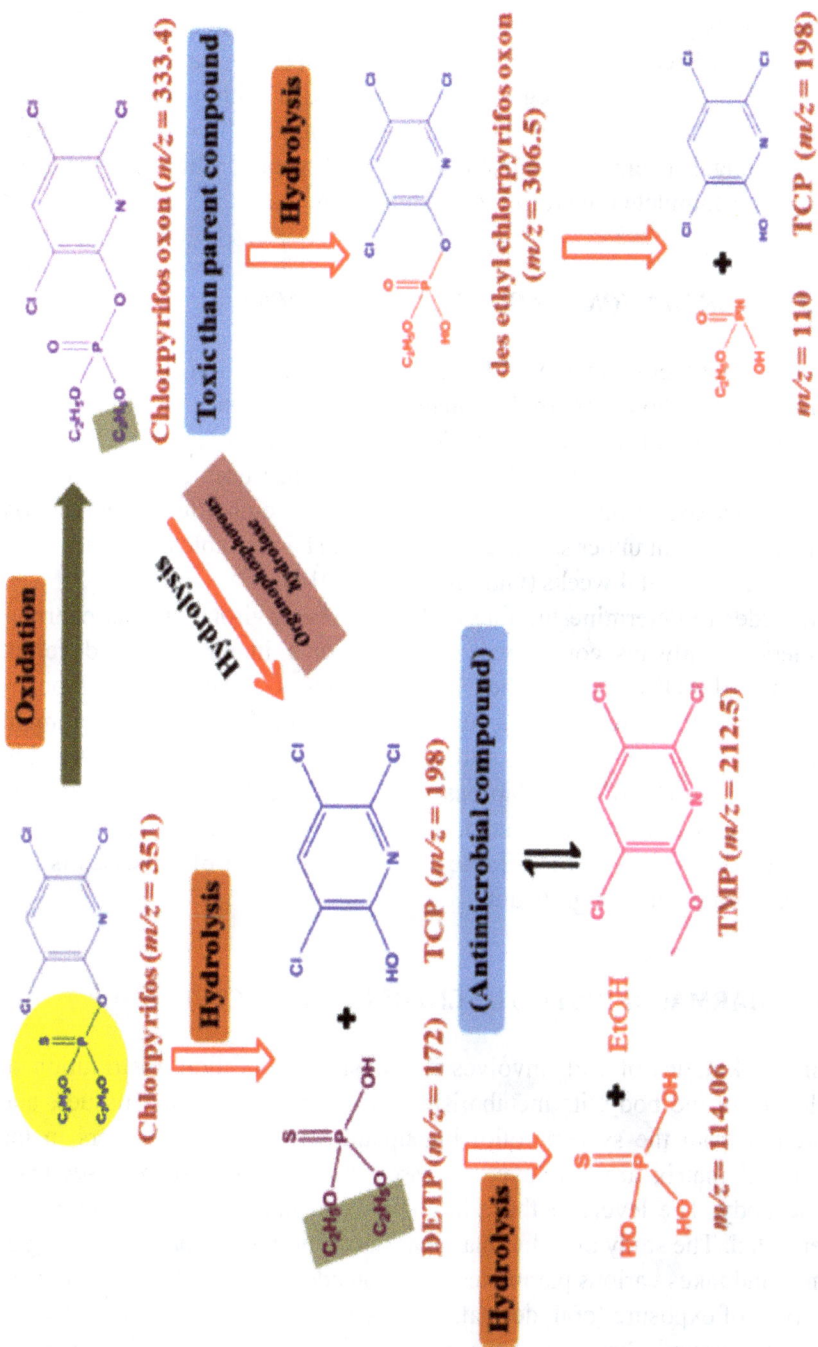

FIGURE 3.3 Biotransformation pathways involved in the degradation of chlorpyrifos (Gilani et al., 2016).

confirmed complete mineralization within 1 week of incubation. Another strain, namely *Verticillium* sp. DSP was capable of degrading the compound only in unsterilized soil. In a similar study, a fungal strain, namely *Aspergillus terreus* – JAS1 was able to degrade concentrated CPF within liquid media just within 24 hours (Shamsa et al., 2016). Degradation of CPF is more efficiently done by fungal species as compared to bacterial species, which are unable to achieve the complete mineralization of the compound.

3.4.3 DEGRADATION IN AQUATIC ENVIRONMENT

In the aqueous environment, CPF undergoes degradation through abiotic hydrolysis or photosensitized oxidation. In the surface waters, abiotic degradation of CPF by hydrolytic cleavage in the aquatic system has been reported to occur. The half-life of the molecule has been seen to decrease with the increase in the pH (Eaton et al., 2008). The photolytic degradation by natural sunlight under sterile conditions, at pH 7 was able to degrade CPF with the half-life of 4 weeks (Giddings et al., 2014).

In order to determine the fate of CPF in the environment, its pharmacokinetics is always considered. An organism's integration of different processes of ADME is described by some factors, which include bioconcentration factors (BCFs), bioaccumulation factors (BAFs), biota/sediment accumulation factor (BSAFs), and in case of movement in food web bio magnifications (BMFs). In lab-based studies, the fate of CPF in some aquatic organisms was evaluated. Different organisms showed different values of BAFs, as shown in Table 3.1, which in most of the cases is less and toxicologically insignificant.

3.5 PHARMACOKINETICS OF CHLORPYRIFOS (CPF)

Pharmacokinetics of CPF involves its physical absorption, distribution of molecule in the body, its metabolism in different organs and tissues, and excretion from the system following uptake. Different biomarkers in the biological matrix are employed in assessing the exposure of insecticide in the body. The levels of these biomarkers in humans are measured and interpreted. The study of individual pharmacokinetics of a particular drug or compound takes various parameters into consideration. Such parameters are the route of exposure (oral, dermal, or inhalation), the dose or concentration of the compound, the form of the compound, and the methodology used in

TABLE 3.1 Bioaccumulation Factors for Chlorpyrifos in Various Biota Forms

Taxon	Species	BCF/BAF/BSAF	BSAF (Apx)	Exposure Time	Exposure Concentration	References
Mollusks	*C. virginica*	BCF ± SD 565 ± 172		4 weeks	0.6 microgram per liter	Woodburn et al. (2003)
Oligochaete	*Lumbriculus variagatus*	BAF(Apx)	BSAF (Apx)	1 week and 5 days	385	Jantumen et al. (2008)
		2	67		54	
		14	99		25.5	
		17	35		30 microgram per kg	
		57	6		Dwt in sediment	
Fish	*Aphanius iberus*	BCF 3.1		3 days	3 microgram per liter	Varo I et al. (2000)
Fish	*Daniorerio eleutheroembryos*	BCF: 3,548 6,918		2 days	1 microgram per liter 10 microgram per liter	El-Amrani et al. (2012)
Fish	*Aphanius iberus*	BMF±SD: 0.30±0.2		4 weeks and 4 days	94 ± 41 nanogram per liter lipid weight	Varo I et al. (2000)

selection of species. The systematic procedures employed for sampling and processing of data are also taken in consideration.

3.5.1   ABSORPTION

Depending upon the route of entry or absorption, the chemical like CPF has been studied in three main categories: Oral absorption, dermal absorption and absorption via inhalation.

3.5.1.1   ORAL ABSORPTION

Most of the studies conducted on oral absorption of CPF involved high dosage treatment of CPF to rats. This is most commonly used approach to characterize the absorption of CPF where anticholinesterase effects are correlated with the high exposure dose. Following treatment, the blood or urinary excretion levels of DETP, DEP, TCP, TCPy, determines both the absorption as well as the vanishing of insecticide at the entry site. Based on oral single-dose gavage studies absorption of CPF in rats ranged from 84–90% (Nitika et al., 2017).

Further CPF uptake studies with relatively low dose (140 umol/kg body weight) had been carried out on rats orally treated with TCPy, DEP, or DETP. It has been identified that peak levels of all three metabolites in the blood occurred just after 1–3 hours of administration which shows that the compounds are absorbed well. In order to evaluate the uptake of CPF in gut, a single-pass intestinal perfusion method with a 100-fold concentration range was employed and it was revealed that the molecule was absorbed well throughout the whole extent of the small intestine (Timchalk et al., 2007).

A study conducted by Jameson et al. (2007) showed that higher doses of CPF inhibited the cholinesterase enzyme. The inhibition was seen just after 2 hours of dose and reached a peak of inhibition at 6 hours after dosing. This inhibition was found to be continued for a few days and then eventually started decreasing, and after almost one month, no major change was seen between negative control and experimental animals. This indicates that adaptive response by animals may be responsible for inducing carboxylesterase biosynthesis.

3.5.1.2 DERMAL ABSORPTION

In order to demonstrate the dermal uptake of the compound, studies on several animal species had been carried out. In one of the studies, a combination of this insecticide and cypermethrin injected into rats apparently caused pyknosis of brain neurons and inhibition of cholinesterase enzyme. This suggests that dermal absorption in animals is much higher (Ray et al., 2007). In human's dermal uptake of CPF has been reported much lower than experimental animals. To further demonstrate the dermal absorption in human's different concentrations of CPF (ranging from 5 to 15 mg) were given to human volunteers through a subcutaneous route at forearms, and it was found that almost 33% to 58% of dosage was well metabolized in the body. It was seen that in both cases that are volunteers with a low dosage of CPF (i.e., 5 mg) and subjects with high dosage (i.e., 15 mg) passed out a similar amount that is 5% of the given dose in urinary excretion. Such results signify that the absorption through dermal exposure is independent of the amount of dose. It was also found that Clearance remained incomplete even after five days of exposure. This suggests that some amount of CPF is subjected either to cutaneous retention or is held in the body (Meuling et al., 2005).

3.5.1.3 ABSORPTION VIA INHALATION

Lung has been considered as the main site of absorption of CPF following the inhalation exposure. Studies on experimental animals have revealed that low exposure levels of insecticide (through the nose only) show considerable systemic absorption but did not show significant affect erythrocyte or plasma cholinesterase activity. Although, the actual reason behind this is not known, but it is hypothesized that a low airborne concentration of pesticide is unable to inhibit blood AChE (Bukowska et al., 2018).

3.5.2 METABOLISM

The chief characteristics of CPF metabolism (Buratti et al., 2003; Tang et al., 2001) are discussed in subsections.

3.5.2.1   DESULFURATION

This process occurs in the liver wherein CPF is activated by hepatic enzyme cytochrome P450 and carries out removal of the sulfur group through oxidation of P=S moiety to P=O, leading to the formation of lethal metabolite, CPF-oxon. These activated sulfur atoms formed during the reaction are able to bind cytochrome P-450 molecule irreversibly. In this way, CPF acts as a "suicide substrate" by inhibiting an important enzyme and may prove fatal (Sultatos et al., 1994).

3.5.2.2   DEARYLATION AND HYDROLYSIS

Hydrolysis of the phosphate ester bonds of CPF and CPF-oxon to form TCPy, DETP, or DEP is carried out by various classes of enzymes, mainly oxonases (particularly some A-esterases such as paraoxonase). In a similar way, di-arylation of CPF to diethyl thiophosphate and 3,5,6-trichloropyridinol is catalyzed by hepatic cytochromes P-450 (Lucio et al., 2013).

3.5.2.3   CONJUGATION WITH GLUCURONIC ACID AND SULFATE

Hydrolysis of TCPy or thioTCPy results in the formation of a free OH group. The enzymes like glucuronyl transferase and sulfurtransferase mediate the conjugation of free OH with glucuronic acid leading to the formation of glucuronide and sulfate conjugates. These reactions produce polar products that are subjected to urinary and biliary excretion (Fujioka and Casida, 2007).

Overall it is summed up that metabolism of CPF is quick and widespread. The molecule is metabolically activated and detoxified predominantly in the liver. The parental molecule and transformed oxon either remain undetected or are detected in small concentrations in blood or urine. The TCP is the key circulatory form of insecticide, and it primarily undergoes urinary excretion (Nolan et al., 1984).

3.5.3   DISTRIBUTION

As mentioned above, some part of CPF is stored in adipose tissue. Therefore, in animals, CPF mainly resides in fatty tissues forming the complex with various types of lipids. The free form of this chemical in the blood is either low or absent because of its increased tendency of binding with plasma protein albumin (Karen et al., 2012). A detailed study of the distribution of CPF in various tissues, showed that higher concentrations of CPF are found in fat followed by brain and liver, and lowest concentrations were found in the kidney. One of the major metabolites of CPF: CPF-oxon, is found to be less membrane-permeable than the parental molecule. In humans, considerable studies on the tissue distribution of CPF and its transformed products are lacking. However, in one of the epidemiology studies on indoor exposure of CPF to pregnant mothers and newborn infants, it has been reported that CPF passes through the placenta quickly (Carol et al., 2013).

3.5.4 EXCRETION

3.5.4.1 EXCRETION OF CHLORPYRIFOS (CPF) THROUGH URINE

In the urine of experimental rats, significant levels of CPF products have been detected after 24 hours of oral exposure. It has been found that DEP and TCPy were detected even after 48 hours of exposure. Maximum excretion in urine was detected after 12 hours; however, measurable excretion continued even at 72 hours (Timchalk et al., 2007). In human studies of CPF excretion, the ingestion of a lethal dose of CPF or its derived conjugates in a subject resulted in urinary excretion of transformed products (Tessema et al., 2019).

3.5.4.2 FECAL AND BILIARY ELIMINATION OF CHLORPYRIFOS (CPF)

In rats, biliary excretion of glutathione (GSH) and its derived conjugates have been reported to a significant extent (Eaton et al., 2008). However, in humans, no such studies have been addressed so far.

3.5.4.3 EXCRETION OF CHLORPYRIFOS (CPF) IN MILK

Experimental studies done on animals have shown that insecticide and its metabolites appear in the milk. Following the oral administration of different concentrations of CPF in experimental rat, milk was checked, and it has been found that levels of CPF were 200 times more than levels in blood found during the perinatal period (Mattsson et al., 2000). In India, detectable concentrations of CPF in the milk of feeding mothers have been reported (Sanghi et al., 2003).

3.5.5   SUMMARY OF PHARMACOKINETICS OF CHLORPYRIFOS (CPF)

Experimental studies on animals and humans indicate that CPF is well metabolized from the intestine and lung. Dermal exposure contributes very little uptake (less than 5%). This indicates that the root of exposure has got a major role in absorption and, therefore void volume and target organ dose. The organ dose and effective dose are influenced by the pharmacokinetics of CPF through its distribution, metabolite production, and enzyme function (Barr et al., 2006). CPF is rapidly metabolized. In the liver, it undergoes desulfuration with the aid of cytochrome P450 and is activated to CPF-oxon, which gets quickly metabolized. This activated metabolite is almost completely excreted in the urine (Tang et al., 2001), but some studies have shown that little storage of it occurs in adipose tissue. Overall levels of parent molecule are very less in blood compared with transformed products (Smegal et al., 2000). The major identified metabolites of CPF shown by many studies are TCPy, DEP, DETP, CPF-oxon, GSH conjugates, and conjugates of TCPy. Besides these metabolites, a DAP metabolite and an organic metabolite are formed from both the oxon form as well as the unconverted form of CPF by hydrolysis, which may be either enzymatic or spontaneous (Barr et al., 2006). Urinary excretion of these metabolites from the system occurs, which are then subsequently measured. In a comparison of taking blood samples, detection through such measurement is more feasible for young children and newborns. In order to test the fetal exposure of CPF, a biomarker, namely meconium has been found to be most reliable and sensitive (Ortega Garcia et al., 2006). The estimation of activity of certain plasma enzymes like butyrylcholinesterase and erythrocyte acetylcholinesterase is done to determine CPF exposure in the blood (Albers et al., 2007). Moreover, it has been seen that the activities of individual enzymes differ significantly between species. Differences can also be seen with age and gender (i.e., between fetus, newborn, child, and adult). Therefore, it may be concluded that inter-individual and interspecies differences do occur in the overall metabolism and toxicokinetics of CPF.

3.6   TOXICITY ASSESSMENT OF CHLORPYRIFOS (CPF)

When CPF enters in body of mammals, there is the generation of three modes of action viz. Blockage of acetylcholinesterase, induction of oxidative stress, and hormonal imbalance. Acetylcholine (ACh) is a well-known neurotransmitter that is responsible for the stimulation of receptors and is important in nerve impulse transmission. Its action is terminated with its degradation by enzyme acetylcholinesterase. During CPF poisoning, CPF binds on the active site of acetylcholinesterase and inhibits it competitively. This competitive inhibition of an important enzyme-acetylcholinesterase by chemical CPF prevents normal neurotransmitter inactivation and retention of ACh molecules in the nerve synapse for a longer time. Increased levels of this neurotransmitter in synaptic vesicles result in stimulation of receptor continuously, which leads to the deregulation of downstream signaling pathways that causes the functional alterations at the tissue or organ level (Pope et al., 2005). Below an established level for extreme cholinesterase inhibition and acute toxicity studies have found that CPF is not able to affect any biological systems (Gibson et al., 1998). *In vitro* studies, as well as *in vivo* studies, have demonstrated that low levels of CPF exposure exert disruptive effects on, DNA replication, gene transcription, cell division, and synaptogenesis leading to defective neural cell development (Crumpton et al., 2000). Some studies have shown that besides ACh esterase CPF is able to inhibit another enzyme, namely NTE (neuropathy target esterase). In such conditions, degeneration of myelin sheath and loss of axon fibers of the PNS and CNS may occur (Nand et al., 2007). In some cases, there occurs permanent inhibition of some enzymes like acetylcholinesterase (AChE) and NTE; such a process is known as aging.

3.7 EFFECTS ON BIOLOGICAL SYSTEMS

3.7.1 NEURAL BEHAVIOR

Even at very low concentrations, CPF has a dominant toxic nature towards aquatic animals (Humphrey et al., 2004). The most vital mechanism behind the CPF toxicity is its capability to restrain acetylcholinesterase (AChE) (Xing et al., 2012). With the drop in AChE activity, there occurs an upsurge in the buildup of ACh in the synapses. Directly or indirectly, this lays a huge negative influence upon general behavior, feeding, and swimming (Glusczak et al., 2006). CPF is able to inhibit the AChE in brain and muscle salmonids *Oncorhynchus kisutch* (Jason et al., 2005). This progresses to the termination of neuronal transmission (paralysis) and overexcitation in *Cyprinus carpio* (Linnaeus). Even under retrieval phases, morphological distortions and impaired behavioral reactions have been reported. Similar kinds of results indicating a decline in AChE upon CPF treatment have been reported in common carp by Xing et al. (2010), *Oryzias latipes* by Khalil et al. (2013), larval zebrafish by Yen et al. (2011). Post CPF exposure, the brain of common carp has exhibited reduced mRNA transcription and AChE enzyme activity Xing et al. (2010).

Another very proficient parameter in ecotoxicology is behavior (Drummond et al., 1990; Scherrer, 1992; Cohn et al., 1996). The occurrence of pollutants in the environment is known to disturb locomotor behavior (Little and Finger, 1990), and fish swimming design is considered an extremely organized species-specific response. The CPF has the potential to instigate irreversible ChE inhibition (Fulton and Key, 2001), which can lead to paralysis and death because of triggering persistent stimulation of the muscles. In sea bass *Dicentrarchus labrax,* 40% to 50% AChE inhibition has been related to behavioral impairment, such as declined mobility, while when the AChE inhibition approaches 80% mortality of fish takes place (Almeida et al., 2010). The muscular AChE inhibition may advance to another type of paralysis, i.e., caudal bending which is headed by sudden contraction of voluntary muscles (Ware, 1989; Habig and DiGiulio, 1991). Because caudal portion is the slimmest structure, the twisting of caudal base due to the AChE activity inhibition can be reviewed as disorientation (Ramesh and David, 2009). In the course of determining lethal concentrations of CPF, when *Gambusia affinis* were exposed to higher concentrations of CPF atypical activities like erratic swimming with spasmodic actions, lack of symmetry and discharge of profuse mucous from entire body. Prior to collapse fish presented with temporary hyperactivity and were exhausted at the time of death. A substantial decrease in swimming speed has been witnessed with effect of toxicants (Venkateswara et al., 2005).

3.7.2 REPRODUCTION

The existence of fish population is promised by reproduction. Insecticide pollution is notorious in producing destructive outcomes on the development, endurance, and breeding of marine fauna. Pesticide exposure is linked to slight fluctuations in behavior and physiology spoiling survival as well as reproduction in doses not high enough to kill fish (Kegley et al., 1999). Longer exposures of pesticides can affect the reproductive abilities of fish (Rice, 1990) which includes deterioration of vitellogenesis performance (Haider and Upadhyaya, 1985), weak fecundity, histological infliction of testicles and ovaries (Duttaa and Meijer, 2003; Banaee et al., 2009), steroidogenesis disturbance (Khan and Law, 2005), lag in buildup of gonads (Skandhan et al., 2011), mutation of reproductive and parental action (Jaensson et al., 2007), worsening of reproductive migrations and olfactory feedback (Scholz et al., 2000), and interruption in synchronized courtship behavior of male and female fish during spawning time (Jaensson et al., 2007).

Studies have revealed reduced growth and reproduction, retarded matura-tion, altered hormone levels, malformed, and shallow populations in marine fauna when subjected to exposure to sub-lethal levels of CPF (Marshall and Roberts, 1978; Jarvinen et al., 1983; Odenkirchen et al., 1988). Another study of the same kind exhibited modifications in fertility, gonadosomatic index, fecundity, gonadal histopathology, and plasma vitellogenin (Daland et al., 2013). Surprisingly, chlorpyrifos at low concentrations under LC50 value deeply terminated male mating patterns. In guppy juvenile's *poecilia reticu-late,* it has been capable of exhibiting signs of paralysis even under minimal concentrations rendering fish immobile and therefore, directly responsible for restriction of mating behavior (Silva and Samayawardhena, 2005).

Vitellogenin (VTG) is a distinct female protein. It is an authentic biomarker of possible reproductive disorders (Folmar et al., 1996, 2001; Le et al., 1999). CPF being an E2 receptor agonist has the capability of sparking estrogenic activity (Bangeppagari and Gundala, 2015). With escalation in exposure time, an increase in vitellogenin levels in male fish has been reported citing estrogenecity of the toxicant. Vacuolization in both the gonads with atretic follicles in ovary and seminiferous tubule expansion in testes has been reported by Bangeppagari and Gundala (2015). In one of the study exposure of CPF for 48 hours showed significant shoot up in the mRNA expression of VTG has been reported (Kaimin et al., 2015). Male and female fish have revealed heavy fall in 17β-estradiol (*E2*) or 11-ketotestosterone (11-KT). These findings are the suggestive of fact that CPF impairs the reproductive performance as these two sex hormones have a very good role to play in gonad gametogenesis, sex differentiation, and determination in zebrafish (Bangeppagari and Gundala, 2015).

3.7.3   RESPIRATION

One of the critical biological indexes of quality of water is fish. The toxicants from the living world obtain their access to the fish predominantly through respiratory complex. The respiratory reaction is therefore the first and foremost to be influenced. This is manifested as a change in respiratory rate and can be detected through the principle of oxygen consumption rate. CPF modifies the rate of respiration in the experimental fish and affects the rate of oxygen consumption at both lethal and sublethal concentrations (Padmanabha et al., 2015). During CPF exposure, respiratory affliction has been reported in *Oreochromis mossambicus* (Ramesh and David, 2009). The possible reason behind this may be the potentiated physiologic response and compromised oxidative metabolism. The unexpected extension of behavioral and respiratory impairments even under retrieval periods can be due to the steady emancipation of CPF from repository tissues (Ramesh and David, 2009). Likewise, a proportional extension in ventilatory frequency to temperature has been exhibited by CPF treated rainbowfish *Melanotaenia duboulayi* and rainbow trout *Oncorhynchus mykiss* on visual monitoring (Ronald et al., 2009). Almost equivalent response has been noted in Zebra mussel *Dreissena polymorpha*. The rate of respiration amplified with increase of the pesticide (CPF) accumulation. This is because of the reason that stress induced by lethal exposure leads to escalation of oxygen requirement in body tissues and organs (Vesela et al., 2017).

Particularly the utmost critical entity for respiration and osmoregulation in fish are gills. Gills have a tremendous role to play in acid-base equilibrium, nitrogenous waste washing, osmoregulation, and most importantly respiration. Because of the reason that the gills are exclusively susceptible to water, the toxicants therefore have remarkable repercussions on respiration. Hence, gill morphology is of appreciable use in screening of ecosystem as an indicator. Nobonita and Suchismita (2013) have reported modification of gill architecture on pesticide exposure. In other study conducted by Sandipan et al. (2012), blood congestion, hypertrophy, and hyperplasia of gill epithelium, marginal channel expansion, epithelial dissipation, lamellar coalition, lamellar derangement, lamellar aneurysm, lamellar epithelium fissure, breach of pillar cells and necrosis have been reported.

In environmental surveillance, histopathological alterations have been adopted as essential biomarkers owing to the fact that they license investigation of distinct target organs. Gills of *chana punctatus* subjected to extreme concentration of CPF have exhibited lamellar epithelia lifting, acute vasodilatation of the lamellar vascular axis and edema. Serious pathological abrasions have been formed in gills under persistent exposure and low doses of CPF (Yogita and Abha, 2013).

3.7.4   IMMUNE SYSTEM

Fishes belong to a unique group of organisms that are having both innate and adaptive immunity (Litman et al., 2005). Of greatest significance is the innate immune system (Magnadottir, 2006; Quiniou et al., 2013; Gao et al., 2012) which comprises of eosinophils, macrophages, and neutrophils (Saurabh and Sahoo, 2008; Boshra et al., 2006; Alvarez et al., 2008) while as defense against repetitive afflictions is overseen by adaptive immunity. The latter reaction being brought about by B and T lymphocytes and antibodies. CPF is one of the most common pesticides that is found in the aquatic ecosystem (Chen et al., 2015). Kai et al. (2004) reported that sub-lethal concentrations of CPF in juvenile Chinook salmon causes change in mediators of immune system at the transcriptional level.

Cytokines like TNF-a, IL-1, IL-6, IL-8 are mediators of inflammatory response (Arican et al., 2005). An escalation in expression of mRNA of IL-1β, IL-1R, and IFN-γ during 24 h (1.16, 11.6, and 116 µg/L) treatment with CPF has been reported in spleen of *C. carpio* (Wang et al., 2011). Almost similar kind of results has been specified by Yuanxiang et al. (2015) in larva of zebrafish where in a remarkable shoot up have been observed in mRNA levels of master cytokines incorporating complement factor 4 (C4), interferon (Ifn), interleukin-1 beta (Il-1b), interleukin 6 (Il6), tumor necrosis factors α (Tnfα) post-exposure to CPF. These findings are suggestive of the fact that CPF has a good hand in deranging innate immunity in zebrafish larva. In contrast, Yuanxiang et al. (2015) reported a noticeable decline in IL-10 and TGF-β mRNA in head, kidney, and spleen of CPF treated carp.

At sub-lethal concentrations of CPF, immunological levels (viz. immuno-globulin M (IgM), lysozyme, nitric oxide, phagocytic activity, total globulin) and interleukin-1β (IL-1β) of *O. niloticus* have exhibited a significant quantum of variance (Zeinab et al., 2016). Significantly declined number of total pronephros has been found in *Oreochromis niloticus* on being subjected to CPF at the concentration levels of 1µg/dl as compared to normal controls. The phagocytic activity of macrophages isolated from pronephros of treated fish has also been found to be reduced relative to control fish (Holladay et al., 1996).

3.7.5 HEMATOLOGY

The health disruption instigated by different agents in the environment can be identified by hematological measures of fish as they can give adequate intimations of stress. *Cyprinus carpio* L. when subjected to higher concen-trations of CPF has revealed a remarkable up shoot in WBC and MCV. The upshot in WBC count maybe because of the immunological reactions to beat the stress instigated by CPF (Ramesh and Saravanan, 2008). In contrast, same species of exposed fish on CPF exposure revealed a significant decline in PCV, MCH, RBC, Hb, and a non-significant decrease in MCHC, thus signifying anemic changes of microcytic and hypochromic type (Noor et al.,

2016). Freshwater fish *Clarias batrachus* has presented with almost similar type of results with expansion of respiratory burst activity, total protein and HSI but hematocrit, plasma glucose, WBC count have exhibited a significant fall (Madhusudan et al., 2012). The reports are in agreement with published reports of Mevlut (2013); Enis Yonar et al. (2012). The possible reason behind the decline in the red blood cells, hematocrit, and hemoglobulin levels can be the osmoregulatory dysfunction, inhibition of RBC formation, hemosynthesis, due to speedy loss of erythrocytes in the hematopoietic organ (Vani et al., 2011).

In innate immunity, leucocytes execute a very serious role. The health of a fish can also be being detailed by leucocyte number/activity (Secombes, 1996). CPF exposed *Cyprinus carpio* reported higher WBC count Enis Yonar et al. (2012). This report is in consensus with previous findings by Ramesh and Saravanan (2008) and Okechukwu et al. (2007). The objective behind this modification could be the immune system stimulation instigated by the presence of contaminant. This may result in production of more effective immune defense response because the response produced due to contaminant is an adaptive response of the organism (Modesto and Martinez, 2010). Leucocyte sedimentation rate is additional and vital element in determining condition of health of an organism. For clinical investigation of fish biology, it is one of the adopted parameter.

On the other hand, other biochemical parameters viz. alkaline and acid phosphatase, lactate dehydrogenase, glycogen content has reportedly been inhibited while as alanine aminotransferase (ALT), aspartate, malate dehydrogenase has exhibited enhancement (Madhusudan et al., 2015; Tripathi and Shasmal, 2011; Narra et al., 2012). The drop in levels of hepatic glycogen is because of the stress caused by CPF and energy consumption.

3.7.6 ENDOCRINE SYSTEM

The haphazard practices of CPF chiefly in the industry and agriculture have greatly influenced the endocrine performance of water organisms (Janner et al., 2005; Minier et al., 2008). It directly interferes with endocrine hormones or hormone action either by competing for receptor binding sites or by disturbing their production from various glands. In fish, it is known to disrupt steroid hormone production (Nobonita and Suchismita, 2013) and its toxicity leads to disturbance in reproductive hormone levels (Baldigo et al., 2006; Hinck et al., 2007a, 2009a). A significant decrease in the levels of estrogen and testosterone has been observed in Tilapia, *Oreochromis niloticus* upon the exposure of CPF. The decline varies with the duration of exposure and concentration of CPF (Oruc et al., 2010). Another finding from the study revealed the decreased blood plasma levels of cortisol, a corticosteroid hormone in *Oreochromis niloticus* in comparison with control (Oruc et al., 2010).

The outcome of CPF toxicity is manifested as dysfunction of thyroid axis (Besselink et al., 1996; Adams et al., 2000; Buckman et al., 2007b). The structural changes in thyroid tissue have been reported in sea bass, *Dicentrarchus labrax* after dietary exposure to CPF (Schnitzler et al., 2011). In various fish species, it has potential effect on hypothalamic pituitary-thyroid axis (HPT) (Brar et al., 2010). Also, there has been a significant decline in blood serum concentrations of tri-iodothyronine (T3), thyroxine (T4), and thyroid-stimulating hormone (TSH) in *Heteropneustes fossilis* after being subjected to sublethal doses of toxicant (Khatun et al., 2014). Decrease in serum calcium levels, degranulation of prolactin cells with rise in nuclear volume has been linked to its long-term use in *Heteropneustes fossilis* (Srivastav et al., 2012). In wild male fish, it has been seen that low estrogen levels are responsible for the abnormal induction of vitellogenin (Miracle et al., 2006; Henry et al., 2009).

ACKNOWLEDGMENT

Special thanks are due to Research Center, College of Pharmacy, King Saud University, Riyadh, and Deanship of Scientific Research, King Saud University, Kingdom of Saudi Arabia.

KEYWORDS

- **chlorpyrifos**
- **environmental degradation and transformation**
- **environmental toxicity**
- **fish toxicology**
- **microbial degradation**
- **pesticides**

REFERENCES

Adams, B. A., Cyr, D. G., & Eales, J. G., (2000). Thyroid hormone deiodination in tissues of American plaice, *Hippoglossoides* platessoides: Characterization and short-term responses to polychlorinated biphenyls (PCBs) 77 and 126. *Comp Biochem. Physiol. C Pharmacol.*, *127*, 367–378.

Albers, J. W., Garabrant, D. H., Mattsson, J. L., Burns, C. J., Cohen, S. S., Sima, C., Garrison, R. P., et al., (2007). Dose-effect analyses of occupational chlorpyrifos exposure and peripheral nerve electrophysiology. *Toxicological Sciences*, *97*(1), 196–204.

Almeida, J. R., Oliviera, C., Gravato, C., & Guilhermino, L., (2010). Linking behavioral alterations with biomarkers responses in the European seabass *Dicentrarchus labrax* L. exposed to the organophosphate pesticide fenitrothion. *Ecotoxicology*, *19*, 1369–1381.

Alvarez-Pellitero, P., (2008). Fish immunity and parasite infections: From innate immunity to immunoprophylactic prospects. *Vet. Immunol. Immunopathol.*, *126*(3/4), 171–198.

Arcury, T. A., Grzywacz, J. G., Barr, D. B., Tapia, J., Chen, H., & Quandt, S. A., (2007). Pesticide urinary metabolite levels of children in eastern North Carolina farm worker households. *Environ Health Prospect*, *115*, 1254–1260.

Arican, O., Aral, M., Sasmaz, S., & Ciragil, P., (2005). Serum levels of TNF-a, IFN-c, IL-6, IL-8, IL-12, IL-17, and IL-18 in patients with active psoriasis and correlation with disease severity. *Mediators Inflamm.*, 273–279.

Axel, M., Christina, R., & Philippe, G., (2018). Safety of safety evaluation of pesticides: Developmental neurotoxicity of chlorpyrifos and chlorpyrifos-methyl. *Environ Health, 17*, 77.

Baldigo, B. P., Sloan, R. J., Smith, S. B., Denslow, N. D., Blazer, V. S., & Gross, T. S., (2006). Polychlorinated biphenyls, mercury, and potential endocrine disruption in fish from the Hudson River. *Aquat. Sci.*, *68*, 206–228.

Banaee, M., Mirvaghefi, A. R., Ahmadi, K., & Ashori, R., (2009). The effect of diazinon on histophatological changes of testis and ovaries of common carp (*Cyprinus carpio*). *Scientific Journal of Marine Biology*, *1*(2), 25–35.

Banerjee, B. D., Seth, V., & Ahmed, R. S., (2001). Pesticide-induced oxidative stress: Perspectives and trends. *Rev. Environ. Health*, *16*, 1–40.

Bangeppagari, M., & Gundala, H. P., (2015). Reproductive toxicity of chlorpyrifos tested in zebrafish (*Danio rerio*), Histological, and hormonal endpoints. *Toxicology and Industrial Health*, 1–9.

Barr, D. B., & Angerer, J., (2006). Potential uses of biomonitoring data: A case study using the organophosphorus pesticides chlorpyrifos and malathion. *Environ. Health Perspect.*, *114*(11), 1763–1769.

Besselink, H. T., Van, B. S., Roex, E., Vethaak, A. D., Koeman, J. H., & Brouwer, A., (1996). Low hepatic 7-ethoxyresorufi n-O-deethylase (EROD) activity and minor alterations in retinoid and thyroid hormone levels in flounder (*Platichthys flesus*) exposed to the polychlorinated biphenyl (PCB) mixture, Clophen A50. *Environ. Poll.*, *92*, 267–274.

Bhatnagar, M. C., & Bana, A. K., (1992). Respiratory distress to *Clarias batrachus* (Linn.) exposed to endosulfan-a histological approach. *J. Environ. Biol.*, *13*, 227–231.

Boshra, H., Li, J., & Sunyer, J. O., (2006). Recent advances on the complement system of teleost fish. *Fish Shellfish Immunol.*, *20*(2), 239–262.

Brar, N. K., Waggoner, C., Reyes, J. A., Fairey, R., & Kelley, K. M., (2010). Evidence for thyroid endocrine disruption in wild fish in San Francisco Bay, California, USA. Relationships to contaminant exposures. *Aquat. Toxicol.*, *96*, 203–215.

Buckman, A. H., Fisk, A. T., Parrott, J. L., Solomon, K. R., & Brown, S. B., (2007b). PCBs can diminish the influence of temperature on thyroid indices in rainbow trout (*Oncorhynchus mykiss*). *Aquat. Toxicol.*, *84*, 366–378.

Bukowska, B., Huras, B., Jarosiewicz, M., Witaszewska, J., Słowińska, M., Mokra, K., Zakrzewski, J., & Michałowicz, J., (2018). The effect of two bromfenvinphos impurities: BDCEE and β-ketophosphonate on oxidative stress induction, acetylcholinesterase activity, and viability of human red blood cells. *J. Environ. Sci. Health a Tox Hazard Subst. Environ. Eng.*, *53*(10), 931–937.

Carol, J. B., Laura, J., Mc, I., Pamela, J. M., Anne, M. J., & Abby, A. L., (2013). Pesticide exposure and neuro-developmental outcomes: Review of the epidemiologic and animal studies. *J. Toxicol. Environ. Health B Crit. Rev.*, *16*(3/4), 127–283.

Chen, D. S. X., Ziwei, Z., Haidong, Y., Yang, L., & Houjuan, X., (2015). Effects of atrazine and chlorpyrifos on oxidative stress-induced autophagy in the immune organs of common carp (*Cyprinus carpio* L.). *Fish and Shellfish Immunology*, 1–9.

Chishti, Z., Hussain, S., Arshad, K. R., Khalid, A., & Arshad, M., (2013). Microbial degradation of chlorpyrifos in liquid media and soil. *J. Environ. Manage, 15*, *114*, 372–380.

Cohn, J., & Mac, P. R. C., (1996). Ethological and experimental approaches to behavior analysis: Implications for ecotoxicology. *Environ. Health Persp.*, *104*, 299–304.

Crumpton, T. L., Seidler, F. J., & Slotkin, T. A., (2000). Is oxidative stress involved in the developmental neurotoxicity of chlorpyrifos? *Brain Res. Dev. Brain Res.*, *121*(2), 189–195.

Daland, R. J., Sean, C. G., Katie, K. C., Matthew, J. L., Vince, J. K., Haitian, L., & Sue, M. M., (2013). Chlorpyrifos: Weight of evidence evaluation of potential interaction with the estrogen, androgen, or thyroid pathways. *Regulatory Toxicology and Pharmacology*, *3*, 249–263.

Das, B. K., & Mukherjee, S. C., (2000). Chronic toxic effects of quinalphos on some biochemical parameters in *Labeo rohita* (Ham.). *Toxicol. Lett.*, *114*, 11–18.

Drevenkar, V., Vasilic, Z., Stengl, B., Frobe, Z., & Rumenjak, V., (1993). Chlorpyrifos metabolites in serum and urine of poisoned persons. *Chem. Biol. Interact.*, *87*(1–3), 315–322.

Drummond, R. A., & Russom, C. L., (1990). Behavioral toxicity syndromes, a promising tool for assessing toxicity mechanisms in juvenile fathead minnows. *Environ. Toxicol. Chem.*, *9*, 37–46.

Dubey, K. K., & Fulekar, M. H., (2012). Chlorpyrifos bioremediation in *Pennisetum rhizosphere* by a novel potential degrader *Stenotrophomonas maltophilia* MHF ENV20. *World J. Microbiol. Biotechnol.*, *28*(4), 1715–1725.

Duttaa, H. M., & Meijer, H. J. M., (2003). Sublethal effects of diazinon on the structure of the testis of bluegill, *Lepomis macrochirus*: A microscopic analysis. *Environ. Pollut.*, *125*, 355–360.

Dwayne, R. J. M., Scott, T. R., Colleen, D. G., Solomon, K. R., & John, P. G., (2014). *Refined Avian Risk Assessment for Chlorpyrifos in the United States* (pp. 163–217). Ecological risk assessment for chlorpyrifos in terrestrial and aquatic systems in the United States. Springer Open.

Dyk, J. S. V., & Pletschke, B., (2011). Review on the use of enzymes for the detection of organochlorine, organophosphate and carbamate pesticides in the environment. *Chemosphere*, *82*, 291–307.

Eaton, D. L., Daroff, R. B., Autrup, H., Bridges, J., Buffler, P., Costa, L. G., Coyle, J., et al., (2008). Review of the toxicology of chlorpyrifos with an emphasis on human exposure and neurodevelopment. *Crit. Rev. Toxicol., 38*, 1–125.

El-Amrani, S., Pena-Abaurrea, M., Sanz-Landaluze, J., Ramos, L., Guinea, J., & Camara, C., (2012). Bioconcentration of pesticides in zebrafish eleutheroembryos (*Danio rerio*). *Sci. Tot. Environ., 425*, 184–190.

Elia, A. C., Waller, W. T., & Norton, S. J., (2002). Biochemical responses of bluegill sunfish (*Lepomis macrochirus*, rafinesque) to atrazine induced oxidative stress. *Bull. Environ. Contam. Toxicol., 68*, 809–816.

Enis, Y. M., Yonar, S. M., Ural, M. S., Silici, S., & Duşukcan, M., (2012). Protective role of propolis in chlorpyrifos-induced changes in the hematological parameters and the oxidative/antioxidative status of *Cyprinus carpio carpio. Food Chem. Toxicol., 50*(8), 2703–2708.

Ensminger, M., Bergin, R., Spurlock, F., & Goh, K. S., (2011). Pesticide concentrations in water and sediment and associated invertebrate toxicity in Del Puerto and Orestimba Creeks, California, 2007–2008. *Environ. Monit. Assess, 175*, 573–587.

Farkas, A., Salanki, J., & Specziar, A., (2002). Relation between growth and the heavy metal concentration in organs of bream *Abramis brama* L. Populating Lake Balaton. *Arch. Environ. Contam. Toxicol., 43*(2), 236–243.

Firdous, A. M., (2012). Assessment of total protein concentration in liver of fresh water fish, channa punctuates (bloch.) With special reference to an organophosphate insecticide, chlorpyrifos. *IJPBS, 3*, 2.

Fogg, P., & Boxall, A. B. A., (2004). Leaching of pesticides from biobeds: Effect of biobed depth and water loading. *J. Agric. Food Chem., 52*, 6217–6227.

Folmar, L. C., Denslow, N. D., Kroll, K., Orlando, E. F., Enblom, J., Marcino, J., Metcalfe, C., & Guillette, L. J., (2001). Altered serum sex steroids and vitellogenin induction in walleye (*Stizostedion vitreum*) collected near a metropolitan sewage treatment plant. *Arch. Environ. Contam. Toxicol., 40*(3), 392–398.

Folmar, L. C., Denslow, N. D., Rao, V., Chow, M., Crain, D. A., Enblom, J., Marcino, J., & Guillette, L. J., (1996). Vitellogenin induction and reduced serum testosterone concentrations in feral male carp (*Cyprinus carpio*) captured near a major metropolitan sewage treatment plant. *Environ. Health Perspect, 104*(10), 1096–1101.

Fujioka, K., & Casida, J. E., (2007). Glutathione S-transferase conjugation of organophosphorus pesticides yields S-phospho-, S-aryl-, and S-alkyl glutathione derivatives. *Chem Res Toxicol., 20*(8), 1211–1217.

Fulton, M. H., & Key, P. B., (2001). Acetylcholinesterase inhibition in estuarine fish and invertebrates as an indicator of organophosphorus insecticide exposure and effects. *Environ. Toxicol. Chem., 20*, 37–45.

Gao, L., He, C., & Liu, X., (2012). The innate immune-related genes in catfish. *International Journal of Molecular Sciences, 13*, 4172–14202.

Gibson, J. E., Peterson, R. K. D., & Shurdut, B. A., (1998). Human exposure and risk from indoor use of chlorpyrifos. *Environ. Health Prospect, 106*, 303–306.

Giddings, J. M., Williams, W. M., Solomon, K. R., & Giesy, J. P., (2014). Risks to aquatic organisms from use of chlorpyrifos in the United States. *Rev. Environ. Contam. Toxicol., 231*, 119–162.

Gilani, R. A., Rafique, M., Rehman, A., Munis, M. F., Rehman, S. U., & Chaudhary, H. J., (2016). Biodegradation of chlorpyrifos by bacterial genus pseudomonas. *J. Basic Microbiol., 56*(2), 105–119.

Glusczak, L., Dos, S. M. D., Crestani, M., Braga, D. F. M., De, A. P. F., Duarte, M. F., & Vieira, V. L., (2006). Effect of glyphosate herbicide on acetylcholinesterase activity and metabolic and hematological parameters in piava (*Leporinus obtusidens*). *Ecotoxicol. Environ. Saf.*, *65*, 237–241.

Habig, C., & DiGiulio, R. D., (1991). Biochemical characteristics of cholinesterases in aquatic organisms. In: Mineau, P., (ed.), *Cholinesterase Inhibiting Insecticides: Their Impact on Wildlife and the Environment-Chemicals in Agriculture* (pp. 19–34). Elsevier, New York.

Haider, S., & Upadhyaya, N., (1985). Effect of commercial formulation of four organophosphate insecticides on the ovaries of a freshwater teleost, *Mystus* vitattus (Bloch), a histological and histochemical study. *J. Environ. Sci. Health.*, *20*, 321–340.

Haohua, C., Chenglan, L., Chuyan, P., Hongmei, L., Meiying, H., & Guohua, Z., (2012). Biodegradation of chlorpyrifos and its hydrolysis product 3,5,6-Trichloro-2-Pyridinol by a new fungal strain *Cladosporium* cladosporioides Hu-01. *Plos One*, *7*(10).

Henry, T. B., Mc Pherson, J. T., Rogers, E. D., Heah, T. P., Hawkins, S. A., Layton, A. C., & Sayler, G. S., (2009). Changes in the relative expression pattern of multiple vitellogenin genes in adult male and larval zebrafish exposed to exogenous estrogens. *Comp. Biochem. Physiol. A Mol. Integr. Physiol.*, *154*, 119–126.

Hinck, J. E., Blazer, V. S., Denslow, N. D., Echols, K. R., Gross, T. S., & May, T. W., (2007a). Chemical contaminants, health indicators, and reproductive biomarker responses in fish from the Colorado River and its tributaries. *Sci. Total Environ.*, *378*, 376–402.

Hinck, J. E., Blazer, V. S., Schmitt, C. J., Papoulias, D. M., & Tillitt, D. E., (2009a). Widespread occurrence of intersex in black basses (*Micropterus* spp.) from US rivers, 1995–2004. *Aquat. Toxicol.*, *95*, 60–70.

Holladay, S. D., Smith, S. A., Habback, H. E., & Caceci, T., (1996). Influence of chlorpyrifos, an organophosphate insecticide, on the immune system of Nile tilapia. *J. Aquat. Anim. Health*, *8*, 104–110.

Humphrey, C. A., Klumpp, D. W., & Raethke, N., (2004). Ambon damsel (*Pomancentrus amboinensis*) as a bioindicator organism for the Great Barrier Reef: Responses to chlorpyrifos. *Bull. Environ. Contam. Toxicol.*, *72*, 888–895.

Jaensson, A., Scott, A. P., Moore, A., Kylin, H., & Olsen, K. H., (2007). Effects of a pyrethroid pesticide on endocrine responses to female odors and reproductive behavior in male parr of brown trout (*Salmo trutta* L.). *Aquat. Toxicol.*, *81*, 1–9.

Jameson, R. R., Seidler, F. J., & Slotkin, T. A., (2007). Nonenzymatic functions of acetylcholinesterase splice variants in the developmental neurotoxicity of organophosphates: Chlorpyrifos, chlorpyrifos oxon, and diazinon. *Environ. Health Perspect*, *115*(1), 65–70.

Janner, G., Ternberg, R. M., Leblanc, G. A., & Porte, C., (2005). Testosterone conjugating activities in invertebrates: Are they targets for endocrine disruptors, *Aquat. Toxicol*, *71*, 273–282.

Jantunen, A. P., Tuikka, A., Akkanen, J., & Kukkonen, J. V., (2008). Bioaccumulation of atrazine and chlorpyrifos to *Lumbriculus variegatus* from lake sediments. *Ecotoxicol. Environ. Saf.*, *71*, 860–868.

Jarvinen, A. W., Nordling, B. R., & Henry, M. E., (1983). Chronic toxicity of dursban (chlorpyrifos) to the fathead minnow (*Pimephales promelas*) and the resultant acetylcholinesterase inhibition. *Ecotoxicol. Environ. Saf.*, *7*, 423–434.

Jason, F., David, H. B., Jeffrey, J. J., & Nathaniel, L. S., (2005). Comparative thresholds for acetylcholinesterase inhibition and behavioral impairment in Coho salmon exposed to chlorpyrifos. *Environ. Toxicol. Chem.*, *24*(1), 36–145.

Jindal, R., & Kaur, M., (2014). Acetylcholinesterase inhibition and assessment of its recovery response in some organs of *Ctenopharyngodon idellus* induced by chlorpyrifos. *International Journal of Science, Environment, 2*, 473–480.

Kai, J. E., Christian, M. L., Barry, W. W., & Ingeborg, W., (2004). Molecular and cellular biomarker responses to pesticide exposure in juvenile Chinook salmon (*Oncorhynchus tshawytscha*). *Mar. Environ. Res., 58*, 809–813.

Kaimin, Y., Guochao, L., Weimin, F., Lili, L., Jiayu, Z., Wei, W., Lei, X., & Yanchun, Y., (2015). Chlorpyrifos is estrogenic and alters embryonic hatching, cell proliferation, and apoptosis in zebrafish. *Chem. Biol. Interact., 239*, 26–33.

Kammerbauer, J., & Moncada, J., (1998). Pesticide residue assessment in three selected agricultural production systems in the Choluteca river basin of Honduras. *Environ. Pollut., 103*, 171–181.

Kane, A. S., Salierno, J. D., Gipson, G. T., Molteno, T. C. A., & Hunter, C. A., (2004). Video based movement analysis system to quantify behavioral stress response of fish. *Water Res., 38*, 3993–4001.

Karen, H., Asa, B., Kim, H., Paul, Y., Dana, B. B., Brenda, E., & Nina, H., (2012). Organophosphate pesticide levels in blood and urine of women and newborns living in an agricultural community. *Environ. Res., 117*, 8–16.

Kegley, S., Neumeister, L., & Martin, T., (1999). *Ecological Impacts of Pesticides in California* (p. 99). Pesticide Action Network, California, USA.

Khalil, F., Kang, I. J., Undap, S., Tasmin, R., Qiu, X., Shimasaki, Y., & Oshima, Y., (2013). Alterations in social behavior of Japanese medaka (*Oryzias latipes*) in response to sublethal chlorpyrifos exposure. *Chemosphere, 92*, 125–130.

Khatun, N., & Mahanta, R., (2014). A study on the effect of chlorpyrifos (20% EC) on thyroid hormones in freshwater fish, *Heteropneustes fossilis* (Bloch.) by using EIA technique. *The Science Probe, 2*, 8–16.

Kime, D. E., Nash, J. P., & Scott, A. P., (1999). Vitellogenesis as a biomarker of reproductive disruption by xenobiotics. *Aquaculture, 177*(1–4), 345–352.

Kristiansen, T. S., Ferno, A., Holm, J. C., Privitera, L., Bakke, S., & Fosseidengen, J. E., (2004). Swimming behavior as an indicator of low growth rate and impaired welfare in Atlantic halibut (*Hippoglossus hippoglossus* L.) reared at three stocking densities. *Aquaculture, 230*, 137–151.

Litman, G. W., Cannon, J. P., & Dishaw, L. J., (2005). Reconstructing immune phylogeny: New perspectives. *Nat. Rev. Immunol., 5*, 866–879.

Little, E. E., & Finger, S. E., (1990). Swimming behavior as an indicator of sublethal toxicity in fish. *Environ. Toxicol. Chem., 9*, 13–19.

Lucio, G. C., Gennaro, G., Toby, B. C., Judit, M., & Clement, E. F., (2013). Paraoxonase 1 (Pon1) As a genetic determinant of susceptibility to organophosphate toxicity. *Toxicology, 307*, 115–122.

Madhusudan, R. N., Kodimyala, R., Rudra, R. R., Venkateswara, R. J., & Ghousia, B., (2015). The role of vitamin C as antioxidant in protection of biochemical and hematological stress induced by chlorpyrifos in freshwater fish *Clarias batrachu. Chemosphere, 132*, 172–178.

Madhusudan, R. N., Rudra, R. R., & Rajender, K., (2012). Sub-acute toxicity effects of chlorpyrifos on acetylcholinesterase activity and recovery in the freshwater field crab barytelphusa guerini. *International Journal of Environmental Sciences, 3*, 98–107.

Magnadottir, B., (2006). Innate immunity of fish (overview). *Fish and Shellfish Immunology, 20*, 37–151.

Marshall, W. K., & Roberts, J. R., (1999). *Ecotoxicology of Chlorpyrifos* (pp. 1–4). National Research Council of Canada, NRC Associate Committee on Scientific Criteria for Environmental Quality, Subcommittee on Pesticides and Related Compounds. Canadian Environmental Quality Guidelines, Canadian Council of Ministers of the Environment.

Masia, A., Campo, J., Navarro-Ortega, A., Barcelo, D., & Pico, Y., (2015). Pesticide monitoring in the basin of Llobregat River (Catalonia Spain) and comparison with historical data. *Sci. Total Environ.*, 58–68.

Mattsson, J. L., Maurissen, J. P., Nolan, R. J., & Brzak, K. A., (2000). Lack of differential sensitivity to cholinesterase inhibition in fetuses and neonates compared to dams treated perinatally with chlorpyrifos. *Toxicol. Sci.*, *53*(2), 438–446.

Meuling, W. J., Ravensberg, L. C., Roza, L., & Van, H. J. J., (2005). Dermal absorption of chlorpyrifos in human volunteers. *Int. Arch Occup. Environ. Health*, *78*(1), 44–50.

Mevlüt, S. U., (2013). Chlorpyrifos-induced changes in oxidant/antioxidant status and hematological parameters of *Cyprinus carpio carpio*: Ameliorative effect of lycopene. *Chemosphere*, *90*, 2059–2064.

Minier, C., Forget-Leray, J., Bjørnstad, A., & Camus, L., (2008). Multixenobiotic resistance, acetylcholine esterase activity and total oxyradical scavenging capacity of the Arctic spider crab, hyasaraneus, following exposure to bisphenol A, tetra bromo diphenyl ether and diallyl phthalate. *Marine Pollutant Bulletin*, *56*(8), 1410–1415.

Miracle, A., Ankley, G., & Lattier, D., (2006). Expression of two vitellogenin genes (vg1 and vg3) in fathead minnow (*Pimephales promelas*) liver in response to exposure to steroidal estrogens and androgens. *Ecotoxicol. Environ. Safe*, *63*, 337–342.

Modesto, K. A., & Martinez, C. B. R., (2010). Effects of Roundup Transorb on fish: Hematology, antioxidant defenses and acetylcholinesterase activity. *Chemosphere*, *81*, 781–787.

Moore, A., & Waring, C. P., (2001). The effect of a synthetic pyrethroid pesticide on some aspects of reproduction in Atlantic salmon *Salmo salar* L. *Aqu. Toxicol.*, *52*(1), 1–12.

Moore, M. T., Schulz, R., Cooper, C. M., Smith, J. S., & Rodgers, J. H., (2002). Mitigation of chlorpyrifos runoff using constructed wetlands. *Chemosphere*, *46*, 827–835.

Nand, N., Aggarwal, N. K., Komal, B., & Chakrabarti, D., (2007). Organophosphate induced delayed neuropathy. *JAIP*, *55*, 72–73.

Narra, M. R., Begum, G., Rajender, K., & Rao, J. V., (2012). Toxic impact of two organophosphate insecticides on biochemical parameters of a food fish and assessment of recovery response. *Toxicol. Ind. Health*, *28*(4), 343–352.

Nitika, S., Vivek, K. G., Abhishek, K., & Bechan, S., (2017). Synergistic effects of heavy metals and pesticides in living systems. *Front Chem.*, *5*, 70.

Nobonita, D., & Suchismita, D., (2013). Chlorpyrifos toxicity in fish: A review. *Current World Environment*, *8*(1), 77–84.

Nolan, R. J., Rick, D. L., Freshour, N. L., & Saunders, J. H., (1984). Chlorpyrifos: Pharmacokinetics in human volunteers. *Toxicol. Appl. Pharmacol.*, *30*, *73*(1), 8–15.

Noor, S. J., Adel, M. R., & Al-Chalabi, S. M. M., (2016). Biochemical, hematological parameters, and histological alterations in fish *Cyprinus carpio* L. as biomarkers for water pollution with chlorpyrifos. *Human and Ecological Risk Assessment: An International Journal*, *23*(3), 605–616.

Odenkirchen, E. W., & Eisler, R., (1988). *Chlorpyrifos Hazards to Fish, Wildlife, and Invertebrates: A Synoptic Review* (p. 13). U.S. Fish and Wildlife Service Biological Report.

Okechukwu, E. O., Auta, J., & Balogun, J. K., (2007). Effects of acute nominal doses of chlorpyrifos-ethyl on some hematological indices of African catfish *Clarias* gariepinus-teugels. *J. Fish. Int., 2*, 190–194.

Ortega, G. J. A., Gallardo, D. C., Ferris, I., Tortajada, J., Garcia, M. M., & Grimalt, J. O., (2006). Meconium and neurotoxicants: Searching for prenatal exposure timing. *Arch Dis. Child., 91*, 642–646.

Oruç, E. O., (2010). Oxidative stress, steroid hormone concentrations, and acetylcholinesterase activity in *Oreochromis niloticus* exposed to chlorpyrifos. *Pestic. Biochem. Physiol., 96*, 160–166.

Padmanabha, A., Reddy, H. R. V., Muttappa, K., Prabhudeva, K. N., Rajanna, K. B., & Chethan, N., (2015). Acute effects of chlorpyrifos on oxygen consumption and food consumption of freshwater fish, *Oreochromis mossambicus* (Peters). *International Journal of Recent Scientific Research, 6*(4), 3380–3384.

Phillips, P. J., Scoh, W. A., & Nystrom, E. A., (2007). Temporal changes in surface-water insecticide concentrations after the phase-out of diazinon and chlorpyrifos. *Environ. Sci. Technol., 41*, 4246–4251.

Pope, C., Karanth, S., & Liu, J., (2005). Pharmacology and toxicology of cholinesterase inhibitors: Uses and misuses of a common mechanism of action. *Environ. Toxicol. Pharmacol., 19*, 433–446.

Pragati, J., Anjali, B., Tanvi, T., Sameena, M., & Zaidi, M. G. H., (2018). Electrochemical sensor for the detection of pesticides in environmental sample: A review. *International Journal of Chemical Studies, 6*(2), 3199–3205.

Quiniou, S. M., Boudinot, P., & Bengtén, E., (2013). Comprehensive survey and genomic characterization of toll-like receptors (TLRs) in channel catfish, *Ictalurus punctatus*: Identification of novel fish TLRs. *Immunogenetics, 65*(7), 511–530.

Rahman, M. Z., Hossain, Z., Mollah, M. F. A., & Ahmed, G. U., (2002). Effect of diazinon 60 EC on *Anabas testudineus*, *Channa punctatus*, and *Gonionotus naga*. *The ICIARM Quarterly, 25*, 8–121.

Ramesh, H., & David, M., (2009). Respiratory performance and behavioral responses of the freshwater fish, *Cyprinus carpio* (Linnaeus) under sublethal chlorpyrifos exposure. *J. Basic Clin. Physiol. Pharmacol., 20*(2), 127–139.

Ramesh, H., & Muniswamy, D., (2009). Behavioral responses of the freshwater fish, *Cyprinus carpio* (Linnaeus) following sublethal exposure to chlorpyrifos. *Turkish Journal of Fisheries and Aquatic Sciences, 9*, 233–238.

Ramesh, M., & Saravanan, M., (2008). Hematological and biochemical responses in a freshwater fish *Cyprinus carpio* exposed to chlorpyrifos. *Int. J. Integr. Biol., 3*, 80–83.

Ray, A., Liu, J., Karanth, S., Gao, Y., Brimijoin, S., & Pope, C., (2009). Cholinesterase inhibition and acetylcholine accumulation following intracerebral administration of paraoxon in rats. *Toxicology and Applied Pharmacology, 236*(3), 341–347.

Reichenberger, S., Bach, M., Skitschak, A., & Frede, H. G., (2007). Mitigation strategies to reduce pesticide inputs into ground- and surface water and their effectiveness: A review. *Sci. Total Environ., 384*, 1–35.

Rezk, M. R., El-Aleem, A. E. B., A., Khalile, S. M., & El-Naggar, O. K., (2018). Selective determination of diazinon and chlorpyrifos in the presence of their degradation products: Application to environmental samples. *J. AOAC Int., 101*(4), 1191–1197.

Rice, J. A., (1990). Bioenergetics modeling approaches to evaluate stress in fishes. In: *Proceedings of 8th American Fisheries Society Symposium* (pp. 80–92). Bethesda, Maryland.

Ronald, W. P., John, C. C., Richard, P. L., Peter, C. G., & Ramasamy, M. S., (2009). Effects of temperature on ventilatory behavior of fish exposed to sub-lethal concentrations of endosulfan and chlorpyrifos. *Environ. Toxicol. Chem.*, *28*, 2182–2190.

Rusyniak, D. E., & Nanagas, K. A., (2004). Organophosphate poisoning. *Semen. Neurol.*, *24*, 197–204.

Sandipan, P., Emiko, K., Jiro, K., Seiichi, U., & Apurba, R. G., (2012). Histopathological alterations in gill, liver, and kidney of common carp exposed to chlorpyrifos. *Journal of Environmental Science and Health*, *47*, 180–195.

Sanghi, R., Pillai, M. K., Jayalekshmi, T. R., & Nair, A., (2003). Organochlorine and organophosphorus pesticide residues in breast milk from Bhopal, Madhya Pradesh, India. *Hum. Exp. Toxicol.*, *22*(2), 73–76.

Sasikala, C., Jiwal, S., Rout, P., & Ramya, M., (2012). Biodegradation of chlorpyrifos by bacterial consortium isolated from agriculture soil. *World J. Microbiol. Biotechnol.*, *28*(3), 1301–1308.

Saurabh, S., & Sahoo, P. K., (2008). Lysozyme: An important defense molecule of fish innate immune system. *Aquaculture Research*, *39*, 223–239.

Sayeed, I., Parvez, S., Pandey, S., Bin-Hafeez, B., Haque, R., & Raisuddin, S., (2003). Oxidative stress biomarkers of exposure to deltamethrin in freshwater fish *Channa punctatus bloch*. *Ecotoxicol. Environ. Saf.*, *56*, 295–301.

Scherrer, E., (1992). Behavioral responses as indicator of environmental alterations: Approaches, results, developments. *Journal of Applied Ichthyology*, *8*, 122–131.

Schnitzler, J. G., Thome, J. P., Lepage, M., & Das, K., (2011). Organochlorine pesticides, polychlorinated biphenyls, and trace elements in wild European sea bass (*Dicentrarchus labrax*) off European estuaries. *Sci. Total Environ.*, *409*, 3680–3686.

Scholz, N. L., Truelove, N. K., French, B. L., Berejikian, B. A., Quinn, T. P., Casillas, E., & Collier, T. K., (2000). Diazinon disrupts antipredator and homing behaviors in chinook salmon (*Oncorhynchus tshawytscha*). *Can. J. Fish. Aquat. Sci.*, *57*, 1911–1918.

Secombes, C. J., (1996). The nonspecific immune system: Cellular defenses. In: Iwama, G., & Nakanishi, T., (eds.), *The Fish Immune System, Organism, Pathogen and Environment* (pp. 63–103). Academic Press, Toronto.

Shamsa, A., & Sikander, S., (2016). Soil bacteria showing a potential of chlorpyrifos degradation and plant growth enhancement. *Braz. J. Microbiol.*, *47*(3), 563–570.

Silva, D. P. M. C. S., & Samayawardhena, L. A., (2005). Effects of chlorpyrifos on reproductive performances of guppy (*Poecilia reticulata*). *Chemosphere*, *58*, 1293–1299.

Singh, D. P., Khattar, J. I., Nadda, J., Singh, Y., Garg, A., Kaur, N., & Gulati, A., (2011). Chlorpyrifos degradation by the cyanobacterium *Synechocystis* sp. strain PUPCCC 64. *Environ. Sci. Pollut. Res. Int.*, *18*(8), 1351–1359.

Skandhan, K. P., Sahab, K. P., & Sumangala, B., (2011). DDT and male reproductive system. *Research Journal of Environmental Toxicology*, *5*(2), 76–80.

Smegal, D. C., (2000). *Human Health Risk Assessment Chlorpyrifos* (pp. 1–131). US Environmental Protection Agency, Office of Prevention, Pesticides and Toxic Substances, Office of Pesticide Programs, Health Effects Division, US Government Printing Office, Washington DC.

Srivastav, A. K., Srivastava, S. K., Tripathi, S., Mishra, D., & Srivastav, S. K., (2012). Morpho-toxicology of chlorpyrifos to prolactin cells of a freshwater catfish, *Heteropneustes fossilis*, *Acta scientiarum. Biological Sciences*, *34*(4), 443–449.

Susan, A. T., Sobha, K., & Tilak, K. S., (2010). A study on acute toxicity, oxygen consumption, and behavioral changes in the three major carps, *Labeo rohita* (ham), *Catla catla* (ham) and *Cirrhinus* mrigala (ham) exposed to fenvalerate. *Biores. Bull.*, *1*, 33–40.

Tang, J., Cao, Y., Rose, R. L., Brimfield, A. A., Dai, D., Goldstein, J. A., & Hodgson, E., (2001). Metabolism of chlorpyrifos by human cytochrome P450 isoforms and human, mouse, and rat liver microsomes. *Drug Metab. Dispos.*, *29*(9), 1201–1204.

Tessema, F. M., Ulrich, P., & Matthias, K., (2019). Glucosylation and glutathione conjugation of chlorpyrifos and fluopyram metabolites using electrochemistry/mass spectrometry. *Molecules*, *24*, 898.

Timchalk, C., Busby, A., Campbell, J. A., Needham, L. L., & Barr, D. B., (2007). Comparative pharmacokinetics of the organophosphorus insecticide chlorpyrifos and its major metabolites diethylphosphate, diethylthiophosphate and 3,5,6-trichloro-2-pyridinol in the rat. *Toxicology*, *31*, *237*(1–3), 145–157.

Tripathi, G., & Shasmal, J., (2011). Concentration related responses of chlorpyriphos in antioxidant, anaerobic and protein synthesizing machinery of the freshwater fish, *Heteropneustes fossilis*. *Pestic. Biochem. Physiol.*, *99*, 215–220.

Vani, T., Saharan, N., Mukherjee, S. C., Ranjan, R., Kumar, R., & Brahmchari, R. K., (2011). Deltamethrin induced alterations of hematological and biochemical parameters in fingerlings of *Catla catla* (Ham.) and their amelioration by dietary supplement of vitamin C. *Pestic. Biochem. Physiol.*, *101*, 16–20.

Varo, I., Serrano, R., Pitarch, E., Amat, F., Lopez, F. J., & Navarro, J. C., (2002). Bioaccumulation of chlorpyrifos through an experimental food chain: Study of protein HSP70 as biomarker of sublethal stress in fish. *Arch Environ. Contam. Toxicol.*, *42*, 229–235.

Varo, I., Serrano, R., Pitarch, E., Amat, F., Lopez, F. J., & Navarro, J. C., (2000). Toxicity and bioconcentration of chlorpyrifos in aquatic organisms: *Artemia parthenogenetica* (Crustacea), *Gambusia affinis*, and *Aphanius iberus* (Pisces). *Bull. Environ. Contam. Toxicol.*, *65*(5), 623–30.

Venkateswara, R. J., Ghousia, B., Pallela, R., Usman, P. K., & Nageswara, R. R., (2005). Changes in behavior and brain acetylcholinesterase activity in mosquito fish, *Gambusia affinis* in response to the sub-lethal exposure to chlorpyrifos *Int. J. Environ. Res. Public Health*, *2*(3), 478–483.

Venkateswara, R. J., Parvati, K., Kavitha, P., Jakka, N. M., & Pallela, R., (2005). Effect of chlorpyrifos and monocrotophos on locomotor behavior and acetylcholinesterase activity of subterranean termites, *Odontotermes obesus*. *Pest Manag. Sci.*, *61*, 417–421.

Vesela, Y., Ivelin, M., Elenka, G., Stela, S., Vesela, T., & Iliana, V., (2017). Ex situ effects of chlorpyrifos on the lysosomal membrane stability and respiration rate in zebra mussel, *Dreissena polymorpha* (Pallas, 1771). *Acta zool. Bulg.*, *8*, 85–90.

Voltz, M., Louchart, X., Andrieux, P., & Lennartz, B., (2005). Process of water contamination by pesticides at catchment scale in Mediterranean areas. *Geophy. Res.*, *7*, 106–134.

Wang, X., Xing, H., Li, X., Xu, S., & Wang, X., (2011). Effects of atrazine and chlorpyrifos on the mRNA levels of IL-1 and IFN-γ2b in immune organs of common carp. *Fish Shellfish Immunol.*, *31*(1), 126–133.

Ware, G., (1989). *The Pesticide Book* (p. 336). Thomson, Fresno, CA, USA.

Woodburn, K. B., Hansen, S. C., Roth, G. A., & Strauss, K., (2003). The bioconcentration and metabolism of chlorpyrifos by the eastern oyster, *Crassostrea virginica*. *Environ. Toxicol. Chem.*, *22*, 276–284.

Xing, H., Wang, J., Li, J., Fan, Z., Wang, M., & Xu, S., (2010). Effects of atrazine and chlorpyrifos on acetylcholinesterase and carboxylesterase in brain and muscle of common carp. *Environ. Toxicol. Pharmacol.*, *30*, 26–30.

Xing, H., Wang, X., Sun, G., Gao, X., Xu, S., & Wang, X., (2012). Effects of atrazine and chlorpyrifos on activity and transcription of glutathione S-transferase in common carp (*Cyprinus carpio* L.). *Environ. Toxicol. Pharmacol., 33*(2), 233–244.

Xingang, M., Lingzhu, C., Yuping, Z., Deyu, H., & Baoan, S., (2018). Hydrolysis and photolysis kinetics, and identification of degradation products of the novel bactericide 2-(4-fluorobenzyl)-5-(methylsulfonyl)-1,3,4-oxadiazole in water. *Int. J. Environ. Res. Public Health, 15*, 2741.

Xu, W., Jilong, L., Houjuan, X., & Shiwen, X., (2011). Review of toxicology of atrazine and chlorpyrifos on fish. *J. Northeast Agric. Univ., 18*, 88–92.

Yen, J., Donerly, S., Levin, E. D., & Linney, E. A., (2011). Differential acetylcholinesterase inhibition of *Chlorpyrifos diazinon* and parathion in larval zebrafish. *Neurotoxicol. Teratol., 33*, 735–741.

Yogita, D., & Abha, M., (2013). Histopathological alterations in gill and liver anatomy of freshwater, air breathing fish *Channa punctatus* after pesticide hilban® (Chlorpyrifos) treatment. *Adv. Biores., 4*, 57–62.

Yuanxiang, J., Zhenzhen, L., Tao, P., & Zhengwei, F., (2015). The toxicity of chlorpyrifos on the early life stage of zebrafish: A survey on the endpoints at development, locomotor behavior, oxidative stress, and immunotoxicity. *Fish Shellfish Immunol., 43*(2), 405–14.

Zaheer, M. D. K., & Law, F. C. P., (2005). Adverse effects of pesticides and related chemicals on enzyme and hormone systems of fish, amphibians, and reptiles: A review. *Proc. Pakistan Acad. Sci., 42*, 315–323.

Zeinab, M. B., Gamal, E. N., Rasha, M. R., & Rowida, E. I., (2016). Effect of insecticide "chlorpyrifos" on immune response of *Oreochromis niloticus*. *Zagazig Veterinary Journal, 44*, 196–204.

Zhang, X., Starner, K., & Spurlock, F., (2012). Analysis of chlorpyrifos agricultural use in regions of frequent surface water detections in California, USA. *Bull. Environ. Contam. Toxicol., 89*, 978–984.

CHAPTER 4

Neurotoxicity of Heavy Metals in Fishes: A Mechanistic Approach

SANA FAROOQ,[1] ADIL FAROOQ WALI,[2] SABHIYA MAJID,[1] SAIEMA RASOOL,[3] HILAL AHMAD WANI,[4] SHOWKAT AHMAD BHAT,[1] SHAFAT ALI,[1,3] JAVAID AHMAD WANI,[1] RAFIQA ECHIKOTI,[1] SHABHAT RASOOL,[1] AJAZ AHMAD,[5] and MUNEEB U. REHMAN[1,5]

[1]Department of Biochemistry, Government Medical College (GMC), Karan Nagar, Srinagar–190010, Jammu and Kashmir, India

[2]RAK College of Pharmaceutical Sciences, RAK Medical and Health Sciences University, Ras Al Khaimah–11172, United Arab Emirates

[3]Department of School Education, Govt. of Jammu & Kashmir, Srinagar, 190001 India

[4]Department of Biochemistry, Govt. Degree College (Sumbal), Sumbal, Bandipora, Jammu and Kashmir, India

[5]Department of Clinical Pharmacy, College of Pharmacy, King Saud University, Riyadh–11451, Saudi Arabia, E-mail: muneebjh@gmail.com (M. U. Rehman)

ABSTRACT

Due to drastic increase in industrial activities, heavy metal pollution has become a source of grave concern. Heavy metals not only have hostile effects on living organisms, but they also disturb the delicate balance of the ecosystem. Aquatic ecosystems are particularly affected by heavy metal pollution since most of the industrial wastes are ultimately drained into the water bodies. Fish being at the top of the aquatic food chain accumulates all these toxic heavy metals. In this book chapter, we have tried to go into the mechanistic depths of neurotoxicity caused by heavy metals in fishes. Trace amounts of some heavy metals like iron, copper, zinc, and cobalt are essential to maintain the physiological and biochemical functions in living organisms; however, beyond a certain threshold level, they become toxic. Other heavy metals like lead (Pb), mercury (Hg), and cadmium (Cd) have no known essential role in living organisms. These essential and non-essential heavy metals together constitute an important group of neurotoxicants. Toxic effects of heavy metals come into play when detoxification, metabolic, storage, and excretory mechanisms in fish are not able to counter uptake. Most of the heavy metals disturb the delicate balance between reactive oxygen species (ROS) and antioxidant defense mechanisms in cells, resulting in an increase in the concentration of ROS. This creates oxidative stress in the cells leading to apoptosis (Mieiro et al., 2009).

4.1 INTRODUCTION

Metals with specific density more than 5 g/cm^3, high atomic number and atomic mass represent heavy metals. Heavy metals, found in nature, may have their origin both from natural and anthropogenic sources. Weathering, soil formation, and different types of rocks (magmatic, sedimentary, and metamorphic) constitute the natural sources that add heavy metals to the surroundings. The anthropogenic sources consist of agricultural and industrial activities (Bradl, 2005).

Heavy metals not only have adverse effect on humans but they also disturb the delicate balance of the ecosystem. Most of the heavy metals have toxic nature and capability of accumulation in bodies of living organisms, thereby acting as a potential ecological and health hazard (Bolognesi et al., 1999). Traces of the heavy metals such as iron, cobalt, copper, and zinc are vital for maintaining physiological and biochemical functions in living organisms; however, these heavy metals can become highly toxic beyond a certain threshold level (Tchounwou et al., 2012; Jaishankar et al., 2014). The heavy metals including mercury, lead, and cadmium have no known essential function in living organisms and therefore, may be considered non-essential. These can be regarded as a major hazard to all life forms since they exhibit toxic effects even at very low concentrations. Hostile intracellular events like mitochondrial dysfunction, DNA fragmentation, oxidative stress, endoplasmic reticulum (ER) stress, protein misfolding, autophagy dysregulation, and activation of apoptosis are linked with raised heavy metal levels in various organisms.

Heavy metal contamination is the most dangerous chemical pollution in water (Sonmez et al., 2012). Higher concentration of heavy metals become accumulated in the body of fish than surrounding sediments or water since fish occupies the highest trophic level in an aquatic food chain (Gumgum et al., 1994; Mansour and Sidky, 2002; Olaifa et al., 2004; Al-Weher, 2008; Sonmez et al., 2012). Accumulation of heavy metals in fish tissues is based on certain factors such as chemistry of water (e.g., hardness, salinity, and pH), water hydro-dynamics, competition for predation and food (Al-Weher, 2008). Moreover, interaction among between different heavy metals may perhaps affect their accumulation (Pagenkopf, 1983; Cicik, 2003).

4.2 HEAVY METAL TOXICITY AND FISH NERVOUS SYSTEM

The brain of fish consists of five components from anterior to posterior: (i) the telencephalon (fore-brain), primarily linked with smell, (ii) the diencephalon, a correlation center for messages vis-à-vis the endocrine system and homeostasis, (iii) the mesencephalon (midbrain), vital for vision, (iv) the metencephalon (hind-brain), essential for maintaining tone of muscles and balance during swimming and has a large median lobe (cerebellum) which is the largest part of the brain of fish, and (v) the myelencephalon (brain-stem or medulla oblongata), the hind part of brain and enlarged anterior part of spinal cord that relays input for all sensory systems except smell and sight (Helfman et al., 2009).

Gills, skin, and food are the primary sources of heavy metal uptake from the environment in fishes. Heavy metals enter into different fish organs largely via two processes, that is, adsorption, and absorption. Their accumulation rate in different organs relies on the uptake and depuration rates (Annabi et al., 2013; Rajeshkumar et al., 2018). By binding to the metal-binding proteins in the tissues, heavy metals can reach a very high concentration in organs. Heavy metals, being persistent, are internally regulated by storage and active excretion. Toxic effects come into play when detoxification, metabolic, storage, and excretory mechanisms fail to balance uptake (Obasohan et al., 2006), ultimately resulting in histopathological and physiological changes (Ribeiro et al., 2005; Georgieva et al., 2014). Water physico-chemistry also has an effect on these changes (Annabi et al., 2013). In three different studies carried out in three different freshwater bodies: Lake Balaton, Hungary (Farkas et al., 2000); Iskenderun Bay, Turkey (Yilmaz et al., 2005); Three Gorges Reservoir, China (Zhang et al., 2007), considerable variations of heavy metal concentrations was reported and even in different fish species inhabiting the same freshwater body. The water bodies of areas with the increased settlement, agricultural, traffic, and industrial activities show higher levels of toxic heavy metals. The toxic heavy metal levels in fish are crucial since it is an essential foodstuff source for human beings. The tissues obtained from fish species living in freshwater bodies contaminated with industrial effluents have been shown containing toxic levels of heavy metals and thus unsafe for human consumption (Egila et al., 2011; Annabi et al., 2013). Heavy metal concentration in fish species reflects the level of these metals in water and sediments of aquatic milieu where from the fish are obtained as well as the time of exposure. Therefore, certain fish species act as biological indicators of pollution due to heavy metals (Burger et al., 2011; Soto et al., 2011).

4.3   HEAVY METAL BIOACCUMULATION IN FISH

Bioaccumulation is described as a type of bio-sorption where metallic elements become assimilated in the biomass of living organisms. Typically assimilation occurs by active uptake, however, occasionally passive uptake is involved as well (Fomina and Gadd, 2014). Heavy metals form complexes with protein molecules of intracellular origin and as a result become immobilized, for instance, sulfur-containing peptides (phytochelatins and metallothionein). It is actually a protective mechanism to oppose the toxic consequences of metals that are assimilated within the biomass (Harms et al., 2011).

Heavy metallic elements first attach to cell surface and then infiltrate the cells of microorganisms. Microorganisms use chemical reactions to digest their food and chemically change these heavy metals; for instance, methyl-mercury is a modified form of mercury, which is formed by some bacteria. Methylmercury is more easily absorbed by insects and other organisms. Fish which consumes these small organisms accumulates methylmercury in much higher concentration. Further, this toxic chemical remains in fish for a long time period and then is transmitted to bigger life forms through the food chain. The concentration of methylmercury thus increases as the food chain continues. In a similar way, other heavy metals like arsenic, lead, cadmium, manganese, etc., are accumulated in the larger living beings. The heavy metal bioaccumulation in trophic food chain is a huge reason for worry because these metals may produce lethal effects on the health of human beings (Jarup and Akesson, 2009; Ersoy and Celik, 2009). Since all life forms depend upon water for sustenance, therefore fish and seafood form the most important connection between heavy metals in the surroundings and exposure to living organisms (Burger and Gochfeld, 2009).

Heavy metals enter aquatic food chain via two routes-dietary that involves consuming contaminated food by way of alimentary canal and non-dietary which includes absorption across the permeable membranes like gills and muscles. Various factors including age, sex, body mass, physiological conditions, and biological habitat of fish influence the bioaccumulation of heavy metallic elements in tissues of fish. Other factors like chemical state of metal in water, concentration of dissolved oxygen, water temperature, transparency, and pH play a vital part in the process of bioaccumulation (Has-Schön et al., 2006). Fish amass heavy metals in a higher concentration in their tissues than the toxic concentration in their adjoining environment via absorption along the surface of gills, liver, kidneys, and gut tract. Differences in heavy metal aggregation among living organisms may be due to the reason of distinction in the assimilation or egestion process or both. Noxious heavy metallic elements with elevated concentrations have been reported from regions undergoing ever-increasing settlement, farming, and traffic activities.

4.4   NEUROTOXICOLOGICAL EFFECTS AND CORRESPONDING MECHANISM OF ACTION OF HEAVY METALS

4.4.1   LEAD

Lead is a legacy pollutant and its sources in the environment include lead pipes/solders, lead-based paints, leaded gasoline, and industrial activities. Lead has no essential biochemical or physiological function. Exposure to lead causes oxidative damage which is closely associated with fish neurotoxicity and continuing exposure to lead is lethal to fish central nervous system (CNS, leading to dysfunctions in cognition and behavior (Hsu and Guo, 2002; Zhu et al., 2016). Lead is an important neurotoxicant that causes morphological changes in the brain leading to neurodegenerative defects, deregulation of cell signaling and impairments in the nerve transmission as a result of lead-induced damage to brain and cognitive functions (Verstraeten et al., 2008). Lead competes with calcium (a vital ion for the secretion and regulation of neurotransmitter) via imitating the cationic calcium thereby disrupting the flux of calcium and regulatory functions leading to neurotoxicity (Marchetti, 2003). Lead exposure damages the calcium transportation system of nervous system and also adversely affects neurotransmission by disturbing calcium homeostasis. Other mechanisms associated with lead toxicity include disruption of biophysics of cell membranes, oxidative stress induction and necrosis of cells (Westerink and Vijverberg, 2002).

Lead substitutes zinc in zinc finger proteins which form the major group of translation factors. Many brain-specific proteins dependent on zinc finger proteins are adversely affected by lead exposure. Neurological injuries associated with hyperactive movements and hyperventilation in fish is caused by lead (Ziazz et al., 2013). Lead exerts its toxic effects mainly by inhibiting the activity of cholinesterase. Cholinesterase is an important enzyme for neurotransmission since it catalyzes acetylcholine (ACh) subsequent to its secretion at the nerve cleft of cholinergic synapses. Reduced secretion of cholinesterase causes ACh accumulation that leads to overstimulation. Acute neurotoxicity by lead may ultimately lead to fatality. Cholinesterase is therefore, widely considered as an important, reliable, and sensitive marker to evaluate lead neurotoxicity (Nunes et al., 2014; Chaing et al., 2016).

Lead exposure produces a number of behavioral and neurological problems in fish. This is because lead damages synapses and alters the concentrations of neurotransmitter. A close relationship has been observed between the structure of neurotransmitter and adenosine triphosphate (ATP). In the wake of discharging extracellular nucleotide, the ATP molecule, a basic emissary, is catabolized to produce ADP (adenosine diphosphate), AMP (adenosine monophosphate), and adenosine sequentially. Among these remnants, adenosine is the key neuroprotective factor. Lead presentation influences the enzymatic pathway of ectoendonuclases meant for the regulation of purinergic signaling (Senger et al., 2006). Consequently, lead introduction may modify the structure or function of proteins, alter expression of genes and disturb the processes such as transduction and repairing of DNA (Richetti et al., 2011). Rice et al. (2011) conducted a study on zebrafish and reported that increasing concentrations of lead were associated with reduced velocity of head turns and period of swimming. Moreover, the number of zebrafish with a startle response was also decreased. In another study, Kim and Kang (2017) found that dietary lead in *Sebastes schlegelii* greatly inhibited the activity of acetylcholinesterase (AchE) initiating neurotoxicity (Figure 4.1).

4.4.2 MERCURY

Mercury (Hg) is amongst the highly noxious heavy metals present in the surroundings. A progression of composite chemical transformations gives rise to three oxidation states of Hg cycle in the environment (Barbosa et al., 2001). Mercury exists in its metallic form with the oxidation state of zero (Hg0); vapor being the most abundant form (98%). The loss of one or two electrons from Hg turns it into two higher oxidation states, viz. mercurous and mercuric, respectively.

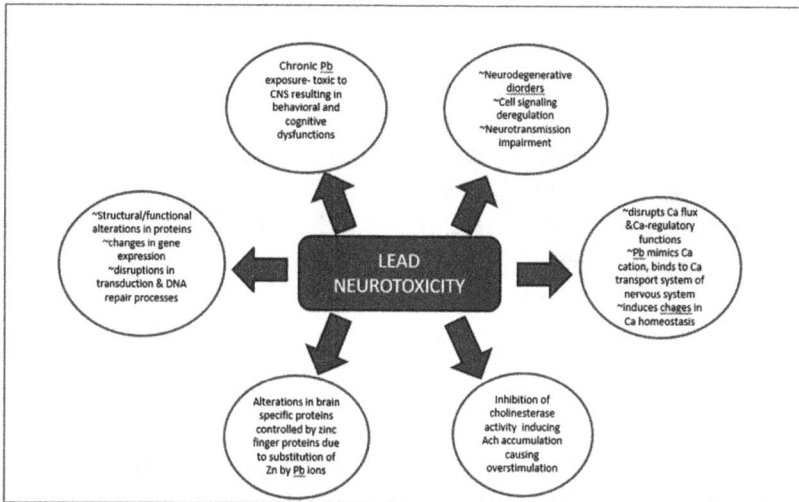

FIGURE 4.1   Summary of toxic effects of lead.

After the disaster in Minamata, Japan (1956), due to the consumption of fish contaminated with methyl-mercury, the studies on effects of heavy metals on fish have gained tremendous importance. The environment receives mercury from three main sources-natural, anthropogenic, and re-emitted source. The anthropogenic sources of mercury include mining, industrial discharges, urban discharges and agricultural activities (Jackson, 1997; Zhang and Wong, 2007).

Methyl mercury is the most toxic form of mercury in terms of health consequences from environmental exposures (Jackson, 1997; Goyer and Clarsksom, 2001). Human beings, phytoplankton, fish, zooplankton, atmosphere, soil, water, and industry are all involved in a cyclin order following contamination path of mercury (Kadar et al., 2000). Inhibition of translation, increase in the levels of intracellular calcium and disruption of microtubules are the major molecular mechanisms associated with the toxicity of methyl mercury. The increase in intracellular Ca^{2+} concentration by methyl mercury disturbs the neurotransmitter function by binding to sulfhydryl/thiol groups of proteins (Sanfeliu et al., 2003; Bridges and Zalups, 2005). Further, this binding results in the activation of sulfur and obstructs associated hormones, enzymes, and cofactors (Mathieson, 1995). The mercury sulfhydryl bond, although by itself is both stable and divided among neighboring sulfhydryl containing ligands, but it forms the basis for chelation by heavy metals by contributing free sulfhydryl groups encouraging mobility of metals within ligands (Bernhoft, 2011).

Furthermore, methyl mercury causes damage to mitochondria, per-oxidation of lipids and also results in accumulation of neuro-toxic molecules such as glutamate, serotonin, and aspartate in cells (Patrick, 2002). This affects the integrity of cells leading to free radical formation. Overproduction of free radical may also be a result of indirect interaction of methyl mercury at essential cell sites or that of inhibition of defensive mechanisms. Following the administration of methyl-mercuric chloride to *in vitro* cells, rodents, and fish, reactive oxygen species (ROS) are produced in brain, kidneys, and liver (Berntssen et al., 2003; Costa et al., 2007; Mori et al., 2007).

4.4.3   CADMIUM

The sources of cadmium in the environment include natural sources (e.g., volcanic eruptions, weathering, and river transport) and various anthropogenic activities (e.g., manufacture of fertilizers, incineration of municipal wastes, smelting, mining, and tobacco smoking). Cadmium is not degraded in the environment and, as such, once released it stays in circulation (Jarup et al., 2009). The cadmium levels in fish are of particular interest since it is the main source of cadmium intake for human populace. In fish, cadmium is absorbed as cadmium chloride ($CdCl_2$). Toxicological and physiological effects of cadmium in aquatic organisms rely on various factors like water hardness and concentration of selenium, calcium, and zinc in water (Davies et al., 1993; Mayers, 1999; Messaoudi et al., 2010b).

Accumulation of cadmium in brain appears to rely on the route of administration. There was no significant rise in cadmium concentration in the brain of trout on receiving oral treatment (Melgar et al., 1997), but cadmium taken up through olfactory epithelium has been shown to be transferred to the brain (Tallkvist et al., 2002), a route in sync with the waterborne acquisition. The accumulation of cadmium in brain augments the expression of detoxifying genes (mt1 and mt2) and apoptotic gene (c-jun) (Gonzalez et al., 2006). The blood-brain barrier (BBB) is an anatomically defined barrier which prevents the entry of many potential toxicants in the nervous system. During the neonatal development, when the BBB is not adequately developed, the CNS is mostly vulnerable to toxic effects of Cadmium. Toxic consequences of cadmium in different freshwater fishes are summarized in Table 4.1. In a study carried out on silver catfish brain and muscle, acetylcholinesterase (AChE) enzyme activity was reported to be reduced after cadmium exposure (Pretto et al., 2010). A decreased AChE activity leads to ACh accumulation

TABLE 4.1 Toxic Effects of Cadmium in Fish

Species	Age at Exposure	Chemical	Potential Toxic Effects	References
Fathead minnows	Adult	$CdCd(NO_3)_2$	• Increase in auditory threshold	Low and Higgs (2015)
Rainbow trout	2–5 dpf	$CdCd(NO_3)_2$	• Higher growth rate, altered social behavior and olfactory accumulation	Sloman et al. (2003)
Zebrafish	Adult	$CdCl_2$	• Induction of nrf2 antioxidant genes (increased olfactory neuron cell death)	Wang and Gallagher (2013); Avallone et al. (2015a, b)
			• Changes in retinal morphology and ultrastructure, increased light sensitivity (at higher concentrations)	
Zebrafish	Adult (males)	$CdCl_2$	• Increased apoptotic (c-jun) and detoxifying genes (mt1 & mt2)	Gonzalez et al. (2006)
Zebrafish	4–24 hpf	$CdCl_2$	• Decreased head size, unclear brain divisions, reduced proneuronal gene expression	Chow et al. (2008)
Zebrafish	Adult (brain)	Cd acetate	• Necleotide hydrolysis	Sonnack et al. (2015)
Zebrafish	0–72 hpf Juvenile	Cd	• Neuromast damage	Zheng et al. (2017)
			• Increased ROS and immunotoxicity	

within synapses damaging important behavioral functions like feeding and swimming (Glusczak et al., 2006). Furthermore, cadmium is well-known to cause oxidative stress and its exposure may cause pathophysiological damages including reduction of growth rate in fishes and various other aquatic organisms (Kaviraj and Ghosal, 1997; Hassan et al., 2002; Das and Khangarot, 2010) (Figure 4.2).

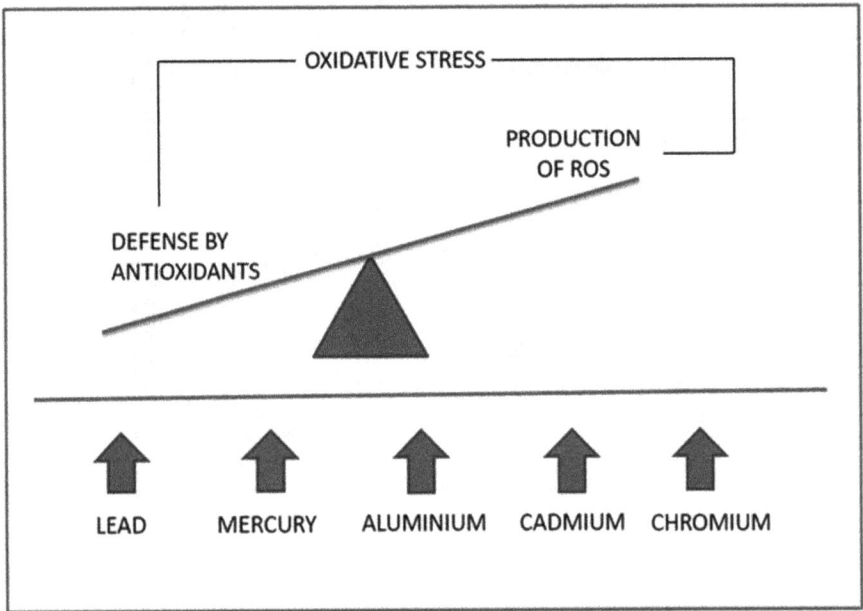

FIGURE 4.2   Mechanism of toxicity by heavy metals in fishes. Imbalance between the production of ROS and defense by antioxidants causes oxidative stress, which ultimately leads to death.

4.4.4   ALUMINUM

Aluminum, one of the most common metals in nature, makes up about 82% of the crust of earth and is present in almost all natural water resources (Gupta et al., 2013). It enters into these water areas by various ways: with partial dissolution of clays and alumino-silicates, with atmospheric precipitation, sewage, and industrial wastes (Walton et al., 2010; Liu et al., 2014; Aksu, 2015). Water quality characteristics like pH, content of organic carbon and hardness determine the biological effects of aluminum (Trenfield et al., 2012; Cardwell et al., 2017).

One important factor for aluminum toxicity is mobility of Al^{3+} ions. Aluminum ions induced double-strand breaks in DNA of embryonic zebrafish cells ZF4 after a few hours of exposure indicating its genotoxic potential (Pereira et al., 2013). The toxic effect of aluminum is associated with its ability to accumulate in animal tissues due to easy integration into the structure of proteins. The degree of aluminum accumulation in living organisms is limited by the insolubility of some of its natural compounds in water. However, due to the fact that insoluble aluminum compounds are solubilized in the acidic environment of vertebrate stomach, the Al^{3+} content in the body rises. Higher than normal concentrations of aluminum ions in the organism provoke the development of various pathologies, caused mainly by a disorder of the oxidation-reduction balance of tissues (Nedzvetskii et al., 2006; Fernandez-Davila et al., 2012). The excessive aluminum levels in the body of aquatic organisms affect the target organs such as kidneys, bones, and the CNS. Neurotoxicity of aluminum may be accompanied by impaired glucose metabolism, cytoskeleton changes and modulation of signal transduction mechanisms (Singla et al., 2015). Different studies have reported that exposure to aluminum in higher concentrations may lead to the induction of oxidative stress, stimulation of the formation of cellular ROS (Li et al., 2006; Sinha et al., 2007), induction of lipid peroxidation (LPX), alteration in the activities of various antioxidant enzymes such as glutathione peroxidase (GPX), superoxide dismutase (SOD) and catalase (CAT) and promotion of protein oxidation (Parvez and Raisuddin, 2005; Almroth et al., 2005; Vlahogianni et al., 2007). The neural system is specifically vulnerable to damage caused by oxidation because brain consumes oxygen at high rate, contains rich amount of extremely oxidizable polyunsaturated fatty acids in its membrane, less active antioxidant enzymes (e.g., GPX, SOD, and CAT) as compared to those present in other body tissues and higher amount of iron (Fe). In addition, numerous neurotransmitter molecules such as norepinephrine and dopamine are auto-oxidizable molecules as they produce active quinines and ROS on reacting with oxygen (O_2) (Verstraeten et al., 2008). In fish, aluminum may deposit on gills and damage them, alter osmoregulation and causes oxidative stress in WBC's (Garcia-Medina et al., 2010). The formation of inositol phosphate is affected by aluminum has been reported by *in vitro* and *in vivo* studies. Some components of the inositol phosphate signal system, including proteins (phosphatidylinositol-specific phospholipase C, protein kinase C and G protein) are extremely sensitive to the content of aluminum ions, which inhibit hydrolysis of phosphoinositol, resulting in the decreased level of inositol triphosphate (IP3) and release of Ca^{2+} ions from intracellular membrane storages (Sun et al., 2008). Glial cell reactivity is one of the widest used biomarkers of neural tissue disorders (Tykhomyrov et al., 2016). The upregulated expression of glial fibrillary acidic protein (GFAP) in astrocytes is an obligatory molecular event that accompanies cell response in neural tissue against toxic chemicals and different damages (Nedzvetskii et al., 2012). This phenomenon is termed astrogliosis and occurs in the brains of all kinds of vertebrates. Several kinds of fish could be more sensitive to aluminum ions toxicity. It has been presumed that aluminum ions activate in the fish brain intracellular proteases which have a capacity to destroy the proteins of intermediate filaments. Aluminum ions induce integrated changes, the more important of which are a significant increase in final LPO products, an increase in antioxidant enzyme activity, a reactivation of brain glial cells. Integrated determination of the content and polypeptide fragments of specific astrocyte proteins in fishes brains coupled with oxidative stress data may be used as valid biomarkers of toxic pollutant effects in aquatic environments.

4.4.5 CHROMIUM

Chromium (Cr) is largely present in the natural environment in its trivalent insoluble form [Cr (III)]. As a result of growing industrialization there is a worldwide increase in the occurrence of soluble Cr (VI). Cr (VI), being soluble gets immediately seeped into groundwater or gets mixed with surface water from soil in concentrations much higher than the permissible levels. Most heavy metals including chromium exhibit their toxic effects on various organ systems of aquatic organisms including the CNS (Lionetto et al., 2003).

Heavy metals exert their lethal effects by penetrating the blood-brain barrier that leads to oxidative stress and changes the metabolic path of protein molecules connected with degeneration of nerves (Senger et al., 2006). Chromium suppresses the activity of AChE, causing ACh accumulation, thereby interfering with the nerve function (Bhattacharya, 1993). ACh stimulates the muscarinic and cholinergic receptors, and mostly deals with cognitive functions of the brain (Richetti et al., 2011). When accumulated at synaptic junctions, ACh can cause severe cholinergic syndrome which can even lead to death (Hsieh et al., 2001). AChE is a vital and reliable molecular marker for evaluating contaminations of environment.

AChE activity after chromium exposure in rockfish *S. schlegelii* was considerably repressed in the brain at a concentration of 240 mg/kg after two weeks and over 60 mg/kg after weeks. Among different contaminants, hexavalent chromium (Cr VI) is one of the potent inducers to produce ROS as it is converted into pentavalent chromium (Cr V) in liver and, additionally, this process causes lipid peroxidation (LPO) (Gunaratnam et al., 2002). Hexavalent chromium is relatively more lethal to fish species of freshwater. Toxicity due to Cr (VI) causes severe damage to nervous tissue by forming ROS (Kumari et al., 2011). AChE was significantly declined in *Mytilus galloprovincialis* when exposed to Cr VI (Ciacci et al., 2012). Constant exposure to Chromium affects various enzyme activities such as succinate dehydrogenase, pyruvate dehydrogenase, and lactate dehydrogenase in liver, kidneys, and brain. Toxicological pathology induced by Chromium is fundamentally influenced by specific factors like age, type of species, conditions of environment, introduction time, and focus.

4.4.6 COPPER

Copper, an essential micronutrient becomes toxic at higher concentrations to water dwellers (Upadhyay and Panda, 2010). Brain, being the focus of CNS in all vertebrates, its homeostasis is of supreme importance for the survival of fish (Mieiro et al., 2011; Mustafa et al., 2012). Fish brain is a vital tool for assessing the environmental risk caused by increased copper exposure since it has a vital role in the bio-magnification process (Mieiro et al., 2009). Excessive copper causes toxicity by generating ROS which attack the polyunsaturated fatty acids in brain (Gumuslu, 2004; Monteiro, 2009). Under oxidative stress GSH scavenges singlet oxygen and hydroxyl radicals to intact cells (Halliwell, 1993). There is a decrease in the brain GSH and AHR (anti-hydroxyl radical) under the exposure of copper and this phenomenon can be explained by the following three reasons. Firstly, increased amounts of GSH are consumed due to an increase in the production of ROS under the exposure of copper. Secondly, the GR activity of brain is declined as a result of copper exposure in fish resulting in a concomitant decrease in the GSH content. This is because the regeneration of GSH is dependent on GR that catalyzes the reduction of GSSG to GSH (Schmidt and Dringen, 2012). GSH is oxidized to GSSG after scavenging ROS (Wu et al., 2004). Thirdly, it has been seen that under copper exposure, brain injury occurred after the lipids and proteins of the brain endured severe oxidative damage. This damage causes GSH leak from the brain. In addition to this, behavioral abnormalities characterized by impairment in swimming patterns are also associated with copper toxicity (Sandahl et al., 2004; De Boeck et al., 2006; McIntyre et al., 2008). Other mechanisms associated with copper-induced neurotoxicity include olfactory transduction signaling and olfactory sensor call disruption (Linbo et al., 2006; De Boeck et al., 2006; Tilton et al., 2008).

4.4.7 ZINC

Zinc is a very important trace element needed for maintaining functions of various organ systems. It is also vital for growth, development, and repair processes in living organisms. Besides, the divalent cation of zinc (Zn^{2+}) acts as a cofactor for a range of enzymes. Nevertheless, zinc beyond a certain level can have deleterious pathological and fatal toxicological effects. Zinc, when in excess, acts as a neurotoxicant, and causes degeneration of the CNS (Mugia et al., 2006). Cytosolic Zn^{2+}, at elevated levels gets transported through calcium channels and triggers downstream processes culminating in cell death of neurons (Cia et al., 2006). Different *in vitro* studies conducted on neurotoxicity of zinc have shown that intracellular free Zn^{2+} causes an increase in the concentration of intracellular protons that affect the role of nerve cells (Kiedrowski et al., 2014).

Studies have associated numerous degenerative diseases like Alzheimer's disease, vascular dementia and Parkinson's disease with heavy metal toxicity. In a study carried out on zebrafish, it was seen that zinc chloride caused damage to brain areas associated with memory. Zinc chloride further affected the behavioral profile of the fish causing decreased locomotor activity, reduced aggression, impaired short-term memory and increased anxiety levels. In the same study, ACh was seen inhibited significantly with $ZnCl_2$ exposure dosage greater than 1 ppm (Sarasamma et al., 2018).

Excess zinc is commonly noticed in neurodegenerative diseases. Zinc finger proteins are vital transcriptional factors that define the response of cells to various stimuli including toxic levels of heavy metals. Zinc gets accumulated in toxic concentrations either through trans-synaptic movement or it may get mobilized through receptor-mediated calcium channels or voltage-gated calcium channels or through membrane receptors sensitive to zinc. Zinc wields its neurotoxicity by different mechanisms including the production of ROS and interference with metabolic enzymes, eventually triggering apoptotic processes.

4.4.8 NICKEL

Nickel is a universal, trace element found in nature which forms approximately 0.0086% of earth's crust. Water bodies contain significantly higher amount of nickel (Kienle et al., 2008). The concentration of nickel generally varies from less than 10 μg/L to about 1000 μg/L in highly polluted waters (Cempel and Nikel, 2006). Nickel has exceptional physical and chemical properties and therefore finds use in many industrial processes like welding, electroplating, and manufacture of stainless steel. It is also utilized as a raw material for in the manufacturing of batteries and catalysts (Miller, 1996). The significant neurotoxic effects of nickel are summarized in Figure 4.3. Topal et al. (2015) reported nickel accumulation in the fish brain after studying the biochemical, immunohistochemical, and histopathological consequences of nickel chloride in the brain of Rainbow trout. The trout was treated with a concentration of 1–2 mg/L of nickel chloride for 21 days. Treatment with nickel brought about a substantial increase in the activity of SOD and LPO in brain. The glutathione levels were also increased in the brain. However, CAT and acetylcholinesterase (AChE) enzyme activities were significantly decreased. Necrotic changes and demyelination of the brain tissues was observed. The results were further indicative of histopathological damage and oxidative stress in the fish brain. Adverse effects such as decrease in locomotor activity, inhibition of antioxidant enzymes and apoptosis in fishes have been linked to nickel chloride (Kienle et al., 2008; Zheng et al., 2014). Additionally, nickel is capable to alter protein function by substituting metals such as zinc in metal-dependent enzymes (Chen et al., 2017; Rana, 2008). It competes with calcium for specific receptors by easily traversing the calcium channels of cellular membranes (Funakoshi et al., 1997). Furthermore, nickel causes formation of ROS by cross-linking amino acids with DNA (Chakrabarti et al., 2001) (Figures 4.3).

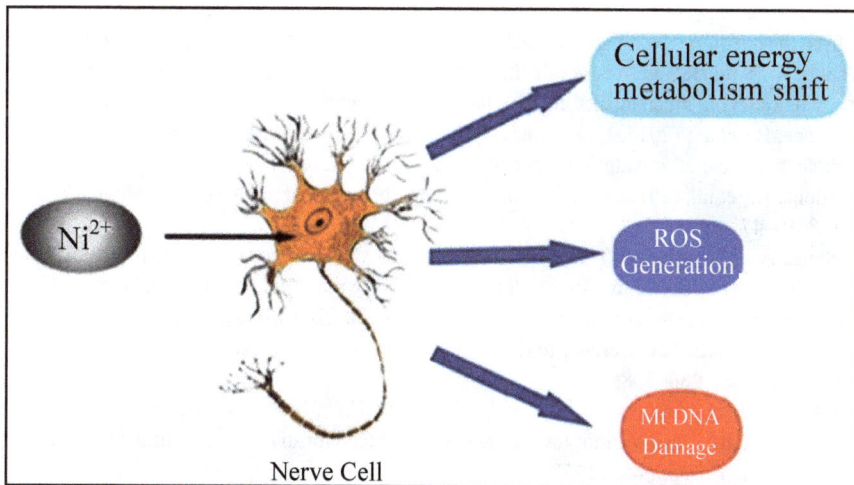

FIGURE 4.3   Neurotoxicity by nickel.

ACKNOWLEDGMENT

Special thanks are due to Research Center, College of Pharmacy, King Saud University, Riyadh, and Deanship of Scientific Research, King Saud University, Kingdom of Saudi Arabia.

KEYWORDS

- bioaccumulation
- bioaccumulation
- heavy metal toxicity
- heavy metals
- neurotoxcity
- neurotoxicological effects

REFERENCES

Aksu, A. J., (2015). Sources of metal pollution in the urban atmosphere (A case study: Tuzla, Istabul). _Journal of Environmental Health Science and Engineering, 13_, 79.

Almroth, B. C., Sturve, J., Berglund, A., & Forlin, L., (2005). Oxidative damage in eelpout (_Zoarces viviparous_), measured as protein carbonyls and TBARS, as biomarkers. _Aquat. Toxicol., 73_, 171–180.

Al-Weher, S. M., (2008). Levels of heavy metal Cd, Cu and Zn in three fish species collected from the northern Jordan valley, Jordan. _Jordan Journal of Biological Sciences, 1_(1), 41–46.

Annab, A., Said, K., & Messaoudi, I., (2013). Cadmium: Bioaccumulation, histopathology and detoxifying mechanisms in fish. _Am. J. Res. Commun., 1_, 60–79.

Avallone, B., et al., (2015a). Structural and functional changes in the zebrafish (_Danio rerio_) skeletal muscle after cadmium exposure _Cell Biol. Toxicol._

Avallone, B., et al., (2015b). Cadmium effects on the retina of adult _Danio rerio. C. R. Biol., 338_, 40–47.

Barbosa, A. C., Jardim, W., Dorea, J. G., Fosberg, B., & Souza, J., (2001). Air mercury speciation as a functioning of gender, age, and body mass index in habitants of the Negro river basin, Amazon. brazil. _Arch. Environ. Contamin. Toxicol., 40_, 439–444.

Bernhoft, R. A., (2012). Mercury toxicity and treatment: A review of the literature. _J. Environ. Public Health,_ 460–508.

Berntssen, M. H., Aatland, A., & Handy, R. D., (2012). Chronic dietary mercury exposure causes oxidative stress, brain lesions, and altered behavior in Atlantic salmon (_Salmo salar_) parr. _Aquat. Toxicol., 65_, 55–72.

Bhattacharya, S., (1993). Target and non-target effects of anti-cholinesterase pesticides in fish. *Sci. Total Environ., 134*, 859–866.

Bolognesi, C., Landini, E., Roggieri, P., Fabbri, R., & Viarengo, A., (1999). Genotoxicity biomarkers in the assessment of heavy metal effects in mussels. *Experimental Studies. Environmental and Molecular Mutagenesis, 33*, 287–292.

Bradl, H. B., (2005). Chapter 1: Sources and origins of heavy metals. *Interface Science and Technology, 6*, 1–27.

Bridges, C. C., & Zalups, R. K., (2005). Molecular and ionic mimicry and the transport to toxic metals. *Toxicol. Appl. Pharmacol., 204*, 274–308.

Burger, J., & Gochfeld, M., (2009). Perceptions of the risks and benefits of fish consumption: Individual choices to reduce risk and increase health benefits. *Environ. Res., 109*(3), 343–349.

Burger, J., & Gochfeld, M., (2011). Mercury and selenium levels in 19 species of salt water fish from New Jersey as a function of species size, and season. *Sci. Total Environ., 409*, 1418–1429.

Cai, A. L., Zipfel, G. J., & Sheline, C. T., (2006). Zinc neurotoxicity is dependent on intracellular NAD+ levels and the sirtuin pathway. *Eur. J. Neurosci., 24*, 2169–2176.

Cardwell, A. S., Adams, W. J., Gensemer, R. W., Nordheim, E., Santore, R. C., Ryan, A. C., et al., (2018). Chronic toxicity of aluminum, at a pH of 6, to freshwater organisms: Empirical data for the development of international regulatory standards/criteria. *Environ. Toxicol. Chem., 37*(1), 36–48.

Cempel, M., & Nikel, G., (2006). Nickel: A review of its sources and environmental toxicology. *Polish J. Environ. Stud., 15*(3), 375–382.

Chiang, C. W., Ng, D. Q., Lin, Y. P., & Chen, P. J., (2016). Dissolved organic matter or salts change the bioavailability processes and toxicity of the nanoscale tetravalent lead corrosion product PbO_2 to medaka fish. *Environmental Science and Technology, 50*, 11292–11301.

Chow, E. S. H., Hui, M. N. Y., Lin, C. C., & Cheng, S. H., (2008). Cadmium inhibits neurogenesis in zebrafish embryonic brain development. *Aquat. Toxicol., 87*, 157–169.

Ciacci, C., Barmo, C., Gallo, G., Maisano, M., Cappello, T., D'Agata, A., et al., (2012). Effects of sub-lethal, environmentally relevant concentrations of hexavalent chromium in the gills of *Mythilus galloprovincialis. Aquat. Toxicol., 120, 121*, 109–118.

Cicik, B., (2003). Bakır-çinko etkileşiminin sazan (*Cyprinus carpio*)'nın karaciğer, solungaç ve kas dokularındaki metal birikimi üzerine etkileri. *Ekoloji Çevre Dergisi, Cilt., 12, Sayı, 48*, 32–36.

Costa, L. G., Fattori, V., Giordano, G., & Vitalone, A., (2007). An *in vitro* approach to assess the toxicity of certain food contaminants methyl mercury and polychlorinated biphenyls. *Toxicology, 237*, 65–76.

Das, S., & Khangarot, B. S., (2010). Bioaccumulation and toxic effects of cadmium on feeding and growth of an Indian pond snail *Lymnaea luteola* L. under laboratory conditions. *J. Hazard. Mat., 15*, 763–770.

Davies, P. H., Gorman, W. C., Carlson, C. A., & Brinkman, S. R., (1993). Effect of hardness on bioavailability and toxicity of cadmium to rainbow trout. *Chem. Speciat. Bioavailab., 5*, 67–77.

Egila, J. N., & Daniel, V. N., (2011). Trace metals accumulation in freshwater and sediment insects of liberty dam, Plateau State Nigeria. *Int. J. Basic Appl. Sci.*, *11*, 128–140.

Ersoy, B., & Çelik, M., (2009). Essential elements and contaminants in tissues of commercial pelagic fish from the eastern Mediterranean sea. *Journal of the Science of Food and Agriculture*, *89*(9), 1615–1621.

Farkas, A., Salánki, J., & Varanka, I., (2000). Heavy metal concentrations in fish of Lake Balaton. *Res. Manag.*, *5*, 271–279.

Fernandez-Davila, M. L., Razo-Estrada, A. C., García-Medina, S., Gomez-Olivan, L. M., Pinon-Lopez, M. J., Ibarra, R. G., & Galar-Martinez, M., (2012). Aluminum-induced oxidative stress and neurotoxicity in grass carp (Cyprinidae-*Ctenopharingodon idella*). *Ecotoxicology and Environmental Safety, 76*(2), 87–92.

Fomina, M., & Gadd, G. M., (2014). Biosorption: Current perspectives on concept, definition, and application. *Bioresour. Technol.*, *160*, 3–14.

Galar-Martinez, M., Gomez-Olivan, L. M., Amaya-Chavez, A., Razo-Estrada, C., & Garcia-Medina, S., (2010). Oxidative stress induced on *Cyprinus carpio* by contaminants present in the water and sediment of Madin reservoir. *J. Environ. Sci. Health, Part A, 45*, 875–882.

García-Medina, S., Razo-Estrada, A. C., Gómez-Oliván, L. M., Amaya-Chávez, A., Madrigal-Bujaidar, E., & Galar-Martínez, M., (2009). Aluminum-induced oxidative stress in lymphocytes of common carp (*Cyprinus carpio*). *Fish Physiol. Biochem.*, *36*, 875–882.

Georgieva, E., Velcheva, I., Yancheva, V., & Stoyanova, S., (2014). Trace metal effects on gill epithelium of common carp *Cyprinus carpio* L. (cyprinidae) *Acta Zool. Bulgarica.*, *66*, 277–282.

Gonzalez, P., Baudrimont, M., Boudou, A., & Bourdineaud, J. P., (2006). Comparative effects of direct cadmium contamination on gene expression in gills, liver, skeletal muscles, and brain of the zebrafish (Danio rerio). *Biometals*, *19*, 225–235.

Goyer, R. A., & Clarsksom, W. T., (2001). Toxic effects of metals. In: Klaassen, C. D., (ed.), *Casarett and Doull's Toxicology* (pp. 811–867). The basic Science of Poisons. McGraw-Hill, New York.

Gumgum, B., Unlu, E., Tez, Z., & Gulsun, N., (1994). Heavy metal pollution in water, sediment and fish from the Tigris River in Turkey. *Chemosphere*, *290*(1), 111–116.

Gunaratnam, M., Pohlscheidt, M., & Grant, M. H., (2002). Pre-treatment of rats with the inducing agents phenobarbitone and 3-methylcholanthrene ameliorates the toxicity of chromium (VI) in hepatocytes. *Toxicol. Vitr.*, *16*, 509–516.

Gupta, N., Gaurav, S. S., & Kumar, A., (2013). Molecular basis of aluminum toxicity in plants: A review. *Am. J. of Plant Sci.*, *4*, 21–37.

Halliwell, B., (1993). The role of oxygen radicals in human disease, with particular reference to the vascular system. *Pathophysiol. Haemostasis Thromb.*, *23*, 118–126.

Hansen, J. A., Welsh, P. G., Lipton, J., & Suedkamp, M. J., (2002). The effects of long-term cadmium exposure on the growth and survival of juvenile bull trout (*Salvelinus confleuentus*). *Aquat. Toxicol.*, *58*, 165–174.

Harms, H., Schlosser, D., & Wick, L. Y., (2011). Untapped potential: Exploiting fungi in bioremediation of hazardous chemicals. *Nat. Rev. Microbiol.*, *9*(3), 177–192.

Helfman, S. G., Collette, B. B., Facey, D. E., & Bowen, B. W., (2009). The diversity of fishes. *Biology, Evolution and Ecology* (2nd edn.). Wiley-Blackwell.

Hsieh, B. H., Deng, J. F., Ger, J., & Tsai, W. J., (2001). Acetylcholinesterase inhibition and the extrapyramidal syndrome: A review of the neurotoxicity of organophosphate. *Neuro Toxicology, 22*, 423–427.

Hsu, P. C., & Guo, Y. L., (2002). Antioxidant nutrients and lead toxicity. *Toxicology, 180*(1), 33–44.

Jackson, T. A., (1997). Long-range atmospheric transport of mercury to ecosystems, and the importance of anthropogenic emissions: A critical review and evaluation of the published evidence. *Environ. Rev., 5*, 99–120.

Jaishankar, M., Tseten, T., Anbalagan, N., Blessy, B. M., & Krishnamurthy, N. B., (2014). Toxicity, mechanism, and health effects of some heavy metals. *Interdiscip. Toxicol., 7*(2), 60–72.

Järup, L., (2009). Current status of cadmium as an environmental health problem. *Toxicol. Appl. Pharmacol., 238*(3), 201–208.

Kadar, I., Koncz, J., & Fekete, S., (2000). Experimental study of Cd, Hg, Mo, Pb and Se movement in soil-plant-animal systems. In: *Kniva, International Conference Proceedings* (pp. 72–76). *Patija, Croatia.*

Kaviraj, A., & Ghosal, T. K., (1997). Effects of poultry litter on the chronic toxicity of cadmium to common carp (*Cyprinus carpio*). *Bioresource Technol., 60*, 239–243.

Kiedrowski, L., (2014). Proton-dependent zinc release from intracellular ligands. *Journal of Neurochemistry, 130*, 87–96.

Kienle, C., Kohler, H. R., Filser, J., & Gerhardt, A., (2008). Effects of nickel chloride and oxygen depletion on behavior and vitality of zebrafish (*Danio rerio*, Hamilton, 1822) (Pisces, Cypriniformes) embryos and larvae. *Environ. Pollut., 152*(3), 612–620.

Kim, J. H., & Kang, J. C., (2017a). Effects of sub-chronic exposure to lead (Pb) and ascorbic acid in juvenile rockfish: Antioxidant responses, MT gene expression, and neurotransmitters. *Chemosphere, 171*, 520–527.

Kumari, K., Ranjan, N., & Sinha, R. C., (2011). Multiple biomarker response in the fish, *Labeo rohita* due to hexavalent chromium. In: *Proceedings of the 2nd International Conference on Biotechnology and Food Science, 7*, 155–158.

Li, M., Hu, C., Zhu, Q., Chen, L., Kong, Z., & Liu, Z., (2006). Copper and zinc induction of lipid peroxidation and effects on antioxidant enzyme activities in the microalga *Pavlova viridis* (Prymnesiophyceae). *Chemosphere, 62*, 565–572.

Lionetto, M. G., Caricato, R., Giordano, M. E., Pascariello, M. F., Marinosci, L., & Schettino, T., (2003). Integrated use of biomarkers (acetylcholinesterase and antioxidant enzymes activities) in *Mytilus galloprovincialis* and *Mullus barbatus* in an Italian coastal marine area. *Mar. Pollut. Bull., 46*, 324–330.

Liu, S., Noth, E. M., Dixon-Ernst, C., Eisen, E. A., Cullen, M. R., & Hammond, S. K., (2014). Particle size distribution in aluminum manufacturing facilities. *Environmental Pollution, 4*(3), 79–88.

Low, J., & Higgs, D. M., (2015). Sublethal effects of cadmium on auditory structure and function in fathead minnows (*Pimephales promelas*) *Fish Physiol. Biochem., 41*, 357–369.

Mansour, S. A., & Sidky, M. M., (2002). Ecotoxicological studies. 3. Heavy metals contaminating water and fish from Fayoum Governorate, Egypt. *Food Chem., 78*, 15–22.

Marchetti, C., (2003). Molecular targets of lead in brain neurotoxicity. *Neurotox. Res.*, *5*(3), 221–236.

Mathieson, P. W., (1995). Mercury: God of TH2 cells. *Clin. Exp. Immunol.*, *102*, 229–230.

Melgar, M. J., Perez, M., Garcia, M. A., Alonso, J., & Miguez, B., (1997). The toxic and accumulative effects of short-term exposure to cadmium in rainbow trout (*Oncorhynchus mykiss*). *Vet. Hum. Toxicol.*, *39*, 79–83.

Messaoudi, I., Hammouda, F., Heni, J. E., Baati, T., Saïd, K., & Kerkeni, A., (2010b). Reversal of cadmium-induced oxidative stress in rat erythrocytes by selenium, zinc or their combination. *Exp. Toxicol. Pathol.*, *62*(3), 281–288.

Meyers, J. S., (1999). A mechanistic explanation for the ln (LC50) vs. ln (hardness) adjustment equation for metals. *Environ. Sci. Technol.*, *33*, 908–912.

Mieiro, C. L., Pacheco, M., Pereira, M. E., & Duarte, A. C., (2009). Mercury distribution in key tissues of fish (*Liza aurata*) inhabiting a contaminated estuary—implications for human and ecosystem health risk assessment. *J. Environ. Monit.*, *11*, 1004–1012.

Mieiro, C. L., Pereira, M. E., Duarte, A. C., & Pacheco, M., (2011). Brain as a critical target of mercury in environmentally exposed fish (*Dicentrarchus labrax*)—bioaccumulation and oxidative stress profiles. *Aquat. Toxicol.*, *103*, 233–240.

Monteiro, S. M., Dos, S. N., Calejo, M., Fontainhas-Fernandes, A., & Sousa, M., (2009). Copper toxicity in gills of the teleost fish, *Oreochromis niloticus*: Effects in apoptosis induction and cell proliferation. *Aquat. Toxicol.*, *94*, 219–228.

Mori, N., Yasutake, A., & Hirayama, K., (2007). Comparative study of activities in reactive oxygen species production/defense systemin mitochondria of rat brain and liver, and their susceptibility to methylmercury toxicity. *Arch. Toxicol.*, *81*, 769–776.

Murgia, C., Lang, C., Truong-Tran, A., Grosser, D., Jayaram, L., Ruffin, R., et al., (2006). Zinc and its specific transporters as potential targets in airway disease. *Curr. Drug Targets.*, *7*, 607–627.

Mustafa, S. A., Davies, S. J., & Jha, A. N., (2012). Determination of hypoxia and dietary copper mediated sub-lethal toxicity in carp, *Cyprinus carpio*, at different levels of biological organization. *Chemosphere*, *87*, 413–422.

Nedzvetskii, V. S., Kirichenko, S. V., Baydas, G., & Nerush, O. P., (2012). Effects of melatonin on memory and learning deficits induced by exposure to thinner. *Neurophysiology/ Neirofiziologiya*, *44*(1), 42–48.

Nedzvetskii, V. S., Tuzcu, M., Yasar, A., Tikhomirov, A. A., & Baydas, G., (2006). Effects of vitamin E against aluminum neurotoxisity in rats. *Biochemistry* (*Moscow*), *71*(3), 239–244.

Nunes, B., Brandão, F., Sérgio, T., Rodrigues, S., Gonçalves, F., Correia, et al., (2014). Effects of environmentally relevant concentrations of metallic compounds on the flatfish *Scophthalmus maximus*: Biomarkers of neurotoxicity, oxidative stress, and metabolism. *Environmental Science and Pollution Research International, Environmental Science and Pollution Research*, *21*(12).

Obasohan, E. E., Oronsaye, J. A. O., & Eguavoen, O. I., (2008). A comparative assessment of the heavy metal loads in the tissues of a common catfish (*Clarias gariepinus*) from Ikpoba and Ogba Rivers in Benin City Nigeria. *Afr. Sci.*, (*9*), 13–23.

Olaifa, F. E., Olaifa, A. K., Adelaja, A. A., & Owolabi, A. G., (2004). Heavy metal contamination of *Clarias garpinus* from a lake and fish farm in Ibadan. Nigeria. *Afr. J. Biomed. Res.*, *7*, 145–148.

Oliveira, R. C. A., Vollaire, Y., Sanchez-Chardi, A., & Roche, H., (2005). Bioaccumulation and the effects of organochlorine pesticides PAH and heavy metals in the eel (*Anguilla anguilla*) at the Camargue nature reserve, France. *Aqua Toxicol., 74*, 53–69.

Pagenkopf, G. K., (1983). Gill surface interaction model for trace-metal toxicity to fishes. Role of complexation, pH and water hardness. *Environ. Sci. Technol., 17*, 342–347.

Parvez, S., & Raisuddin, S., (2005). Protein carbonyls: Novel biomarkers of exposure to oxidative stress-inducing pesticides in freshwater fish *Channa punctate* (Bloch). *Environ. Toxicol. Pharmacol., 20*, 112–117.

Patrick, L., (2003). Toxic metals and antioxidants. Part II. The role of antioxidants in arsenic and cadmium toxicity. *Altern. Med. Rev., 8*, 106128.

Pereira, S., Cavalie, I., Camilleri, V., Gilbin, R., & Adam-Guillermin, C., (2013). Comparative genotoxicity of aluminum and cadmium in embryonic zebrafish cells. *Mutation Research, 750*(1/2), 19–26.

Rajeshkumar, S., & Xiaoyu, L., (2018). Bioaccumulation of heavy metals in fish species from the Meiliang Bay, Taihu Lake, China. *Toxicology Reports, 5*, 288–295.

Rice, C., Ghorai, J. K., Zalewski, K., & Weber, D. N., (2011). Development lead exposure causes startle response deficits in zebrafish. *Aquatic Toxicology, 105*, 600–608.

Richetti, S., Rosemberg, D., Ventura-Lima, J., Monserrat, J., Bogo, M. R., & Bonan, C., (2011). Acetylcholinesterase activity and antioxidant capacity of zebrafish brain is altered by heavy metal exposure. *Neurotoxicology, 32*, 116–122.

Sahin, E., & Gümüslü, S., (2004). Alterations in brain antioxidant status, protein oxidation and lipid peroxidation in response to different stress models. *Behav. Brain Res., 155*, 241–248.

Sanfeliu, C., Sebastià, J., Cristòfol, R., & Rodríguez-Farré, E., (2003). Neurotoxicity of organomercurial compounds. *Neurotox. Res., 5*, 283–305.

Sarasamma, S., Audira, G., Juniardi, S., Sampurna, B. P., Liang, S., Hao, E., et al., (2018). Zinc chloride exposure inhibits brain acetylcholine levels, produces neurotoxic signatures, and diminishes memory and motor activities in Adult Zebrafish. *Int. J. Mol. Sci., 19*(10), 3195.

Schmidt, M. M., & Dringen, R., (2012). *Glutathione (GSH) Synthesis and Metabolism Neural Metabolism In Vivo* (pp. 1029–1050). Springer, US. (Reprinted).

Senger, M. R., Rico, E. P., De, B., Arizi, M., Frazzon, A. P., Dias, R. D., Bogo, M. R., et al., (2006). Exposure to Hg^{2+} and Pb^{2+} changes NTPDase and ecto-5'-nucleotidase activities in central nervous system of zebrafish (*Danio rerio*). *Toxicology, 226*, (2/3), 229–237.

Singla, N., & Dhawan, D. K., (2015). Modulation of (14) C-labeled glucose metabolism by zinc during aluminum-induced neurodegeneration. *Neuroscience Research, 93*(9), 1434–1441.

Sinha, S., Mallick, S., Misra, R. K., Singh, S., Basant, A., & Gupta, A. K., (2007). Uptake and translocation of metals in *Spinacia oleracea* L. grown on tannery sludge-amended and contaminated soils: Effect on lipid peroxidation, morpho-anatomical changes, and antioxidants. *Chemosphere, 67*, 176–187.

Sloman, K. A., et al., (2003). Cadmium affects the social behavior of rainbow trout, *Oncorhynchus mykiss*. *Aquat. Toxicol., 65*, 171–185.

Sönmez, A. Y., Hisar, O., & Yanık, T., (2013). A comparative analysis of water quality assessment methods for heavy metal pollution in Karasu stream, Turkey. *Fresenius Environmental Bulletin, 22*(2), 579–583.

Sonnack, L., et al., (2015). Effects of metal exposure on motor neuron development, neuromasts and the escape response of zebrafish embryos. *Neurotoxicol. Teratol., 50,* 33–42.

Soto, D. X., Roig, R., Gacia, E., & Catalan, J., (2011). Differential accumulation of mercury and other trace metals in the food web components of a reservoir impacted by the chlo-alkali plant (Flix Ebro River, Spain), implications for biomonitoring. *Environ. Pollut., 159,* 1481–1489.

Sun, J. J., Liu, Y., & Ye, Z. R., (2008). Effects of P2Y1 receptor on glial fibrillary acidic protein and glial cell line-derived neurotrophic factor production of astrocytes under the ischemic condition and the related signaling pathways. *Neuroscience Bulletin, 24*(4), 231–243.

Tallkvist, J., Persson, E., Henriksson, J., & Tjalve, H., (2002). Cadmium-metallothionein interactions in the olfactory pathways of rats and pikes. *Toxicol. Sci., 67,* 108–113.

Tchounwou, P. B., Yedjou, C. G., Patlolla, A. K., & Sutton, D. J., (2012). *Heavy Metals Toxicity and the Environment, 101,* 133–164.

Trenfield, M. A., Markich, S. J., Ng, J. C., Noller, B., & Van, D. R. A., (2012). Dissolved organic carbon reduces the toxicity of aluminum to three tropical freshwater organisms. *Environmental Toxicology and Chemistry, 31*(2), 427–436.

Tykhomyrov, A. A., Pavlova, A. S., & Nedzvetsky, V. S., (2016). Glial fibrillary acidic protein (GFAP), on the 45th anniversary of discovery. *Neurophysiology (Springer), 48*(1), 54–71.

Upadhyay, R., (2010). Zinc reduces copper toxicity induced oxidative stress by promoting antioxidant defense in freshly grown aquatic duckweed *Spirodela polyrhiza* L. *J. Hazard Mater., 175*(1–3), 1081–1084.

Verstraeten, S. V., Lucila, A., & Oteiza, P. I., (2008). Aluminum and lead: molecular mechanisms of brain toxicity. *Archives of Toxicology, 82*(11), 789–802.

Vlahogianni, T., Dassenakis, M., Scoullos, M. J., & Valavanidis, A., (2007). Integrated use of biomarkers (superoxide dismutase, catalase and lipid peroxidation) in mussels *Mytilus galloprovincialis* for assessing heavy metals' pollution in coastal areas from the Saronikos Gulf of Greece. *Mar. Pollut. Bull., 54,* 1361–1371.

Walton, R. C., McCrohan, C. R., Livens, F., & White, K. N., (2010). Trophic transfer of aluminum through an aquatic grazer-omnivore food chain. *Aquatic Toxicology, 99*(1), 93–99.

Wang, L., & Gallagher, E. P., (2013). Role of Nrf2 antioxidant defense in mitigating cadmium-induced oxidative stress in the olfactory system of zebrafish. *Toxicol. App. Pharmacol., 266,* 177–186.

Westerink, R. H. S., & Vijverberg, H. P. M., (2002). Ca^{2+} independent vesicular catecholamine release in PC12 cells by nanomolar concentrations of Pb^{2+}. *J. Neurochem., 80,* 861–873.

Wu, G., Fang, Y., Yang, S., Lupton, J. R., & Turner, N. D., (2004). Glutathione metabolism and its implications for health. *J. Nutr., 134,* 489–492.

Yilmaz, A. B., (2005). Comparison of heavy metal levels of grey mullet (*Mugil cephalus* L.) and sea bream (*Sparus aurata* L.) caught inIskenderun Bay (Turkey) Turk. *J. Vet. Animal Sci., 29,* 257–262.

Zhang, L., & Wong, M., (2007). Environmental mercury contamination in China: Sources and impacts. *Environment International, 33*(1), 108–121.

Zhang, Z., He, L., Li, J., & Wu, Z., (2007). Analysis of heavy metals of muscle and intestine tissue in fish in banan section of Chongqing from three gorges reservoir China. *Pol. J. Environ. Stud., 16,* 949–958.

Zheng, G. H., Liu, C. M., Sun, J. M., Feng, Z. J., & Cheng, C., (2014). Nickel induced oxidative stress and apoptosis in *Carassius auratus* liver by JNK pathway. *Aquat. Toxicol.*, *147*, 105–111.

Zheng, J. L., Yuan, S. S., Wu, C. W., Lv, Z. M., & Zhu, A. Y., (2017). Circadian time-dependent antioxidant and inflammatory responses to acute cadmium exposure in the brain of zebrafish. *Aquat. Toxicol.*, *182*, 113–119.

Zhu, B., Wang, Q., Shi, X., Guo, Y., Xu, T., & Zhou, B., (2006). Effect of combined exposure to lead and decabromodiphenyl ether on neurodevelopment of zebrafish larvae. *Chemosphere*, *144*, 1646–1654.

Ziaaz, M., Giusi, G., Crudo, M., Canonaco, M., & Facciolo, R. M., (2013). Lead-induced neurodegenerative events and abnormal behaviors occur via ORXRergic/GABAA Rergic mechanisms in a marine teleost. *Aquatic Toxicology*, *126*, 231–241.

CHAPTER 5

Mechanism of Toxicity in Aquatic Life

HILAL AHMAD GANAIE

*Department of Zoology, Government Degree College (Boys),
Pulwama–192301, Jammu and Kashmir, India,
E-mail: hilalganie@hotmail.com*

ABSTRACT

The term toxicity can be defined as the level of the chemical substance that poses damage to a living organism. The aquatic environments are facing the problems of the toxic chemicals much more as we are directly or indirectly associated with the same environment. There are various fields that are integrated in the aquatic toxicology to find the activity of the toxicant that can lead to the signs and even death of the organism. The aquatic environments are contaminated with different kinds of pesticides which are having different levels of toxicity. There is various mechanism of the mutagenicity of these pesticides, but the most important one is the use of oxidative stress as biomarkers because there occurs a positive correlation between the genotoxicity and oxidative stress. The exposure of these pesticides resulted in the increase in the oxidative stress, which may reflect increase in the lipid peroxidation in erythrocytes. The various anti-oxidative enzymes like superoxide, glutathione, dismutase, and catalase (CAT) also fluctuate significantly in the blood of these exposed organisms. Thus, we can say that the genotoxic effects of the pesticides are accompanied by the change in the concentrations of these enzymes. In a nutshell, we can say that the findings suggest that genotoxic effects are not alone responsible but are related with changes in the antioxidant enzymes and oxidative stress.

5.1   INTRODUCTION

The term toxicity can be defined as the extent of chemical substance or a pesticide to which it can cause damage to an individual or organism. The chemical substance or pesticide can cause damage to the whole individual, such as a bacterium, an animal, or a plant. It can also damage a substructure of the animal or plant such as a cell of a plant called cytotoxicity or the organ of an animal such as liver known as hepatotoxicity. The word toxicity in broader sense can be used to describe the effects of the organisms in a large scale such as the family or society and at times it can be used in day-to-day poisoning.

The main concept of toxicology is to describe the effects of the toxin in a dose-dependent manner. This means that if any substance which is not a toxic can be taken in a very high dose may prove to be a toxic whereas an established toxin if taken in a very low dose may not prove to be fatal. E.g., if water is taken in a very high dose, it can lead to intoxication whereas snake venom is in low dose is not toxic to the individual whom the snake bites. The concept of novel drug toxicity index (DTI) has been proposed recently in order to consider the dose-dependent relationship of the toxicant. The functions of the DTI is to define the toxicity of the drugs, to identify the drugs that are toxic to the liver, to provide insights about the mechanism of action of the drugs, to predict the outcomes and also acts as a potential tool for screening the drugs. The toxicity of a particular substance is species-specific, and hence the analysis of cross-species is problematic. The new concepts are emerging in which the process of animal testing is bypassed while the concepts of endpoints of toxicity are maintained.

5.2   AQUATIC TOXICOLOGY

The term "aquatic toxicology" is a broad term which includes various disciplines such as aquatic ecology, toxicology, and water chemistry. The aquatic toxicology deals not only with the study of the effects of artificial or man-made chemicals but also with natural substances and their action on aquatic animals at different levels. These levels vary from the sub-cellular of an individual to communities and to ecosystems. The field of aquatic toxicology includes all the aquatic systems like marine as well as freshwaters. The common tests used in the determination of aquatic toxicity include the standard acute and the chronic tests. The acute tests last from 24 to 96 hours while as the chronic tests last for 7 or more days. These tests are employed to measure the various endpoints of the model organism with reference to the control group. The endpoints, growth, survival, and reproduction, are measured at each concentration in a dose-dependent manner. The model organisms used to study the effects of the toxic chemical must have relevance with the ecosystem and ample literature must be available for the same. Last but not least, these model organisms can be acquired easily or cultured in laboratory and must be easy to handle.

5.3 GENETIC TOXICOLOGY

The term toxicology can be defined as the field of biology dealing with the study of the effects of a toxicant. The special branch of toxicology dealing with the assessment and study of the effects of chemical and physical toxicants on the hereditary material (DNA and RNA) of living cells is called genetic toxicology. The substances or the agents that cause changes in the DNA or RNA and their associated components at low toxic levels thereby changing the characteristics of the hereditary material or inactivate the DNA are known as genotoxic substances or mutagens. These mutagens are having some physical or chemical properties in common that facilitate their interaction with the hereditary material (DNA or RNA). Thus the key fact of the discipline of genetic toxicology is to study the target molecule. Genetic toxicologists primarily develop and validate assay systems that detect agents that alter the ability of DNA to replicate or indicate that agent has interacted with DNA. This branch has evolved when Muller (1972) first demonstrated that genes of an organism can be mutated. It was recognized as a separate branch when the "Environmental Mutagen Society" (EMS) was established by the geneticists under the leadership of Alexander Hollaender in 1971 in order to find the effects of the various artificial chemicals on the environment. This branch has an important role in evaluating the safety of the environment. Among the various functions of this branch, an important function is to implement the methods of testing and risk assessment in order to define the effects of the mutagenic substances that may be found in the environment or whose presence may change the gene-pool of the humans. The second function of this discipline is to apply the technologies to detect the carcinogens and their mechanisms. During the evolution of genetic toxicology, more than 200 *in vitro* and *in vivo* systems have been validated to a significant extent. These systems range from simple single-cell organism to complex multicellular organisms, and include systems that putatively identify somatic cell and/or germ cell geno-toxins. The US Environmental

Protection Agency (EPA) under its "Gene-Tox" program has developed an approach to determine the maturity for various genetic bioassays. The function of this branch is to develop new test systems, evaluate them and then to incorporate them in this science.

The presence of the genotoxins or mutagens in the environment and their interaction with other such substances may add up the effects of these toxic chemicals, as a result of which future generations of the species are affected. Genotoxicity and mutagenicity are two different terms and should not be confused. One should keep in mind about these mutagens and genotoxins that "all mutagens are genotoxic but not all genotoxic substances are mutagenic." All the three processes, i.e., the mutation induction, activation of mistimed event and DNA damage directly, are responsible for the indirect or direct effects of DNA. The main aim of the genotoxicity is to analyze and detect the harmful effects of the mutagens or geno-toxicants. These genotoxins not only brought about changes in the hereditary material even at low levels, but their interaction is also specific with DNA. Hence, genotoxicity has two main aims:

- To depict the results that are produced by the toxicants to various model organisms that must be based only from the viewpoint of genetics and later to put forth the conclusions that must be applied to humans; and
- To investigate the method(s) by which these toxicants act on the model organisms and then assess the risks associated with the environment that is based on this knowledge (Moutschen, 1985).

This field of genotoxicity came from the studies of Alexander-Hollaender (1960s), in whose leadership the "EMS" was founded in the USA. Muller (1927) first studied genotoxicity in Drosophila. He observed that X-rays induced sex-linked recessive lethal mutation in *Drosophila*. Auerbach (1940) had done an extensive work on the mutagenicity and carcinogenicity of various chemicals in relation to the production of mutations in *Drosophila melanogaster*. Auerbach and Robson (1944) tested allyl isothiocyanate and found that this compound is also a carcinogen. Brusick and Auletta (1985) quoted that "the EMS has a profound positive impact on many disciplines in toxicology and environmental risk assessment." After the organization of "International Association of Environmental Mutagen Societies" (IAEMS) in 1973, the active research in the field of genotoxicity starts to begin. The various International and National organizations such as the National Toxicology Program (NTP), the National Institute of Health (NIH), Food and Drug Administration (FDA), and EPA became not only interested in mutagenesis but also supported the basic research in mutagenesis. These agencies recognize the importance of genotoxicology in order to understand the genetic diseases and the carcinogenesis (Moutschen, 1985).

Genotoxins are the substances that can induce changes in DNA or RNA even at very low level. These substances or genotoxins are having some common physical or chemical characteristics which facilitate their reactions with DAN or RNA. Much of these mutagens or genotoxins can damage the hereditary material when present at high levels, thus seems to have acute and non-specific effects in various processes that take place at sub-cellular level. Hence, this branch has been set up to have a dual role in evaluation the safety programs. The first one being to implement the test for mutagenicity and assessment of the risk associated with the method in order to evaluate the effects of these mutagenic substances present in the environment, especially for those whose presence changes the entire gene-pool. The second one is to apply this test in order to understand and detect the mechanisms that are related with cancer-causing agents (Depledge, 1994). An extensive data can be collected with regard to its history as various online resources are available on the internet like GENE-TOX, TOXNET, PUBMED, MEDLINE, TOXLINE, CHEMIDPLUS at NLM, etc. In the recent because of the advances in molecular biology and various computational tools, novel fields like the toxico-genomics, databases of mutational spectra and structural activity relationship analysis are the areas which emerged in genotoxicity with ample resources online (Young, 2002).

5.4   TOXICITY TESTING WITH FISH

However, this is obvious that the advancement in research in any branch of science depends upon the methodology of that research. Based on this perspective, we can say that with the development in the tools and techniques one can easily understand the interactions between the organisms and their environment at different levels, and considerable development has been achieved in order to examine the effects of genotoxins in the aquatic systems (Jha, 1998). The effects of these genotoxic substances to the aquatic organisms mainly fish is the main concern of the research in determining the mutagenesis. Moreover, stress is being laid to develop the *in vivo* assays with the fish species in order to assess the potential effects to other higher organisms (Dar et al., 2015). Since, the fish is one of the sources of food to the humans and a rich source of proteins; hence the field of aquaculture is expanding as food industry (Tolla et al., 1995; Nagel and Isberner, 1998). In contrast, the water resources have become the dumping site for large number of chemicals which is evident from the high levels of pollution; the fish has become the target for these chemical pollutants (Schwarzenbach et al., 2006). The pollution level of any aquatic system and human health indirectly are monitored by checking the water quality of the ecosystem in which the fish lives. The fishes which are the main source of proteins are often having high levels of the methyl mercury. The damage in the chromosomes has been found in the lymphocytes of the individuals that have consumed the fish in which high levels of methyl mercury were present (WHO, 1990).

The fish is being regarded very crucial model organism in case of the testing systems that are mainly using the aquatic ecosystems. The present guidelines provided at OECD level are the best examples that cover the target organism with acute toxicity (OECD, 1992a), during the early stages of life (OECD, 1992b) and the tests at juvenile growth (OECD, 2000). Thus we say that the acute fish toxicity tests are playing a vital role in assessment of the risk to the environment and the classification of hazardous chemicals, as the fishes are the first organisms that give us the relative toxicity of different substances in different species (Wedekind et al., 2007). According to Hutchinson et al. (2003), the fish as model organism is of vital importance as it is used to assess the hazards of the aquatic systems and to assess the synthetic chemicals for their evaluation (Hutchinson et al., 2003). Hence, we can say that the fish are imported in tons per year; it has become mandatory for us to have a "base set" of data that may check the acute toxicity of these fishes (LC_{50}-96h; OECD, 1992a). Besides, a single fish species may not be sensitive to all the toxicants universally (Isomaa and Lilius, 1995). The LC_{50} of a species may be questioned with respect to its accuracy and its relevance with the toxicity (Braunbeck et al., 2005). Hence, we can say that no such species is available that can be useful in the testing approach (Fent, 2003). If the tests are performed with one of the most sensitive species of fish, it means that we are ignoring the complexity of the aquatic ecosystem. Thus, it is very hard to predict the standard tests of acute toxicity from a species which may be susceptible and tolerant. One should perform several tests on several fish species in order to have an approximate estimate of their susceptibilities to these chemical toxins. The fish as a model of genotoxicity studies has entered into a golden era because this approach is being used in various fields of applied sciences like that of monitoring of the mutagens or genotoxic substances (Rishi and Grewal, 1995). As we know after the pesticides and other chemicals to our crops, they finally reach to the aquatic ecosystems in which fish are the first that may get affected and it should be seen whether they may pose some genetic damage (Ondarza et al., 2015).

The damage that may be induced in the cells of the fish by these genotoxic agents either in the lab or in the natural ecosystem gives us the information about their presence in the environment from which the model organism has been taken (Al-Sabti and Metcalfe, 1995). In the past various protocols were developed for testing that allow us to evaluate the mutagenic properties of different physical and chemical agents (De Serres and Sheridan, 1973). The short protocols that were developed include to check the micronuclei formation, changes in the chromosomes, analysis of sister chromatids, acute toxicity and effects on hereditary material. The long protocols are used to assess the mutagenicity and the toxicity in general (Sharma, 1984). There are various ways by which the chemical can be given to the model organism like may be injected intraperitoneally, may be given orally or may be provided with the aqueous environment (Kligerman, 1982). The fishes are exposed to these chemicals either through the digestive route or through the respiratory route after they absorb the water in which the chemical has been well mixed (De Flora et al., 1993). The fish have the ability to convert these chemicals into some metabolites. This ability of the fish has successfully been used to show the presence of the pesticides or chemicals in the water ecosystems (Al-Sabti, 1994; Farah et al., 2003; Kumar et al., 2010).

5.5   MUTAGENICITY ASSESSMENT PROTOCOLS

The different chemical substances have been used for their mutagenicity or genotoxicity by various methods in both fish and mice. The various methods have differences in their endpoint and each have some specificity in exploring the mutagenic effects of different chemicals at various levels in the biological systems, though most of these methods are having some of the limitations. The various methods used to assess the mutagenicity or genotoxicity of a chemical substance are listed below:

5.5.1   *MICRONUCLEUS TEST (MNT)*

Micronucleus test (MNT) is regarded as one of the reliable, easy, and successful tests that are employed *in vivo* for evaluating the genotoxic effects of the chemicals. About a century ago, this test was first done in the erythrocyte cytoplasm and Howell called them as "fragment of nuclear material," and Jolly termed them "intraglobularies corpuscles" late in the eighteenth century and in early 19th century. The hematologists called these structures as "Howell-Jolly bodies" (Kirsch-Volders et al., 2003). Currently, the research is well established that the micronuclei originated from the acentric fragments of chromosome. Further the fragments of chromatids are also acentric to the whole chromosome. Simultaneously chromatids that failed to enter in the daughter cell during the end of telophase in mitosis may not get attached to the spindle at the time of process of segregation during anaphase (Hartwell et al., 2000; Fenech, 2007, 2010). These fragments or full chromosomes then get enclosed by their nuclear membrane (Figure 5.1). These are only having small size but are similar to the nuclei morphologically when nuclear stain is applied to them. It is easier to count the micronuclei during the interphase than in metaphase stage (Al-Sabti and Metcalfe, 1995). According to Salyadori et al. (2003), damage to the DNA can be of any type, micronucleus formation occurs at some stage in the cell division. The damage to the genetic material by exposing it to the genotoxic chemicals gets expressed in the formation of MN soon after first cycle of the division of that cell and the number of the MN depends upon the number of cells that are going to divide. The interests in this branch go on increasing with the progress of finding large number of tests for evaluating the mutagenic effects of different chemicals in different organisms. MNT has remained as a popular and promising method of genotoxicity testing and an indicator of the damage to DNA for more than three decades (Fenech et al., 2003).

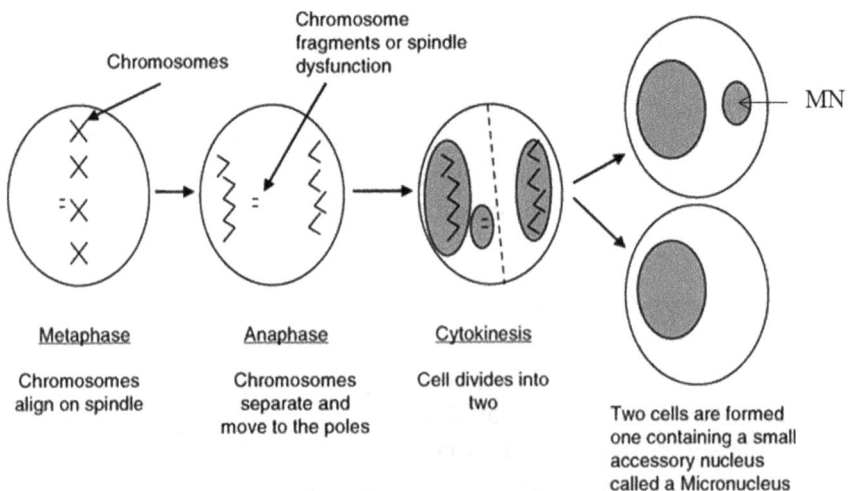

FIGURE 5.1   Formation of micronucleus (MN) in a cell.

It has been found that various drugs, alkylating agents and other chemicals cause mutagenesis as they induce aberrations in the chromosomes (Fishbein et al., 1970; Cattanach et al., 1968; Avakaki and Schmid, 1970; Goetz et al., 1975; Kirkland and Venitt, 1976; Ishidate and Odashima, 1977). Ethyl methanesulphonate (EMS), an alkylating agent causes mutations to different organisms. This alkylating agent induces point mutations in maize (Neuffer and Fiscor, 1963; Amano and Smith, 1965) and barley. EMS induces dominant lethal mutations and/or translocations in the offspring of male rats. Lim and Synder (1968) found that EMS is very less effective in case of Drosophila where the alkylating agent can produce few translocations only, though it can cause breaks in the chromosomes. Magee and Barnes (1956) reported hepatic cancer after making intensive studies using dimethylnitrosamine (DMN) as a mutagen in mice. Ishidate and Odashima (1977) described that dialkylnitrosamines induce fewer or no chromosomal aberrations in cultured hamster cells, when they were directly treated for 24 or 48 hrs. On the other hand, with a dose of 150 mg/kgbw of EMS, there was a significant increase in the MN (Heddle, 1973). Montesano et al. (1973) induced kidney tumors in rats with DMN and EMS and they reported that rats treated with single dose of 30 mg/kgbw of DMN or with a single dose of 100, 200, or 300 mg/kgbw of EMS, tumors of kidney developed in fewer animals receiving EMS. According to Hahn and Sikkim (1979), the dose of 150 mg/kgbw of EMS induced only 0.27% of micronuclei and the frequency of micronucleus increased to 2.10% when a dose of 250 mg/kgbw was given. The frequencies of chromosomal aberrations markedly increased with dose of UV as well as with concentration of methyl methanesulphonate (MMS) and EMS in eggs of mice fertilized *in vitro* with sperms exposed to UV light, EMS, and MMS (Matsuda and Tobari, 1988).

According to Bhagavathy et al. (2011), the mutagens are playing an important role in the initiation of various diseases to humans, being cancer also. The machinery of the process of mutagenesis is very complex because most of the carcinogens and mutagens act by the formation of ROS (reactive oxygen species). It has been found that these reactive species are the main culprits of these processes that lead to the mutations and damage of DNA which is ultimately related with the cancer, aging, and diseases of heart (Maryam et al., 2010).

1. **Reagents and Solutions:**
   - Chloroform.
   - Methanol (absolute).
   - May-Grunwald stain: It is readymade and used as such.

- **Giemsa Stain:** Stock solution of Giemsa stain is prepared by dissolving 1 g of Giemsa powder in 66 mL of glycerine. This solution is incubated for 2 h at 60°C. After cooling the contents, 66 mL of methanol is added. Thereafter, the working solution of Giemsa is prepared by diluting the stock solution with the help of distilled water or phosphate buffer.
- **Phosphate Buffer (pH 7):** Prepared by mixing disodium hydrogen phosphate (Na_2HPO_4; anhydrous, 1.42 g) and sodium dihydrogen phosphate dehydrate ($NaH_2PO_4.2H_2O$, 1.56 g) in 500 mL of distilled water.
- Ethanol (absolute).

2. **Procedure:** Blood samples are withdrawn by puncturing the caudal peduncle, and peripheral blood smears are immediately made by applying two micro drops of blood on pre-cleaned, grease-free slides, using the standard micronucleated erythrocytes method of MacGregor et al. (1987). The smeared slides are air-dried at room temperature in a dust and moisture-free environment. After drying, the slides are fixed by dipping in cold absolute methanol (4°C) for 15 min and again left to air dry at room temperature for 1 h. Finally, the slides are stained in May-Grunwald stain for 5–10 min followed with 6% Giemsa in phosphate buffer for 30 min. The slides are then washed thoroughly in double distilled water, dried, and studied under microscope.

5.5.2   CHROMOSOMAL ABERRATION TEST

Chromosomal aberration is the most important result of the biological systems that are exposed to mutagenic or genotoxic agents. Recently, the research on the chromosomal aberration has been given attention keeping in view that this method is also used to evaluate the mutagenic or genotoxic potential of toxicants present in the environment. The first study on the chromosomal aberrations that lead to the mutations was carried out by de Vries (1918) in *Oenothera*. The mechanism of the chromosomal aberrations with their fate and behavior was carried out in Drosophila and in many plant species like the *Tradescantia* and *Allium*. The protocol gradually developed after colchicine was discovered that arrests the cells during the metaphase (Nagpure et al., 2005).

Chromosomal aberrations are defined when a chromosome may be missing or an extra chromosome or a portion of chromosome may be present in the cells. When normal set of chromosomes is taken from the metaphase cell of an organism and arranged in the series we called this as karyotype. This karyotype is used to compare the chromosome sets of different organisms and we are able to see the abnormalities in the chromosomes. Chromosomal aberrations usually get formed from the error during cell division followed by mitosis or meiosis. Te effects in the cells can be viewed in the organism itself or in some of the cells that were grown in the culture. In both the situations, the cell or the organism is to be exposed to the mutagenic or genotoxic substance prior treating with colchicines that arrests the cells at the metaphase (Nagpure et al., 2005). After proper staining procedure, these chromosomes are visualized under high power microscope for the presence of any aberrations. The chromosomal aberrations are of many types but all of them can be grouped into two main types, the numerical, and the structural aberrations.

5.5.2.1   PROCEDURE

1. **Reagents and Solutions:**
   - **Colchicine (0.05%):** It is prepared by mixing 0.05 mg colchicine powder in 100 mL of distilled water.
   - **Hypotonic Solution (0.56% KCl):** It is prepared by adding 0.56 mg of KCl in 100 mL of distilled water.
   - **Cornoy's Fixative:** It is prepared by mixing methanol and glacial acetic acid in the ratio of 3:1.
   - **Sorenson's Buffer (pH 6.8):**
     Solution A: $Na_2HPO_4.2H_2O$ 5.938 g/100 mL
     Solution B: KH_2PO_4 4.539 g/100 mL
   25 mL of solution A is mixed with 25 mL of solution B to make 50 mL of Sorenson's buffer.
   - **Giemsa Stain:** Procedural details on Giemsa stain preparation are same as described under Section 5.5.1.
   - Ethanol (absolute).
2. **Slide Preparation for Chromosomal Aberration Test:** Blood samples are withdrawn by puncturing the caudal peduncle after applying 0.05% aqueous solution of colchicines 2 h prior to dissection to arrest the metaphase stage. The tube was then incubated at 370 C for 20 minutes. After incubation, centrifugation at 800 rpm for 4 minutes was carried out. Supernatant was discarded and fresh Carnoy's fixative was added (3:1 methanol: acetic acid). The

process of centrifugation was repeated three times. Then slides were prepared, stained with 4% Giemsa prepared in Sorensen's buffer (pH 6.8), air-dried, and studied under compound microscope.

5.5.3   BIO-INDICATORS OF OXIDATIVE STRESS

5.5.3.1   LIPID PEROXIDATION (LPO)

1. **Reagents and Solutions:**
   - Trichloroacetic acid (30%).
   - Butylated hydroxytoluene (10 mmol L^{-1}): It was prepared in 95% ethanol.
   - o-phosphoric acid (1%).
   - 2-thiobarbituric acid (0.67%).
2. **Procedure:** LPO in fish blood, obtained by cardiac puncture, is determined as described (Uchiyama and Mihara, 1978) by the thiobarbituric acid (TBA) method, with some modifications. A 500 μL aliquot of blood is mixed with 1 mL of 30% trichloroacetic acid and centrifuged for 10 min at 5000 rpm. An aliquot of 25 μL of supernatant is mixed with 25 μL of an aqueous solution 10 mmol L^{-1} of butylated hydroxytoluene, 3 mL of 1% aqueous solution of phosphoric acid and 1 mL of 0.67% aqueous solution of 2-thiobarbituric acid. The mixture is incubated for 45 min at 90°C. The concentration of TBA reactive substances (TBARS) is calculated from the absorption at 535 nm and a molar extinction coefficient of 1.56×10^5 L mol^{-1} cm^{-1}.

5.5.3.2   ANTI-OXIDATIVE ENZYME ASSAYS

1. **Glutathione (GSH):**
   - ➤ **Reagents and Solutions:**
     - Precipitating reagent: It is prepared by mixing 5 mL each of phosphoric acid, ethylene diamine tetraacetic acid (EDTA) and sodium chloride.
     - Disodium hydrogen phosphate (0.3 mol L^{-1}).
     - Dithio-bis-2-nitrobenzoic acid: It is prepared in 1% sodium citrate.
   - ➤ **Procedure:** The assay mixture for the estimation of reduced glutathione (Beutler et al., 1963) consisted of 100 μL blood,

900 µL double distilled water, and 1.5 mL precipitating reagent (m-phosphoric acid, EDTA, and sodium chloride). The mixture is incubated at room temperature for 5 min and then centrifuged at 4000 rpm for 15 min at 4°C. An aliquot of 1 mL of the supernatant is taken and mixed with 4 mL of 0.3 mol L^{-1} disodium hydrogen phosphate solution and 0.5 mL of dithiobis-2-nitrobenzoic acid prepared in 1% sodium citrate. The color intensity of reaction product is measured at 412 nm using a UV-Vis spectrophotometer (Shimadzu, Kyoto, Japan). The GSH content is expressed as nmol GSH/mg protein.

2. **Superoxide Dismutase (SOD):**
   ➤ **Reagents and Solutions:**
   - Normal saline: It is prepared by dissolving 0.9 g of NaCl in 100 mL of distilled water.
   - Phosphate buffer (pH 8.24).
   - Pyrogallol.
   - HCl (10 mmol L^{-1}).
   ➤ **Procedure:** SOD activity is determined by modifying the method of Marklund and Marklund (1974), which depends on the auto-oxidation of pyrogallol. A sample of 200 µL blood is centrifuged at 3000 rpm for 15 min at 4°C and the supernatant aspirated. The erythrocyte rich precipitate is washed thrice-using normal saline (3:1 v/v); lysed by distilled water and lysate is used for measuring the enzyme activity. The assay mixture consisted of 8.7 mL of 50 mmol L^{-1} phosphate buffer (pH 8.24), 100 µL lysate and 300 µL of pyrogallol dissolved in 10 mmol L^{-1} HCl. The rate of pyrogallol auto-oxidation is read at 420 nm. The SOD activity is expressed as units/mg protein. One unit of SOD activity is defined as the amount of the enzyme that inhibits 50% of the oxidation rate of 0.1 mmol L^{-1} pyrogallol in 1 mL of solution at 25°C.

3. **Catalase (CAT):**
   ➤ **Reagents and Solutions:**
   - Normal saline: It is prepared by dissolving 0.9 g of NaCl in 100 mL of distilled water.
   - Potassium phosphate (pH 7; 100 mmol L^{-1}).
   - Hydrogen peroxide (35 mmol L^{-1}).
   - Potassium hydroxide (10 mol L^{-1}).
   - 4-amino-3-hydrazino-5-mercapto-1, 2, 4-triazole.
   - Potassium periodate.

> **Procedure:** The method described by Johansson and Borg (1988) is used for the estimation of CAT activity. The erythrocyte rich precipitate obtained by centrifuging 500 μL blood at 3000 rpm for 15 min at 4°C is washed thrice using normal saline (3:1 v/v) and lysed by distilled water. Briefly, three replicates of 20 μL lysate are mixed with 100 mmol L^{-1} potassium phosphate (pH 7.0) and 30 μL methanol. The reaction is initiated by adding 20 μL of 35 mmol L^{-1} hydrogen peroxide (H_2O_2) and the mixture is incubated for 20 min on shaker at room temperature. The reaction is terminated by adding 30 μL of the 10 mol L^{-1} potassium hydroxide. The triplicates of each sample are added with 30 μL of 4-amino-3-hydrazino-5mercapto-1, 2, 4-triazole, a chromogen, and incubated for 10 min at room temperature on a shaker. After incubation 10 μL of potassium periodate is added to the mixture and again incubated for 5 min. Finally, the absorbance of assay mixture is read at 540 nm using a UV-Vis spectrophotometer. Enzyme activity is expressed as nmol L^{-1} H_2O_2 decomposed/min/mg protein.

5.6 MECHANISM OF THE TOXICITY

The living systems come across various stresses throughout their interaction with the environment. This stress-induced by the environment results in the activation of various ROS, commonly known as ROS, the most of which are produced as by-products of the respiration of tissues. Thus, if the organisms like the aquatic ones, are exposed to these stress molecules or pesticides, it will enhance the production of ROS mediated oxidative damage to the tissues. The agricultural wastes ultimately find their way into the aquatic ecosystem. These wastes are then taken up the organisms like fish which can induce multiple changes in these organisms. Some of these changes may directly increase the production of ROS formation while some may have indirect effect. The indirect effect includes their binding with the cellular thiols thereby reducing antioxidant potential. The fish population is mainly threatened by these aquatic pollutants. As fishes are used in biomonitoring, their vulnerability to these aquatic pollutants needs to be discussed.

Lushchak (2011) defined oxidative stress as "a situation when steady-state ROS concentration is transiently or chronically enhanced, disturbing cellular metabolism and its regulation and damaging cellular constituents."

When the oxidative processes are going in the body, it leads to the antioxidant response, activates the expression of those genes related with antioxidant enzymes, and increases the concentration of ROS scavengers. It has been observed in field studies having different activities of antioxidant enzymes that in polluted waters, there occur serious oxidative damage as compared to the clean waters.

There are various studies that have proven that various pesticides especially organochlorines such as hexachlorocyclohexane (HCH), endrin, endosulfan, lindane, and others induce the oxidative stress in various organs of fish and mammalian species (Sahoo et al., 2000; Bachowski et al., 1998; Bagchi et al., 1995; Bagchi and Stohs, 1993). It has been reported that endo-sulfan, a pesticide, induces oxidative stress by increasing LPO and changes in glutathione (GSH) redox cycle in hepatic and cerebral tissues of rats (Hincal et al., 1995). Pandey et al. (2001) found that after the exposure of fishes to copper, endosulfan induces the oxidative stress. The role of catalase (CAT) and glutathione redox cycle against endosulfan-induced oxidative stress in the adrenocortical cells of rainbow trout (*Oncorhynchus mykiss*) was demonstrated by (Dorval and Hontela, 2003). They found that the glutathione redox cycle is more efficient as compared to CAT in protecting adrenocortical cells against oxidative stress induced by the endosulfan. The toxicity of endosulfan has also been found in the case of human Jurkat T cells (Kannan and Jain, 2003; Kannan et al., 2000) and in rat tissues (Bebe and Panemangalore, 2003). This toxicity has been linked to oxidative damage by ROS and the reduction of GSH. In another fish species, *Cyprinus carpio,* a decrease in CAT activity has been reported by Oruç and Usta (2007) when exposed to diazinon. Tripathi and Shasmal (2010), reported that CAT and superoxide dismutase (SOD) activity reduces in the gill, brain, and liver tissues of *Heteropneustes fossilis* when they were exposed to chlorpyrifos (CPF), thus signifying these as biomarkers of aquatic pollution.

The biochemical constituents such as LPO and antioxidant enzymes are used as the biomarkers of pesticide exposure in various living organisms (Bechard et al., 2008; Regoli et al., 2004; Livingstone, 2003). These biochemical constituents are less variable, more sensitive, conserved, and easier to measure as inducers of stress (Agrahari et al., 2007).

The LPO (Lipid peroxidation), also known as the oxidation of polyun-saturated fatty acids, is calculated mainly by the (thiobarbituric acid reactive substances) levels. These TBARS levels are mainly used to access the effect of various pollutants (Livingstone, 2001; Lushchak, 2011). Sanchez et al. (2007) observed the high LPO levels in case of fish when exposed to polluted waters. The water bodies located near the industries have higher rates of TBARS levels as compared to the clean environments. The increased LPO levels have also been reported in three species of cichlid fish (*Tilapia rendall, Geophagus brasiliensis,* and *Oreochromis niloticus* after their exposure to metals (Ruas et al., 2008).

Another component of the mechanism, the antioxidant enzymes have also been found to protect the tissues against the stress and damage to the tissues. These have also been shown to act as the biomarkers of the ROS in organisms exposed to contaminants and as tool in assessment of the environmental risk (Livingstone, 2001; Kohen and Nyska, 2002). The toxic effect of ROS is minimized by various antioxidant enzymes thereby protecting the bulky molecules like lipids, proteins, and nucleic acids (DNA & RNA) against damage by the toxicants (Lushchak et al., 2001; Ozmen et al., 2004). There are various studies available in the literature that reveals the formation of intracellular ROS that cause damage to the biological systems, when exposed to the contaminated environments (Ercal et al., 2001; Ferreira et al., 2006).

This component, i.e., the antioxidant system can be classified into two groups: enzymatic and non-enzymatic. The enzymatic group includes the SOD and calatase (CAT), while the non-enzymatic group include the glutathione reduced (GSH). SOD brings about the dismutation of superoxide anion to oxygen and hydrogen peroxide while the CAT, an intracellular enzyme present in the peroxisomes, facilitate the removal of the H_2O_2 (hydrogen peroxide) that gets converted into water and molecular oxygen (Livingstone, 2001; Ozmen et al., 2004). Because of the handling procedure and cost-effective of these antioxidant enzyme systems, they are used in various bio-monitoring programs of the fish (Romeo et al., 2000; Viarengo et al., 2007). The recent research suggests that the variations in the antioxidant defense systems proved that these could be used as biomarkers of pollutant-induced oxidative stress in aquatic ecosystems. Sayeed et al. (2003) investigated that the antioxidant enzymes and non-enzymatic antioxidants can be used in the assessment of environmental risks caused by the effects of insecticide, deltamethrin in fish. It was found that there is not balance between various biomarkers that were used to measure the oxidative stress in freshwater fish *Wallago attu*. He found that the fish which was from polluted site possess higher activities of xanthine oxidase, glutathione reductase, and SOD both in the liver and gills, while the activity of CAT was very low when compared with fish from clean waters.

5.7 CONCLUSION

The field of aquatic toxicology aids the other branches of science in finding the ill effects of various chemicals that may prove fatal to the aquatic organisms. In order to find the mechanism of the aquatic toxicology using fish as a model organism, it has been found that the oxidative stress biomarkers play an important role. It has also been found that a positive relationship occurs between genotoxicity and oxidative stress. It has been found in various research studies that with the increase in the oxidative stress, the LPO in the erythrocytes of the fish shows an increase in a dose and time-dependent manner. The various antioxidant enzymes also show differences in their concentrations in the case of polluted environments. It has also been found by various researchers that the formation of micronuclei increased with the increase in the LPO. Now, it can be demonstrated that the LPO and antioxidant enzymes can give us the mechanism of the toxicity in the case of aquatic ecosystems in addition to the conventional methods of MNT and Chromosomal aberration tests.

KEYWORDS

- antioxidant enzymes
- aquatic toxicity
- lipid peroxidation
- micronucleus test
- mutagenicity
- oxidative stress

REFERENCES

Agrahari, S., Pandey, K. C., & Gopal, K., (2007). Biochemical alteration induced by monocrotophos in the blood plasma of fish *Channa punctatus* (Bloch). *Pest Biochem. Physiol., 88,* 268–272.

Alexander, P., & Connell, D. I., (1963). The failure of the potent mutagenic chemical ethyl methane sulphonate to shorten the life-span of mice. In: Harris, R. J. C., (ed.), *Cellular Basis and Aetiology of Late Somatic Effects of Ionizing Radiation* (pp. 259–265). A

Symposium held in London, 1962 Under Auspices of UNESCO and the IAEA, Academic Press, London, New York.

Al-Sabti, K., & Metcalfe, C. D., (1995). Fish micronuclei for assessing genotoxicity in water. *Mutation Research, 343*, 121–135.

Al-Sabti, K., (1994). Micronuclei induced by selenium, mercury, methyl mercury and their mixtures in binucleated blocked fish erythrocyte cell. *Mutation Research, 320* (1/2), 157–163.

Amano, E., & Smith, H. H., (1965). Mutations induced by ethyl methane sulfonate in maize. *Mutation Research/Fundamental and Molecular Mechanisms of Mutagenesis, 2*(4), 344–351.

Auerbach, C., & Robson, J. M., (1944). Production of mutations by allyl isothiocyanate. *Nature, 154*, 81.

Auerbach, C., (1940). Tests of carcinogenic substances in relation to the production of mutations in *Drosophila melanogaster. Proc. R. Soc. Edinb., B60*, 164–173.

Avakaki, D. T., & Schmid, W., (1970). Chemical mutagenesis: Chinese hamster as an *in vivo* test system. II Correlation with *in vitro* results on Chinese hamster fibroblasts and human fibroblasts and lymphocytes. *Humangenetik, 11*, 119–131.

Bachowski, S., Xu, Y., Stevenson, D. E., Walborg, E. F., & Klaunig, J. E., (1998). Role of oxidative stress in the selective toxicity of dieldrin in the mouse liver. *Toxicol. Appl. Pharmacol., 150*, 301–309.

Bagchi, D., Bagchi, M., Hassoun, E. A., & Stohs, S. J., (1995). *In vitro* and *in vivo* generation of reactive oxygen species, DNA damage and lactate dehydrogenase leakage by selected pesticides. *Toxicology, 104*, 129–140.

Bagchi, M., & Stohs, S. J., (1993). *In vitro* induction of reactive oxygen species by 2,3,7,8-tetrachlorodibenzo-p-dioxin, endrin and lindane in rat peritoneal macrophages and hepatic mitochondria and microsomes. *Free Rad. Biol. Med., 14*, 11–18.

Bebe, F. N., & Panemangalore, M., (2003). Exposure to low doses of endosulfan and chlorpyrifos modifies endogenous antioxidants in tissue of rats. *J. Environ. Sci. Health B., 38*, 349–363.

Bechard, K. M., Gillis, P. L., & Wood, C. M., (2008). Trophic transfer of Cd from larval chironomids (*Chironomus riparius*) exposed via sediment or waterborne routes, to zebrafish (*Danio rerio*) tissue-specific and subcellular comparisons. *Aquat. Toxicol., 90*, 310–321.

Bhagavathy, S., Sumathi, P., & Madhushree, M., (2011). Antimutagenic assay of carotenoids from green algae *Chlorococcum humicola* using *Salmonella typhimurium* TA98, TA100 and TA102. *Asian Pacific Journal of Tropical Disease, 1*(4), 308–316.

Boller, K., & Schmid, W., (1970). Chemical mutagenesis in mammals. The Chinese hamster bone marrow as an *in vivo* test system. Hematological findings after treatment with trenimon. *Humangenetik, 11*(1), 35.

Braunbeck, T., Bottcher, M., Hollert, H., Kosmehl, T., Lammer, E., Leist, E., Rudolf, M., & Seitz, N., (2005). Towards an alternative for the fish LC50 test in chemical assessment: The fish embryo toxicity test goes multispecies: An update. *ALTEX, 22*, 87–102.

Brusick, D., & Auletta, A., (1985). Developmental status of bioassays in genetic toxicology: A report of phase II of the US environmental protection agency gene-tox grogram. *Mutation Research/Reviews in Genetic Toxicology, 153*(1), 1–10.

Cattanach, B. M., Pollard, C. E., & Isaacson, J. H., (1968). Ethyl methane sulfonate-induced chromosome breakage in the mouse. *Mutation Research/Fundamental and Molecular Mechanisms of Mutagenesis, 6*(2), 297–307.

Dar, S. A., Yousuf, A. R., Balkhi, M. H., Ganai, F. A., & Bhat, F. A., (2015). Assessment of endosulfan induced genotoxicity and mutagenicity manifested by oxidative stress pathways in freshwater cyprinid fish crucian carp (*Carassius carassius* L.). *Chemosphere, 120,* 273–283.

De Flora, S., Vigano, L., D'Agostini, F., Camoirano, A., Bagnasco, M., Bennicelli, C., Melodia, F., & Arillo, A., (1993). Multiple genotoxicity biomarkers in fish exposed in situ to polluted river water. *Mutation Research, 319*(3), 167–1677.

De Serres, F. J., & Sheridan, W., (1973). The evaluation of chemical mutagenicity data in relation to population risk. *Environ. Health Perspect., 6,* 1–232.

De Vries, H., (1918). Mass mutations and twin hybrids in *Oenothera grandiflora. Ait. Bot. Gaz., 65,* 377–422.

Depledge, M., (1994). Series forward. In: Forbes, V. E., & Forbes, T. L., (eds.), *Ecotoxicology in Theory and Practice* (pp. vi–ix). Chapman and Hall, London.

Dorval, J., & Hontela, A., (2003). Role of glutathione redox cycle and catalase in defense against oxidative stress induced by endosulfan in adrenocortical cells of rainbow trout (*Oncorhynchus mykiss*). *Toxicology and Applied Pharmacology, 192,* 191–200.

Ercal, N., Gurer-Orhan, H., & Aykin-Burns, N., (2001). Toxic metals and oxidative stress part I: Mechanisms involved in metal induced oxidative damage. *Curr. Topics Med. Chem., 1,* 529–539.

Farah, M. A., Ateeq, B., Ali, M. N., & Ahmad, W., (2003). Evaluation of genotoxicity of PCP and 2, 4-D by micronucleus test in freshwater fish *Channa punctatus. Ecotoxicology and Environmental Safety, 54,* 25–29.

Fenech, M., (2007). Cytokinesis-block micronucleus cytome assay. *Nat. Protoc., 2,* 1084–1104.

Fenech, M., (2010). The lymphocyte cytokinesis-block micronucleus cytome assay and its application in radiation biodosimetry. *Health Phys., 98,* 234–243.

Fenech, M., Chang, W. P., Kirsch-Volders, M., Holland, N., Bonassi, S., & Zeiger, E., (2003). HUMN project: Detailed description of the scoring criteria for the cytokinesis-block micronucleus assay using isolated human lymphocyte cultures. *Mutation Research, 534,* 65–75.

Fent, K., (2003). *Okotoxikologie* (p. 332). Georg Thieme Verlag.

Ferreira, M., Moradas-Ferreira, P., & Reis-Henriques, M. A., (2006). The effect of long-term depuration on phase I and phase II biotransformation in mullets (*Mugil cephalus*) chronically exposed to pollutants in River Douro Estuary. *Portugal. Mar. Environ. Res., 61,* 326–338.

Fishbein, L., Flamm, W. G., & Falak, H. L., (1970). *Chemical Mutagen.* Academic Press. New York.

Goetz, P., Sram, R. J., & Dohnalova, J., (1975). Relationship between experimental results in mammals and man: I. Cytogenetic analysis of bone marrow injury induced by a single dose of cyclophosphamide. *Mutation Research/Environmental Mutagenesis and Related Subjects, 31*(4), 247–254.

Hahn, S., & Kim, D. S., (1979). The production of micronuclei from chromosome aberrations by chemical carcinogens in mice. *Yonsei Medical Journal, 20*(2), 105–112.

Hartwell, L. H., Hood, L., Goldberg, M. L., Reynolds, A. E., Silver, L. M., & Veres, R. C., (2000). *Genetics: From Genes to Genomes* (pp. 70–351). McGraw Hill Higher Education.

Heddle, J. A., (1973). A rapid *in vivo* test for chromosomal damage. *Mutation Research/ Fundamental and Molecular Mechanisms of Mutagenesis, 18*(2), 187–190.

Hincal, F., Gurbay, A., & Giray, B., (1995). Induction of lipid peroxidation and alteration of glutathione redox status by endosulfan. *Biol. Trace Elem. Res., 47,* 321–326.

Hutchinson, T. H., Barrett, S., Buzby, M., Constable, D., Hartmann, A., Hayes, E., Huggett, D., et al., (2003). A strategy to reduce the numbers of fish used in acute ecotoxicity testing of pharmaceuticals. *Environ. Toxicol. Chem.*, *22*, 3031–3036.

Ishidate, M., & Odashima, S., (1977). Chromosome tests with 134 compounds on Chinese hamster cells *in vitro*: A screening for chemical carcinogens. *Mutation Research/Fundamental and Molecular Mechanisms of Mutagenesis*, *48*(3), 337–353.

Isomaa, B., & Lilius, H., (1995). The urgent need for *in vitro* tests in ecotoxicology. *Toxciol. In vitro*, *9*, 821–825.

Jha, A. N., (1998). Use of aquatic invertebrates in Geno toxicological studies. *Mutation Research*, *153*, 1–10.

Kannan, K., & Jain, S. K., (2003). Oxygen radical generation and endosulfan toxicity in jurkat T-cells. *Mol. Cell. Biochem.*, *247*, 1–7.

Kannan, K., Holcombe, R. F., Jain, S. K., Alvarez-Hernandez, X., Chervenak, R., Wolf, R. E., & Glass, J., (2000). Evidence for the induction of apoptosis by endosulfan in a human T-cell leukemic line. *Mol. Cell. Biochem.*, *205*, 53–66.

Kirkland, D. J., & Venitt, S., (1976). Cytotoxicity of hair colorant constituents: Chromosome damage induced by two nitro phenylenediamines in cultured Chinese hamster cells. *Mutation Research/Genetic Toxicology*, *40*(1), 47–55.

Kirsch-Volders, M., Sofuni, T., Aardema, M., Albertini, S., Fenech, M., Ishidate, Jr. M., Kirchner, S., et al., (2003). Report from the *in vitro* micronucleus assay-working group. *Mutation Research*, *540*, 153–163.

Kligerman, A. D., (1982). In: Hsu, T. C., (ed.), *Cytogenetic Assays of Environmental Mutagens* (p. 161). Allenheld, Osmum.

Kohen, R., & Nyska, A., (2002). Oxidation of biological systems: Oxidative stress phenomena, antioxidants, redox reactions, and methods for their quantification. *Toxicol. Pathol.*, *30*, 620–650.

Kumar, R., Nagpure, N. S., Kushwaha, B., Srivastava, S. K., & Lakra, W. S., (2010). Investigation of the genotoxicity of malathion to freshwater teleost fish *Channa punctatus* (Bloch) using the micronucleus test and comet assay. *Arch Environ. Contam. Toxicol.*, *58*, 123–130.

Lim, J. K., & Snyder, L. A., (1968). The mutagenic effects of two monofunctional alkylating chemicals of mature spermatozoa of *Drosophila*. *Mutation Research/Fundamental and Molecular Mechanisms of Mutagenesis*, *6*(1), 129–137.

Livingstone, D. R., (2001). Contaminant-stimulated reactive oxygen species production and oxidative damage in aquatic organisms. *Marine Pollution Bulletin*, *42*, 656–666.

Livingstone, D. R., (2003). Oxidative stress in aquatic organisms in relation to pollution and aquaculture. *Revue De Medecine Veterinaire*, *154*, 427–430.

Lushchak, V. I., (2011). Adaptive response to oxidative stress: Bacteria, fungi, plants and animals. *Comparative Biochemistry and Physiology*, *153*, 175–190.

Lushchak, V., Lushchak, L. P., Mota, A. A., & Hermes-Lima, M., (2001). Oxidative stress and antioxidant defenses in goldfish *Carassius auratus* during anoxia and reoxygenation. *Am. J. Physiol. Regul. Integr. Comp. Physiol.*, *280*, 100–107.

Magee, P. N., & Barnes, J. M., (1956). The production of malignant primary hepatic tumors in the rat by feeding dimethylnitrosamine. *British Journal of Cancer*, *10*(1), 114.

Maryam, S., Sasan, M., & Hasan, M., (2010). Cadmium-induced genotoxicity detected by the random amplification of polymorphism DNA in the maize seedling roots. *J. Vet. Sci.*, *7*(2), 181–187.

Matsuda, Y., & Tobari, I., (1988). Chromosomal analysis in mouse eggs fertilized *in vitro* with sperma exposed to ultraviolet light (UV) and methyl and ethyl methanesulfonate (MMS and EMS). *Mutation Research/Fundamental and Molecular Mechanisms of Mutagenesis, 198*(1), 131–144.

Montesano, R., Saint, V. L., & Tomatis, L., (1973). Malignant transformation *in vitro* of rat liver cells by dimethylnitrosamine and N-methyl-N'-nitro-N-nitrosoguanidine. *British Journal of Cancer, 28*(3), 215.

Moutschen, J., (1985). *Introduction to Genetic Toxicology*. Wiley.

Muller, H. J., (1927). Artificial transmutation of the gene. *Science, 66*, 84–87.

Nagel, R., & Isberner, K., (1998). Testing of chemicals with fish-critical evaluation of tests with special regard to zebrafish. In: Braunbeck, T., Hinton, D. E., & Streit, B., (eds.), *Fish Ecotoxicology* (pp. 337–352). Birkhauser.

Nagpure, N. S., Pandey, S., & Sharma, S., (2005). Single cell gel electrophoresis (SCGE) or comet assay. In: Kapour, D., & Nagpour, N. S., (eds.), *Training on Genotoxic Assays in Fishes* (p. 68). National Bureau of Genetic Resources. Dilkusha, Telibagh, India.

Neuffer, M. G., & Ficsor, G., (1963). Mutagenic action of ethyl methanesulfonate in maize. *Science, 139*(3561), 1296–1297.

OECD, (1992a). *Guidelines for Testing of Chemicals*. Section 5.2: Effects on biotic systems test no. 203: Acute toxicity for fish. Paris, France.

OECD, (1992b). *Guidelines for the Testing of Chemicals*. Section 5.2: Effects on biotic systems test no. 210: Fish, early life stage toxicity test. Paris, France.

OECD, (2000). *Guidelines for the Testing of Chemicals*. Section 5.2: Effects on biotic systems test no. 215: Fish, juvenile growth test. Paris, France.

Ondarza, P. M., Gonzalez, M., Fillmann, G., & Miglioranza, K. S. B., (2014). PBDEs, PCBs and organochlorine pesticides distribution in edible fish from Negro River basin, Argentinean Patagonia. *Chemosphere, 94*, 135–142.

Oruç, E. O., & Uner, N., (2000). Combined effects of 2,4-D and azinphosmethyl on antioxidant enzymes and lipid peroxidation in liver of *Oreochromis niloticus. Comp. Biochem. Physiol. Part C, 127*, 291–296.

Ozmen, I., Bayir, A., Cengiz, M., Sirkecioglu, A. N., & Atamanalp, M., (2004). Effects of water reuse system on antioxidant enzymes of rainbow trout (*Oncorhynchus mykiss* W, 1792). *Veterin. Med. Czech., 49*, 373–378.

Pandey, S., Ahmad, I., Parvez, S., Bin-Hafccz, B., Haque, R., & Raisuddin, S., (2001). Effect of endosulfan on antioxidants of freshwater fish *Channa punctatus* Bloch: 1. Protection against lipid peroxidation in liver by copper preexposure. *Arch. Environ. Contam. Toxicol., 41*, 345–352.

Regoli, F., Frenzilli, G., Bochetti, R., Annarumma, F., Scarcelli, V., Fattorini, D., & Nigro, N., (2004). Time-course variations of oxyradical metabolism, DNA integrity, and lysosomal stability in mussels, *Mytilus galloprovincialis*, during a filed translocation experiment. *Aquat. Toxicol., 68*, 167–178.

Rishi, K. K., & Grewal, S., (1995). Chromosome aberration test for the insecticide, dichlorvos, on fish chromosomes. *Mutation Research, 344*, 1–4.

Romeo, M., Bennani, N., Gnassia-Barelli, M., Lafaurie, M., & Girard, J. P., (2000). Cadmium and copper display different responses towards oxidative stress in the kidney of the sea baas *Dicentrarchus labrax. Aquatic Toxicology, 48*, 185–194.

Ruas, C. B. G., Carvalho, C. S., Araujo, H. S. S., Espindola, E. L. G., & Fernandes, M. N., (2008). Oxidative stress biomarkers of exposure in the blood of cichlid species from a metal-contaminated river. *Ecotoxicology and Environmental Safety, 71*, 86–93.

Sahoo, A., Samanta, L., & Chainy, G. B. N., (2000). Mediation of oxidative stress in HCH-induced neurotoxicity in rat. *Arch. Environ. Contam. Toxicol., 39,* 7–12.

Salvadori, D., Ribeiro, L., & Fenech, M., (2003). Micronuclei test in human cells *in vitro.* In: Ribeiro, L., Salvadori, D., & Marques, E., (eds.), *Environmental Mutagenesis* (pp. 201–223). Canoas: ULBRA.

Sanchez, W., Ait-Aissa, S., Palluel, O., Ditche, J. M., & Porcher, J. M., (2007). Preliminary investigation of multi-biomarker responses in three spined stickleback (*Gasterosteus aculeatus* L.) sampled in contaminated streams. *Ecotoxicology, 16,* 279–287.

Sayeed, I., Parvez, S., Pandey, S., Bin-hafeez, B., Haque, R., & Raisuddin, S., (2003). Oxidative stress biomarkers of exposure to deltamethrin in freshwater fish, *Channa punctatus* Bloch. *Ecotoxicology and Environmental Safety, 56,* 295–301.

Schwarzenbach, R. P., Escher, B. I., Fenner, K., Hofstetter, T. B., Johnson, C. A., Von, G. U., & Wehrli, B., (2006). The challenge of micro-pollutants in aquatic systems. *Science, 313,* 1072–1077.

Sharma, A., (1984). *Environmental Chemical Mutagenesis.* Perspective report series 6, Golden Jubilee Publications, Indian National Science Academy, New-Delhi.

Tolla, L. J. D., Srinivas, S., Whitaker, B. R., Andrews, C., Hecker, B., Kane, A. S., & Reimschuessel, R., (1995). Guidelines for the care and use of fish in research. *Ilar. Journal,* 37.

Tripathi, G., & Shasmal, J., (2010). Reparation of chlorpyrifos-induced impairment by thyroxine and vitamin C in fish. *Ecotoxicol. Environ. Saf., 73,* 1397–1401.

Viarengo, A., Lowe, D., Bolognesi, C., Fabbri, E., & Koehler, A., (2007). The use of biomarkers in biomonitoring: A 2-tier approach assessing the level of pollutant-induced stress syndrome in sentinel organisms. *Comparative Biochemistry and Physiology, 146,* 281–300.

Wedekind, C., Von, S. B., & Gingold, R., (2007). The weaker point of fish acute toxicity tests and how tests on embryo can solve some issues. *Environ. Pollut., 148,* 385–389.

WHO, (1990). *International Program on Chemical Safety* (p. 144.). Environmental Health Criteria 101: Methylmercury. Geneva: World Health Organization.

Young, R. R., (2002). Genetic toxicology: Web resources. *Toxicology, 173,* 103–121.

CHAPTER 6

Pesticide Pollution in an Aquatic Environment

UMAIR RIAZ,[1] FAIZAN RAFI,[2] MUHAMMAD NAVEED,[3]
SHEHZADA MUNAWAR MEHDI,[4] GHULAM MURTAZA,[2]
ABDUL GHAFFAR NIAZI,[4] and HASSAN MEHMOOD[5]

[1]Soil and Water Testing Laboratory for Research, Bahawalpur–63100, Pakistan, E-mail: umairbwp3@gmail.com

[2]Institute of Soil and Environmental Sciences, University of Agriculture, Faisalabad, Pakistan

[3]Soil and Water Testing Laboratory for Research, Lahore, Pakistan

[4]Rapid Soil Fertility Research Institute, Lahore, Punjab, Pakistan

[5]Department of Soil Science, The Islamia University of Bahawalpur, Pakistan

ABSTRACT

The contamination of the aquatic ecosystem from pesticides, animal wastes, insecticides, fertilizers, etc., is the most common problem of water nowadays. In the aquatic environment, the pesticide can adsorb or desorb on suspended solids and further settle down in the bottom sediments. Pesticides reach aquatic systems mainly through two processes, namely runoff, and leaching. These two main processes are directly connected with the hydrological cycle. Due to rapid urbanization and industrialization, the use of toxic chemicals in pesticides increased. Pesticides lower the levels of oxygen, particulate organic carbon, and dissolved organic carbon (DOC) from decreased primary production and increased respiration in aquatic plants. When atrazine is exposed to an aquatic environment, it increases calcium in the water. This is due to reduced production of calcium carbonate ($CaCO_3$) precipitation due to inhibition of crystal growth or the decrease in pH. Organochlorine pesticides (OCP), atrazine, etc., increases ammonium (NH^{4+}), nitrate (NO_3^-), sulfate (SO_4^{2-}), and nitrite in the aquatic system. There are several methods to remove pesticide contamination from the aquatic environment like biological (bioaugmentation, natural attenuation, and biostimulation), chemical (denitrification, reverse osmosis, electrodialysis, and catalytic denitrification), and physical (zeolites, activated carbon, clays, and polymer materials).

6.1 INTRODUCTION

One of the biggest problems faced by humanity in the 21st century is either related to water quantity or quality of water. The present condition can become worse due to new challenges like climate change, global warming, melting of glaciers, improper use of pesticides, and many more. A significant portion of renewable freshwater is used for agricultural, industrial, and domestic purposes. This leads to contamination with synthetic chemicals, which pose a threat to human health. An aquatic ecosystem is of two type's freshwater and marine ecosystems. The freshwater ecosystem covers 0.78% of Earth's surface. The most extensive ecosystem is a marine ecosystem covering 71% of the entire earth's surface. This ecosystem has a large amount of dissolved salt. It is home to a large number of plants and animal species. Pollution is the contamination of a system with foreign or naturally occurring substances (contaminants). Aquatic pollution infers to contamination of freshwater and marine ecosystems with contaminants from both point and non-point sources of pollution. Point sources are those sources that can be identified, and their source is known. Its example includes a pipe coming directly from a paint industry. The source of non-point sources is not known because it is a complex of many contaminants that cannot separately be identified. Marine pollution occurs when effluents from agricultural, industrial, and domestic wastes enter the ecosystem. The land is responsible for 80% of marine pollution.

The Mediterranean Sea receives 80% untreated sewage discharges annually. The Arctic Ocean clogged with 300 billion pieces of plastic which can be seen with the naked eye. 1.4 billion pounds of domestic waste is dumped in oceans every year (National Ocean and Atmospheric Administration (NOAA) 2018 report). Every year, 5–13 million tons of plastic are leaked into oceans. Marine pollution is more common in deep waters. When excessive nutrients and agricultural runoff can enter in oceans, it results in several low oxygen (hypoxic) areas. These areas are known as Dead Zones. There are 500 dead zones, and they cover more than 245,000-kilometer square. According to a study conducted in 2017, 233 marine species, 100% marine turtles, 59% seabirds, 36% seals, and 6 invertebrate species had plastic in them. This is responsible for stomach problems, starvation, and the death of these organisms.

According to the team of international scientists led by Nicolas Gruber, professor of Environmental Physics at ETH Zurich, the dissolution of carbon dioxide in oceans lowers the pH making the water acidic. This acidic pH can have many serious consequences, which include the dissolution of calcium carbonate in this acidic environment. This can cause some severe problems for some aquatic organisms. The shells and skeletons of many aquatic animals are composed of calcium carbonate, which would be hazardous for them due to the dissolving of calcium carbonate. Many physiological and chemical processes of aquatic organisms are adversely affected due to change in the chemical composition of oceans. One example includes breathing of fish. This acidification leads to slow growth of coral reefs. There is a reaction between seawater and carbon dioxide coming from emissions making carbonic acid. This carbonic acid causes corrosion of shellfish, corals, and other marine life.

In the last 70–80 years, the oil has been used extensively by humans. The discharge of oil to its environment is worth discussing here. The noteworthy oil spills have taken place at sea, which has affected the aquatic environment in one and many ways. The spills from oil tankers after accidents are read in newspapers most often. These spills play havoc for the marine environment. According to an NRC report on average 5000 tons per year, oil is spilled as a result of tanker accidents. This was way more in the past because of the absence of laws for oil spills in the world. Apart from oil spills from tanker accidents, it is also spilled from pipelines, operational discharges, coastal facilities, atmospheric deposition, and aircraft dumping. There are two categories of oil toxic effects. First is the coating of an organism with oil, and second is any type of disruption in the metabolism of an organism due to ingestion of oil. Low molecular weight aromatics such as toluene and benzene are the most responsible compounds for the second type of toxic effects. These compounds are responsible for the disruption of cell membranes of many organisms resulting in cell death. Other hydrocarbons have adverse effects on proteins and reproduction of organisms. The oil affects the larval stage of some fish. There is a little concern from oil pollution of the aquatic environment for humans because fish and other organisms become tainted in taste, which is mostly thrown away before eating. Legislation of laws has led to the control of oil spills. Mostly three approaches are adopted to control oil spills mechanical collection, burning, and chemical dispersal, but these approaches are not natural to conduct. Sinking with the help of sand has been used in the past and may be used in the future. Bioremediation is also being applied in this field and has led to many positive results (Aquatic Pollution: An introductory text, Laws, 2017).

After the incidents like Chernobyl and Fukushima Daiichi atomic power plants, the whole world is concerned for the safe use of radioactivity, especially near the sea. Radioactivity has a lot of adverse effects on all organisms. When aquatic systems are exposed to it, the situation becomes alarming because no one knows to how the extent and in which ways would it affect plants, animals, and microorganisms present in that aquatic system. Radioactivity imparts direct negative impacts on the organism, especially fish, because the reproduction of fish is mostly affected by radioactivity. The discharges from atomic power plants are also of concern. Many environmentalists even demand to ban atomic power plants due to their environmental issues (NEA report 2016).

The solubility of water of pesticides and organic pollutants is one of the most critical factors in controlling the fate and transport of chemicals in aquatic systems. Their magnitudes determine the specific limiting loads in water and partition constants. Solubility and stability of many hydrophobic organic compounds like those that PCBs are enhanced by low concentrations of dissolved and suspended particulate-bound natural organic matter in water. This has an interaction with organic pesticides in aqueous solution due to hydrophilic functional groups of humic molecules (Chiou, 1986).

6.2   PESTICIDES AS WATER POLLUTANTS

A pesticide is any substance or mixture of substances intended for repelling, destroying, preventing, or mitigating any pest (nematodes, weeds, insects, mites, rats, etc.), including insecticide, herbicide, fungicide, and various other substances used to control pests (EPA, 2009). This definition varies with countries and times. The history of pesticides is divided into the following three phases:

- The first phase includes era before the 1870s in which natural pesticides were used, e.g., Greeks used sulfur to control pests;
- The second phase includes the period from 1870–1945. This is the era of inorganic synthetic pesticides in which inorganic compounds and natural materials were used; and
- The third phase is from 1945 since now. This era is known as the era of organic synthetic pesticides (Zhang et al., 2001).

The human-made pesticides like HCH, dieldrin DDT, etc., have terminated era of inorganic and natural pesticides. One of the marks of human civilization is the use of these chemical pesticides which increase agricultural productivity extensively. In the era of organic synthesized pesticides, only three kinds of insecticides were used, namely organophosphorus insecticide, carbamate insecticide, and organochlorine insecticides.

The largest pesticide consumers, traders, or producers in the world are China, the United States, France, Brazil, and Japan. Mostly these pesticides are sprayed on vegetable crops and fruits. The total production of pesticides is 3 to 7 million tons annually. 0.2 and 2 kg of active substance per hectare of arable land in developing versus developed countries is the estimation of pesticide use approximately. These are not precise estimates. Geographic and climatic conditions, the application technique, type of pesticide used, and the crop treated determine the amount of active chemical required to control pests. The recently developed pesticides operate at low lower doses as compared to established products of the past, but their toxic loads per dose active ingredients are variable among different types of agrochemicals. These pesticides are available with tons of brand names having thousands of active ingredients. The peak concentrations of pesticides and their products when exceeding ecotoxic levels for those aquatic or soil organisms which are considered non-target for pesticides is a significant concern. The contamination of aquatic systems by pesticides mainly depends on the route by which these contaminants reach the aquatic system. The use of pesticides is more in developing countries due to pressure of food security, economic, and political implications. These are used extensively to feed the rapidly growing populations. The pesticide use per hectare has increased rapidly in developing countries to gain maximum yields. The capabilities and resources to monitor pesticide pollution in developing countries in marine systems and environment are often very limited (Jian, 2011).

Does question arise that why pesticides are used so much around the globe? The answer to this is that there are approximately 800,000 species of weeds, 9000 species of insects and mites and 50,000 species of plant pathogens that damage crops. The losses by plant pathogens are 13%, by insects and pests are 14% and by weeds are 13%. That is why pesticides are so dispensable in agricultural systems. If pesticides are not applied the losses due to attacks of pests are 78% in fruits, 32% in cereals, and 54% in vegetables. 35–42% crop loss from pest attack declines after the application of pesticides. Due to rapid growth in world population and decreasing cropland, it is necessary to apply these chemicals for food security. The economic values of fruits, vegetables, and other crops have increased due to the application of weedicides and fungicides (Bo, 2009).

The global amount of spraying of pesticides in the environment is 4.6 million tons. Many of these chemical pesticides contain mercury, arsenic, and lead, which are not only poisonous to the environment but also all organisms. Only 1% of the sprayed pesticides are active, and 99% of pesticides applied released to non-target atmosphere, soils, and water bodies and ultimately absorbed by every organism in them. The Environment Protection Agency (EPA) of the USA states that almost every rural well has at least one of 127 pesticides. Lindane, DDT, and aldrin residues were found in 90 sites from the equator to high altitudes in Greenland by a research panel of Indiana University (Tang and Li, 1998). The Antarctic penguins also had residues of pesticides like DDT, which must be due to ocean currents, atmospheric circulation, ocean currents, and biological enrichment of pesticides. One of the joint reports of WHO and UNEP gives a figure of 26 million human poisonings by pesticides per year. The pesticide-poisoning rate of humans in China is 0.5 million while in the US it is 67 thousand per year. Deaths due to these poisonings are 0.1 million in China. Pesticides are also responsible for many diseases in humans and marine organisms.

The OCPs which are included in persistent organic pollutants (POPs) are heptachlor, chlordane, DDT, endrin, toxaphene, aldrin, dieldrin, etc. Their persistence in the environment makes them different from other pesticides. Nine in 12 POPs are OCPs as stated by "Stockholm Convention on POPs." The persistence of pesticides depends on specific abiotic and biotic degradational purposes. Oxidation, photolysis, and hydrolysis are abiotic processes. Metabolism and biodegradation are biotic processes.

There are certain factors which affect pesticide toxicity in aquatic systems. They are listed below:

- Toxicity;
- Degrades;
- Persistence.

The toxicity in aquatic systems in mammals and non-mammals is expressed in terms of LD_{50}. This has an inverse relationship with that of toxicity lower the value of LD_{50} higher the toxicity. 0–10 are considered extremely toxic values for it. Risk-based assessment is used to check the guidelines for water and food.

The risk is calculated as:

$$Risk = Toxicity \times Exposure \ (duration \ and \ amount)$$

The response to this toxicity, i.e., effect, may be acute or chronic. The acute effect means death while chronic means the effect which does not cause death throughout testing, but observable effects can be seen such as tumors, growth inhibition, teratogenic effects, reproductive failures, cancers, etc.

Degrades are formed by some degradational processes of original pesticide. These degrade may have lesser, equal, or higher toxicity as compared to parent pesticide. For example, DDT degrades to DDE or DDD having varying toxicities.

There is an example which may be called the most crucial regional example of human contamination of pesticides from an aquatic system, i.e., the Aral Sea region. There are two routes of entry of pesticides in the human body by the skin and by inhalation. There is an increased risk for farmers and workers at the farm, which mixes and applies pesticides to crops. However, most people are affected by pesticide-contaminated food. Water contaminated by pesticide has two main human health impacts. First is the consumption of pesticide-contaminated fish and other seafood, and second is the intake of water contaminated by pesticides. The first impact is a significant concern to those economies which rely on kinds of seafood. The harmful effects of pesticides include death, cancers tumors on aquatic animals, physical deformities, damage to cells and DNA of organisms, thinning of eggshell, damage to endocrine glands, low red to white blood cells ratio in fish, production of excessive slime in fish and many more.

6.3 PESTICIDE SOURCES, THEIR FATE, AND DIFFERENT WAYS TO IMPACT AQUATIC ORGANISMS

Pesticides reach aquatic systems mainly through runoff and leaching process. These two main processes are directly connected with the hydrological cycle. The exact mechanism of entry of pesticides in water systems is complicated to determine. Sometimes water itself is responsible for their entry. The pollutant-transporting source of water may be natural like rainfall, or it may be anthropogenic such as diversions and irrigation. Pesticides enter the aquatic system through both points as well as non-point sources. Point sources are easy to identify as they are mostly small and include areas of high pesticide concentration such as spills, tanks, or containers. Non-point source areas are broad and cannot be easily defined and mostly contain pesticide residue.

Mostly, the water quality is deteriorated by surface runoff from agriculture systems and other non-point sources of pollution. Pesticide water pollutes not only drinking water but also lakes, canals, streams, rivers, seas, and estuaries.

This impaired quality of water not only cause damage of billions of dollars to agriculture alone but also to commercial fishing, ponds, and sites of aesthetic value. The nitrogen and phosphorus from fertilizers are responsible for algal bloom which affects estuaries, lakes, and ponds. This process is called eutrophication. These sediments block sunlight damaging photosynthesis and may be toxic to aquatic life. The pesticides from surface water runoff from croplands affect vigor and health of freshwater and marine organisms.

The contamination of groundwater from pesticides, animal wastes, insecticides, fertilizers, etc., is the most common problem of water nowadays.

Some of the residues from these substances remain in the soil and are taken up by plants and then may leach down to subsurface water. Some physical and chemical processes are responsible for the transformation of residues into products. Its example is the transformation of nitrogen either from animal waste or fertilizers first to ammonium and then to nitrates. Nitrates may be changed to nitrites; both are concerns for health.

The fate of pesticides is complicated to predict because of their complex formulations. Some models called Multimedia Mass Balance Box Models were developed. These are one of the adequate forms of explaining the fate of pesticides. The physiochemical properties of the pollutants and features of the medium to write mass balances are used as a principle to explain these models (Mackay, 2001; Scheringer and Wania, 2003). This is the most acceptable thermodynamic criteria to describe the partition and transport of these chemicals. There are different levels of accuracy and complexity in these models. The most straightforward multimedia environmental calculation (level I) is to take in account thermodynamic equilibria and estimation of a contaminant in different parts from partition coefficients without considering their transport and reactions while more complex models (level III and IV) consider reaction, intermedia transport, and advection (Mackay, 2001). The validation of these simple models has been done by comparing the values of concentrations with real ones (Otto et al., 2016; Xia et al., 2011; Muir et al., 2014). The efficiency of them has been proved to explain the fate of pesticides (Zhang et al., 2015) and is best for pesticide screening. Other elaborative models have proved their effectiveness for air pollution or climatic research (Sameena et al., 2005).

For many compounds, a lot of experimental partition data is available in the literature, and sometimes conflict is seen in data (Ma et al., 2005). It is necessary to rationalize the data present in the literature. Due to the presence of a lot of experimental data on the properties used in these models, group contribution method is applied to estimate the values for new pesticides (Sarraute, 2018).

Regarding the movement of pesticide in water, when there is no advection, random molecular motion is due to diffusion which is responsible for mixing and too homogenous dispersion in the medium. Fick's Law gives the calculation of macroscopic fluxes and molecular diffusivity in water. There are very few data in the literature about the diffusion coefficient of pesticides in water (Scott and Phillips, 1973; Raveton et al., 1999). The first reason for this is in many cases; only approximate values are used because advection is essential. Secondly, accurate experimental data is complicated to obtain. A constant value is used by authors in these models, with the same order of magnitude of diffusivity of organic compounds in water (Xie et al., 2011). There are many methods to calculate the molecular diffusion coefficient in liquid with accuracy. The use of a diaphragm cell is a suitable and useful method. It is based on the calculation of diffusion coefficient through a perforated membrane separating the two constituents filled with solutions of different concentration (Stokes, 1950). The accuracy of nuclear magnetic resonance and dynamic light scattering techniques is low, and they require costly equipment. Taylor dispersion technique is the most accurate technique to measure the diffusion coefficient (Ye et al., 2012).

Fluka provided all the materials in the experiment as conducted by Husson and Gomes (2018). Their purity was 98.8%. Distilled water by weight was used to make aqueous solutions of pesticides. To avoid photodegradation, they were kept away from sunlight. Less than 1 mmol/L pesticide was injected into a diffusion tube; this insured infinite dilution conditions during dynamic measurement of diffusivity. Due to its accuracy, Taylor dispersion technique was used. Its principle is injecting a pulse in a laminar flow of water and observes a quasi-Gaussian peak. This curve gives the measurement of diffusivity coefficient of solute in the solvent. Experimental results were obtained using different models (Table 6.1).

The vital parameter for knowing the fate of pesticides in the aquatic system is their diffusion in water. The diffusion constant explained in the above table is a function of defusing species and the defusing solvent. For some pesticides, the values mentioned above may be less precise due to inaccurate values of molar volume.

6.4 ORGANOCHLORINE PESTICIDES (OCPS): A THREAT TO AQUATIC ECOSYSTEMS

Organochlorine pesticides (OCPs) have long persistence, low biodegradability, and chronic adverse effects on human life and aquatic organisms

(Table 6.2). These have been added to the list of persistent toxic substances (UNEP, 2003). Deepwater and high fat solubility, low vapor pressure, and stable behavior to photo-oxidation are those chemical properties which are not only responsible for their efficiency to kill pests but also for their persistence. They are easily transported in aquatic systems due to persistence in combination with physicochemical properties (Singh et al., 2005). They are highly persistent to degradation in the environment which has made them global pollutants. Pesticides enter the aquatic system through discharges of runoff from agriculture, polluted sewage and industrial wastewater and direct discharge of wastes in rivers or oceans. Pesticides have an ability of rapid distribution among the aquatic system. As these are a fat-soluble accumulation of low concentration of these in mammals of marine has been reported. This is a significant threat to marine mammals and responsible for many harmful effects on them (Metcaff, 1997). The residues of these OCPs reside in rivers and seas for many years, owing to their greater persistence. They are now very less in use due to their adverse effects on organisms. They are accumulated in the reproductive and nervous systems of aquatic organisms due to their environmental persistence. These pesticides mostly disrupt the endocrine system. They not only affect aquatic organisms but also are also responsible for breast and liver cancers in humans (Hileman, 1994; Cocco et al., 1997). The carcinogenic activity of DDT has been explained by the UNEP (2003) report.

TABLE 6.1   Experimental and Calculated Diffusion Coefficients for Pesticides under Study

Pesticides	$D/10^{-9}$ m^2 s^{-1}		RD (%)	References
	D_{exp}	D_{cal}		
Atrazine	0.26	0.66	−150.5	Raveton et al. (1997)
Prometone	0.66[a]	0.54	18.3	Scott and Phillips (1973)
Diphenamid	0.57[a]	0.49	14.7	Scott and Phillips (1973)
Sulcotrione	0.70	0.79	−13.3	Sarraute et al. (2018)
Tebuconazole	0.35	0.36	−3.1	Sarraute et al. (2018)
Metazachlor	0.60	0.52	13.3	Sarraute et al. (2018)
Pirimicarb	0.59	0.65	−9.5	Sarraute et al. (2018)
Chlortoluron	0.64	0.71	−10.4	Sarraute et al. (2018)
Cyromazine	0.73	0.56	23.9	Sarraute et al. (2018)

Note: D is the diffusion coefficient, and RD is random deviation (%) = $(D_{exp} - D_{cal})/D_{exp} \times 100$ and [a]Measured at 23°C.

TABLE 6.2 Biochemical Effects of Major Organochlorine Pesticides

Chemical	Organism	Biochemical Effects	References
Aldrin and Dieldrin	Human	Developmental, immunological, vomiting, genotoxic, nausea, developmental, and a plastic anemia.	USEPA (2003)
Chlordane	Human	Mental confusion, tremor, and convulsions.	ATSDR (1997)
	Seals	Meningoencephalitis, cancer, and trauma.	Kajiwara et al. (2000)
DDT	Human	Incoordination, tremors, nausea, anemia, anxiety, anorexia, confusion, headache, and fatigue.	Klaassen et al. (1996)
	Birds	Eggshell thinning.	USEPA (1975)
	Salmons	Impaired behavioral development	USEPA (1975)
	Fish	Affects enzymes and membrane functions.	USEPA (1975)
Diazion	Birds	Prostration, diarrhea, wing beat convulsions, wing spasms, ptosis of the eyelid, wing drop, tenesmus, and hunched back.	Peterson and Talcot (2006)
	Mammals, reptiles, and fishes	Anxiety, hyperactivity, anorexia, vomiting, depression, lacrimation, abdominal pain, bradycardia, and salivation.	
Methoxychlor	Rats	Reduced fertility.	Cummings and Grey (1989)
	Sea urchins	Fertilization and early development of eggs.	Pesando et al. (2004)
PCB	Rats, fishes, mice, and monkey	Decreased size of the thymus gland, cancer Hodgkins, cancer, lymphoma, and decreased birth weight.	Jacobson and Jacobson (1996)

Fish are the most affected aquatic organisms. That is why they are regarded as indicators for toxicity in the aquatic system. The main reason for this that pesticides enter fish through skin and gills directly (Zhang et al., 2014). OCPs also enter the aquatic system through the trophic chain. Persistent OCPs include aldrin, dieldrin, chlordane, and dichlorodiphenyltrichloroethane (DDT). They are also accumulated in food webs of aquatic systems, disturbing them in one and different forms (Chou et al., 2003). OCPs are mainly found in muscles and liver of fish. Sharks had these chemicals in their liver (Muir and Sverko, 2006). Crustaceans are also affected by these chemicals. Another critical factor in the increased amount of pesticides in the aquatic system is the improper dispose of pesticide container (Costillo, 1992).

Environment cost can be very high when pesticides enter aquatic systems. The kills with these OCPs may be substantial, and these include fish, water birds, frogs, muscles, and many more organisms. The endangered and rare fish and other species like osprey, peregrine falcon, and the bald eagle are adversely affected by these pesticides. Pesticide use is one of the essential factors in declining in several many organisms of the aquatic system (Finley et al., 1980). An aquatic ecosystem is disturbed by these pesticides in a way that they alter the habitat and food chains. There is an elimination of insects and plants by these chemicals in an aquatic ecosystem, which is food, and habitat for fish and other aquatic animals. By pesticide application, insect-eating fish lose a significant part of their food. When insects suddenly vanish from the food chain, insect-eating fish must cover long distances in search of food, exposing them to predation. This dramatically disturbs the food web in an ecosystem. When the food web is disturbed, there is an increase in predator or prey population, which disturbs the aquatic ecosystem (Lutz, 1992).

The reproductive success of fish and other aquatic animals is disturbed by herbicides. There are many shallow weedy nurseries which home to young fish and source of food are destroyed by herbicides. The shelter and cover for these fish are essential to get protection from predators. These are refugees of these aquatic plants. Another drastic effect of herbicides in an aquatic ecosystem is that they also kill those aquatic plants which account for providing 80% of dissolved oxygen in the ecosystem. When oxygen level gets low due to the killing of these aquatic plants, it causes suffocation of fish and other aquatic animals. Freshwater ecosystems are mostly affected by herbicide poisoning. When these are sprayed near freshwater systems, food supply, dissolved oxygen, fish habitat, and fish productivity are significantly reduced. Fish populations are decreased when herbicides are used for "beautification" of lakes (Humberg et al., 1989).

A case study is critical to discuss here. This is a study of Manzala Lake, Egypt in Nile Delta. The study is conducted by Azab et al. (2013). This lake is polluted by OCPs. It has been converted to a freshwater system from the marine system due to increased inputs from agriculture and urban wastes. This lake discharges to the Mediterranean Sea. After affecting the lake, they now have affected the Mediterranean Sea, and they have threatened the aquatic ecosystem there (Baur et al., 2010). It has been labeled as the most polluted lake in the country by the Egyptian National Environmental Action

Plan of 1992. The effects of the pollution can be seen. One of the significant sources of water pollution is pesticides due to their persistence and stability. Majority of OCPs get accumulated in the aquatic food web, for example, shrimps, and fish and their concentration may reach 1000 to 10,000 folds. DDT was banned from most parts of the world due to its bioaccumulation (FAO/WHO, 1978; Roberts et al., 2004; Karl et al., 2010). One of the most affected species by these pesticides is mammals and wild birds which are regarded as the best "bio-indicator species" (Akhtar et al., 2009). The use of OCPs was banned in 1980 in Egypt by the Ministry of Agriculture and Manzala Lake has been the main subject of aquatic weeds and fish status in the last 50 years (Malhat, 2010; Barkat et al., 2012). The fish in different eco-compartments of Manzala Lake was affected by OCPs. The pesticides' mobilization efficacy was transferred in fish, sediment, water, and aquatic weeds and recorded long half-life of these pesticides (Azab et al., 2012). Although the ban on OCPs in Egypt in the 1980's many OCPs have been detected in water and fish samples (Mansour et al., 2010). Samples were taken from different dour sites of Manzala Lake. The sampling from these sides was done in the winter of 2012. 2.5 L water samples were taken in clean water bottles from the surface of the water and 50 cm below the water surface. The bottles were immediately transferred to the lab for analysis. Debris and sand were filtered from samples (Wilde, 2005). By the technique given by Berglund and Ralska-Jasiewiczowa (1986) 2 kg sediment samples were collected from selected locations at a 5 cm depth just at the time of water sampling. Decantation removed water from samples and then taken to the lab. Before analysis, they were air-dried for 48 hours in the dark. Water-weeds (*Pennisetum glaucum*) sampling was also done from four selected sites from localities and analyzed according to the method given by Holm et al. (1977). At the same time of water and sediment sampling, 1–2 kg of fish were also caught from different locations. Only a muscle tissue sample of fish was taken from dorsal muscle and kept in a deep freezer at- 20°C. All the pesticides used in the experiment were 98.5% pure and purchased from well-reputed firms.

The extraction of 1 L water was done twice with 60 ml 15% methylene chloride in n-hexane (APHA, 1975). The number of sediments taken was 50 g weighed and homogenized with 20 g anhydrous sodium sulfate in a ceramic mortar and then shake and then the extract was dried by 50 g anhydrous sodium sulfate (Leyva-Cardoso et al., 2003). After removal of sulfur by treating with acid-activated copper, the rotatory evaporator was used for evaporation of extract to about 1 ml. The most challenging task was to homogenize aquatic weeds. The fish samples were homogenized by the method like sediments fraction with 20 g of 0.5% deactivated florisil of fish, aquatic weeds, sediment, and water was done after cleaning. The solvents were then poured in the column. The first fraction eluted by 70 ml and it was discarded. The second fraction, when eluted at 60 ml of 30% methylene chloride in n-hexane, contained P,P'-DDE, HCHs, and P,P'-DDD. 40 ml-eluted column of methylene chloride gave endrin and dieldrin. After this, the evaporation of these columns was done, and the final extract was transferred to the tube for dryness (UNEP, 1988). They were then transferred for gas chromatography with electron capture detector (GC-ECD).

The results from statistical and quantitative experiments showed that P,P'-DDE and P,P'-DDT was present at all tested areas in the range of 0.067–1.629 ng/ml and from 0.19–4.74 ng/ml respectively. Other pesticides were also present in different sites in considerable amounts but not in all sites. Environmental conditions play a crucial role in the persistence of OCPs or their derivatives. Their derivatives are more toxic than the parent compound (Roberts et al., 2004). For example, DDT has a biodegradation half-life of 15 years in soil and is slowly converted to DDD or DDE. Their stability is more in the environment than DDT. Azab et al. (2013) concluded that variation in concentrations of OCPs was seen I different sites of Manzala Lake. The levels of residues of OCPs were more in summer as suggested by published data as compared to that in winter.

6.5 PESTICIDES AND THEIR IMPACT ON AQUATIC MICROORGANISMS

Aquatic microorganisms play a critical role in decomposition, primary productivity, and nutrient cycling and are a vital resident of the aquatic system. Microorganisms are exposed to pesticides by direct or indirect inputs of pesticides in the aquatic environment. Pesticides impart acute and chronic toxic effects on microorganisms. Microbes can metabolize, accumulate, or detoxify pesticides to some extent. Higher trophic levels are affected when microbes are exposed to the detrimental effects of pesticides subsequently. Effect on the growth of zooplankton grazers when changes take place among the macromolecular composition of phytoplankton species or shift in the composition of a community is an example of this (Ahlgren et al., 1990).

Coastal watersheds receive millions of pounds of active pesticide ingredients each year. These directly affect estuaries, which are nursery ground and critical feeding place for many aquatic organisms. They damage shellfish and recreationally important fish. As these serve as a productive ecosystem, so, they are vulnerable to pesticide pollution. Another reason for this is that they also act as repositories for pollutants. Estuaries are also posed of toxicity from agricultural estuaries (Scott et al., 1990).

Data for the toxicity of pesticides in microorganisms is limited. Rather than impacts of pesticides on the microbial population, most studies have focused on microbial degradation of pesticides. Many studies are on the effects of pesticides on soil than those on marine systems. The data which involves estuarine or marine microbes are even scarcer. Based on the mechanism of action, pesticides are classified. For instance, organophosphate, organochlorine, and carbamates disrupt nervous systems, and herbicides mainly affect on photosynthesis. The action of pesticides on microbes is not the same as target organisms. Cell division and growth, biosynthetic reactions, respiration, molecular composition, and photosynthesis are mainly damaged by pesticides in microorganisms.

The light reaction of photosynthesis in microbes is majorly affected by herbicides. Triazines, triazinones, urease, and cyclic urease are the groups that disturb the "Hill reaction" of electron transport chain. Electrons from reducing side of photosystem bipyridinium herbicides such as diquat intercept me. When oxygen is present oxidation of bipyridinium occurs and forms an original ion. This activated species affects cell tissues. There are other compounds like pyridazinones that are involved in the inhibition of carotene formation. They also alter fatty acid composition of chlorophyll lipids (Corbet et al., 1984). A process called bleaching inhibits carotene production. It does not inhibit chlorophyll formation directly, but as a photo-oxidative shield of carotene is missing, so the pigment is destroyed in the presence of light (Sikka and Pramer, 1968). *S. obliquus* is poisoned by diphenyl ether; maintenance of high oxygen concentration is achieved by splitting water by photosynthetic electron transport chain in algal suspension. The herbicide is reduced to radical species by electron transport chain in case of paraquat, which starts lipid peroxidation. Herbicides also influence microtubular systems of some aquatic microorganisms. Tubulin forms the microtubular system. The microtubular formation is stopped when pesticide colchicine binds with tubulin. Separation of chromosome and cell division are affected by the loss of spindle microtubules. A lack of cortical microtubules causes the disruption morphogenesis of cells and tissues. Propyzamide, phosphoric amides, dinitroanilines, terbutol, and carbamates are all those herbicides that destroy microtubules (Fedtke, 1982). Protein, carbohydrate, and chlorophyll production in cyanobacteria is affected by p-nitrophenol. These nitrophenols also hampered the activities of enzymes like nitrate reductase and nitrogenase (Megharaj et al., 1991).

There is a considerable diversity found in the sensitivity of microorganisms to pesticides. There are three kingdoms of microorganisms, including more than 50,000 species of bacteria, protozoa, and algae. They are present in many sizes and morphologies classes, a variety of growth and reproductive strategies, mobile, and nonmobile forms, multiple feeding types, and exist in every habitat. The chlorinated hydrocarbon pesticides like endrin, aldrin, and dieldrin had no significant effect on algal respiration at a concentration up to 1 mg/L. Aldrin and dieldrin lowered adenosine triphosphate levels at higher concentrations (100 mg/L), but population densities were not affected (Vance and Drummond, 1969; Clegg and Koevenig, 1974). DDT inhibited photosynthesis in marine planktonic algae at a concentration below ten microgram/L (Lee et al., 1976). There is a class that is only used in the USA, known as endosulfan. It is chlorinated cyclodiene insecticide with same chemical formula as DDT. It is applied to vegetables, tea, tobacco, cotton, and grains. Detrimental effects on reproduction of green algae are seen by endosulfan. During the four developmental stages of the sexual life cycle delayed meiosis by five days just only by a single exposure. There was no effect on zygote and gamete production. When this was applied to gametogenesis cells, it reduced meiosis by 35%. It chronically impaired tetra sporophytes and female growth in red algae. Lasting reproductive effects were seen on algae when higher concentrations were applied (Netrawali, 1986; Thursby, 1985).

A study was conducted on axenic batch cultures of 12 freshwater algae to check effects of organophosphorus insecticide fenitrothion. There was a reduction in growth at 10 mg/L. It also inhibited normal mitotic division leading to macromolecules accumulation and cell was also increased (Lal and Lal, 1988). Cellular lipid content is strongly related to fenitrothion sensitivity, and it suggests that lipophilic compounds are more toxic to those phytoplankton species that have large lipid fractions. This also observes bleaching of chlorophyll in several blue-green algae (Kent and Currie, 1995). The growth of algae was completed stopped at higher concentrations of dimethoate and Kimpton. They also cause a decrease in algal carbohydrate and protein. Cell density was significantly reduced at a concentration higher than 1%. Carotenoid was also depleted by these chemicals which affected chlorophyll indirectly (Jampani and Kumari, 1995).

Carbofuran had interspecific sensitivity on diatoms, green algae, and cyanobacteria. There was a 50% less uptake of carbon dioxide in these species. Carbaryl reduced the population density of algae by 30% even at low concentrations. This was also lethal to species of phytoplankton (Ukeles,

1962). Even the concentration of 0.3 mg/L of diquat altered a significant ratio of bacterial and algal densities in mainly freshwater ecosystems. This chemical also increased the population density of protozoa, and no recovery was seen even after 21 days. This also inhibits the growth of cyanobacteria (Melendez et al., 1993).

Diuron affects the photosynthesis of marine periphyton. It also inhibits reproduction processes in other microorganisms. There is an indirect negative impact on communities of microorganisms to which soluble organic matter excreted by phytoplankton is essential (Pesce et al., 2006). There are facultative anaerobic gram-negative bacteria called Aeromonads which possess which have a wide range of exoenzymes that are active in bio-degradation of many organic components. That is why they act as an indicator of the hygienic status of water. Some of the microbial species may also be grown in culture media (Sumpono et al., 2003). A reduction in phototrophic organisms also decreased the feed to some grazers. They cause a decrease in productivity and biomass of many photographic organisms. This affects the quality and quantity of nutrients which are circulated in microbial food web at higher trophic levels (Ahlgren et al., 1990). In microcosm, some pesticides change the concentration of those algae and bacteria that originate from natural ecosystems even at deficient concentrations. However, the population of green algae was increased considerably at diquat concentrations of 10 and 30 mg/L. This leads to eutrophication of freshwater systems depleting oxygen in them.

The growth of some bacteria was also seen to increase the application of some herbicides (Cragg and Fry, 1984). Aldrin shows negative impacts on phosphate-solubilizing and heterotrophic bacteria. It also decreased nitri-fying activity (Lopez et al., 2002). The strains *Pseudomonas, Rhodococcus, Bacillus,* and *Xanthomonas,* have shown high degradation capacity of many pesticides. This has led to becoming a key research field. Some fungi have also shown biodegradable properties. There is no effect of chlorpyrifos (CPF) on phytoplankton, but a small decrease in photosynthetic activity was indicated by atrazine (Van den Brink et al., 1995).

There is a significant effect of pyrethroid insecticides on phytoplankton. When these were tested at various concentrations in many artificial and natural ponds. There is a decline in zooplankton abundance (Heimbach, 1992). Marine diatoms were poisoned by tributyltin oxide and tributyltin chloride. Many algal species were unable to take up carbon efficiently in the presence of aldehyde herbicides. The Carbon uptake in Diatoms and nitrogen-fixing cyanobacteria was significantly reduced by glycine derivative herbicide (Goldsborough and Brown, 1986). From this detailed discussion, it is concluded that pesticides are responsible for many negative impacts on microorganisms of aquatic systems. However, some herbicides have shown positive effects on the growth of some microbes.

6.6   IMPACT OF PESTICIDES ON AQUATIC LIFE

As discussed, earlier pesticides have a lot of lethal effects on aquatic life. This is due mainly to persistence, bioaccumulation, and chemical formulation of these compounds. From the above discussion, it has been clear that not only plants and animals but also microorganisms are adversely affected by these compounds. The reasons for this are the direct entry of polluted water from industry or agricultural runoff from fields. The freshwater systems are majorly affected by pesticide pollution because these are present near to those sites where pesticides are sprayed in abundance. Apart from runoff these also enter freshwater systems by air currents or disposal of cans of pesticides directly in lakes, canals, or freshwater ponds. There are many examples in which you can observe that many freshwater lakes are subjected to eutrophication only due to the application of fertilizers and pesticides. The residues or daughter compounds of pesticides are even more dangerous. These have a half-life of many years. Another problem with residues is that only a few microorganisms can biodegrade them. These poses even more danger to aquatic life. Much attention has been given these days to these compounds.

The annual production and sale along with total acres covered of herbicides have surpassed all other chemicals like insecticides, pesticides, or fungicides. There is a growing awareness of adverse effects of herbicides on human, terrestrial, and aquatic organisms in recent years (Pèrez and Miranda, 2011). Herbicides are formulated to kill unwanted plants by inhibiting cell division, cell metabolism, pigment synthesis, or by directly stopping photosynthesis. Some of their chemical properties which are used in their mode of action are a threat to aquatic life. These chemical properties are:

- Volatility;
- Water solubility and polarity;
- Chemical structure;
- Formulations.

Volatility is the essential factor in determining the fate of herbicide in environmental or aquatic environments. Those herbicides which have high vapor pressure are readily volatile while those having low vapor pressure are non-volatile. The solubility of herbicides in water being manufactured as salts is quite well. These are manufactured in this way to be easily dissolved in water and sprayed quickly. However, this property also makes the threat to aquatic life when they enter aquatic systems. This characteristic also determines their method of application and movement in the soil profile.

The portion of herbicide that is biologically active is called active ingredient. This active ingredient remains persistent in ecosystems and causes toxicity. These ingredients are dangerous chemicals for every form of life. It also contains an inert ingredient which is used for dispersal and dilution of herbicides. It may act as surfactant, thickener, activator, emulsifier, additive, carrier, or other adjuvants. Latest studies show that these carriers and adjuvant are responsible for increased toxicity of herbicides and sometimes even more toxic than the active ingredient. The endocrine disruption, developmental abnormalities, mutagenicity, biomagnification, and carcinogenicity are widely studied (Radosevich, Holt, and Ghersa, 2007).

There are some sublethal effects imparted by pesticides. They do not cause immediate death in animals. Fish present in freshwater streams receive repeated low doses of pesticides from surroundings. They may reduce the production of eggs in fish. The most common sublethal effects in aquatic life include:

1. Weight loss;
2. Loss of attention;
3. Sterility;
4. Low predator avoidance;
5. Low disease resistance;
6. Reduced egg production;
7. Rate of growth;
8. Inefficient feed utilization (Hudson et al., 1984).

Now several routes of entry of pesticides in fish will be discussed. There are three main ways of pesticides to enter in fish:

a. By breathing;
b. Orally; and
c. Dermally.

Fish mainly respire through gills and pesticides also enter through them during respiration. The oral pathway is by drinking pollutant pesticide water or by preying on pesticide-poisoned organisms. When excess pesticides contaminate water, it starts to enter in fish even by the skin. This, in turn, is a threat to humans when they eat these contaminated fish. When pesticides enter in fish, they mainly disrupt the endocrine system. It results in the death of fish. The most fish kill by pesticides goes unreported. The estimation of fish killed by pesticides is a complicated task. The underwater conditions like size and camouflage by young fish, depth, and clarity of water make it even more challenging to study. The stressed or adversely affected fish may hide in dense cover or may leave the area thoroughly. Scavenging micro-organisms quickly remove carcasses from the site. Fear of liability or not considering fish kill important are essential factors in not reporting of the fish kill. Some aquatic systems are very remote.

The coastal waters and estuaries receive fewer amounts of pesticides than rivers and due to degradation and dilution. The number of pesticides like atrazine varies with the weather. For example, the highest amount of atrazine was found in summer and spring months in the USA, as reported by Solomon et al. (1996). A little information is available for the half-life of pesticides and their persistence in aquatic ecosystems. Davies et al. (1994); Detenbeck et al. (1996) reported that atrazine in estuarine sediments resides for two weeks approximately and three months in streams after spraying. The degradation of atrazine in water like soil's results from both chemical and biological processes. Splitting of ethyl groups from triazine ring provides energy to fungi and bacteria. There is chemical hydrolysis of this ring which yields chloride ions. 2- and 4-dihydroxy derivatives which are primary metabolites are released on subsequent dealkylation and hydroxylation reactions. Very less degradation is done through photochemical processes (Hamilton et al., 1989; Winkelmann and Klaine, 1991).

Pesticides lower levels of oxygen, particulate organic carbon, and dissolved organic carbon from decreased primary production and increased respiration in aquatic plants. When atrazine is exposed to an aquatic environment, it increases calcium in the water. This is due to reduced production of calcium carbonate precipitation due to inhibition of crystal growth or the decrease in pH. Atrazine also increases ammonium, nitrate, sulfate, and nitrite in the aquatic system (Hamilton et al., 1989). There is a direct influence of pesticides on phytoplankton, macrophytes, and periphyton. Indirect effects were seen on growth or abundance of fish, invertebrates, and tadpoles (Dewey, 1986). Both periphyton and phytoplankton communities are affected by even low concentrations of pesticides. The main effect is on the photosynthesis of these communities (Lakshminarayana et al., 1992). Now some toxicity tables (Table 6.3) will be discussed to show different levels of toxicity to fish. These pesticides are conventional in or near aquatic systems. The effects which cause occurs rapidly in animals are called acute effects. A single dose of pesticides is necessary to have symptoms on organisms. Death is an example of acute effects. These effects are stated as an LC_{50} value in Table 6.4. The amount of concentration of chemicals in water usually in parts per million or milligrams per liter that can kill 50% of test animals in a given time. Lower the LC_{50} value more the toxicity of the pesticide. There are many factors which affect the toxicity of pesticides on aquatic animals. Condition of an animal, species, water temperature, pesticide formulation, sex, and age are the main factors. Dilution and formulation of pesticides are vital factors of toxicity.

TABLE 6.3   Toxicity to Fish of Commonly Used Herbicides

Toxicity Classification	LC_{50} (mg/L)
Super	<0.01
Extreme	0.01–0.10
High	0.11–1.0
Moderate	1.1–10
Slight	11–100
Minimal-Nontoxic	>100

TABLE 6.4   Toxicity to Fish of Commonly Used Insecticides, Miticides, and Nematicides

Toxicity Classification	LC_{50} (mg/L)
Super	<0.01
Extreme	0.01–0.10
High	0.11–1.0
Moderate	1.1–10
Slight	11–100
Minimal-Nontoxic	>100

Source: Mayer, Foster, and Mark (1986); Ware and George (1994).

6.7   IMPACT OF PESTICIDES ON INVERTEBRATES IN AQUATIC ECOSYSTEM

The effects of pesticides on aquatic life have been discussed in much detail in previous sections, but effects on pesticides were not discussed in detail. This section focuses on impacts only on invertebrates. The significant phyla included are mollusk, coelenterate, annelid, etc. 95% of animal species are invertebrates with the highest number of phyla. That's why it is essential to discuss the effects of pesticides on invertebrates. From the above discussion, pesticides would impart adverse effects only. Therefore, it is not necessary to discuss those factors or formulations that make them toxic.

There is an increased risk of aquatic invertebrates to pesticide poisoning. It also affects other vertebrates because they largely depend on invertebrates for their nutritional requirements (Gibbons et al., 2014). The most affected species by pesticides like neonicotinoids are some crustaceans and many members of phylum arthropods. These insecticides disturb the central nervous systems (CNS) of these invertebrates. Mammals and other vertebrates are not affected by them due to less affinity for the compounds found in these insecticides (Yamamoto et al., 1995). These insecticides bind to receptors of invertebrates, which show irreversible damages, and they become more cumulative with exposure time (Tennekes, 2010). There is no surprise in stating that neonicotinoids show lethal and sublethal effects on the population of a high number of invertebrates. Aquatic arthropods are the most affected organisms of the aquatic system with varying toxicities. The most sensitive class of arthropods is class Insecta (Alexander et al., 2008). The most tested global species of invertebrates for industry standards is *Daphnia magna*. Almost 82 chemicals have been tested on it globally (Sanchez-Bayo, 2006). However, this is least sensitive for chronic and acute effects of neonicotinoids as reported by Jemec et al. (2007).

Other pesticides like atrazine are responsible for inhibiting growth and other metabolic processes. Atrazine reduces succession and growth of phytoplankton. Some species are tolerant of these pesticides under lab conditions. However, in the natural environment, their abundance decreased due to a reduction in the number of edible phytoplankton species (Dewey, 1986). The composition of species in zooplankton communities has been altered by the application of atrazine. The dominance of species varies in ecosystems (DeNoyelles et al., 1982). The change in community structure is mainly because of a change in the food source. The direct physiological effects have not been observed in zooplankton species. Transient behavioral response is shown by some invertebrates in the drift of streams. Gastropods are those invertebrates which graze on algae. When the algal population is decreased by any means, it leads to a reduction in the gastropod population and change in the structure of the community. When pesticides reach sediments, gastropods are the most affected species because they live there (Davies et al., 1994). Macrophytes death by pesticides has a direct effect on invertebrates. They are used for grazing by herbivorous gastropods and are the substrate for periphyton growth. When macrophytes are lost, it causes destruction of habitat which has effect diversity and abundance of gastropods.

There is no effect of pesticides on crustaceans in a series of species test. However, some species are impacted at high concentrations. It is complicated for fiddle crabs to survive or escape from the area polluted by pesticides. Adult amphipod's survival is also tricky at higher amounts of pesticides, but low amounts don't affect them (Dewey, 1986). There is a need for more testing of some crustacean species that mainly feed on algae to have knowledge of pesticide poisoning on them. There may be some indirect impacts on these species. The most affected species by atrazine are those who have a benthic lifestyle. There is an open research field for researchers to research on changes in community structure and recovery of crustaceans from atrazine application.

The concentrations between 20 and 100 micrograms per liter of atrazine impact growth development, and survival of many aquatic insects. The atrazine concentration was more significant than 230 micrograms per liter cause reduction of several larvae pupating, hatching success, and promote abnormal larval developments (Dewey, 1986; Macek et al., 1976). In the experiment conducted by Dewey (1986), adverse effects were seen of atrazine exposure were seen on many other chironomid species. For example, the population of new chironomid *Labrundinia Pilosella* was reduced even at low concentrations of pesticide atrazine. Many other species had decreased emergence rates from its application in the aquatic ecosystem. The benthic insects were affected mainly by pesticides in changes to species richness, diversity, and abundance as pesticide concentrations increased. An apparent reduction in species richness and abundance was seen in benthic non-predatory insects as compared to predator species. This was mainly due to food unavailability and habitat destruction. The sensitivity to atrazine of herbivore species is highest in comparison to other aquatic species of invertebrates. When food is not available to predatory species, the switch to other prey which includes tubificid worms and other benthic microcrustaceans because these species are less sensitive to pesticides. The earlier emergence period is seen in many species of herbivores in response to atrazine. Other groups are adversely affected by this change in the emergency period as this leads to a limited number of hatches that occur in that season, causing a reduction in the availability of food for predators (Gruessner and Watzin, 1996).

6.8 METHODS OF REMOVAL OF PESTICIDES FROM AQUATIC MEDIUM

Before discussing methods of pesticide removal, it is essential to discuss some facts about the use of pesticides in developing countries. Developed countries have controlled the application of very toxic pesticides such as DDT. However, due to some factors, many banned pesticides are still being used in third world countries, mainly due to their less cost. Many laws have been legislated in these countries, but no one applies these laws. Due to more significant pressure of population, governments of these countries neglect the adverse effects of these dangerous pesticides. From the above-detailed discussion, it has been clear that pesticides are not the only threat to aquatic life but also to humans and other terrestrial organisms. There are clear signs of pesticide pollution in soils which damage agriculture. The most straightforward approaches to reduce pesticide pollution are use pesticides only when it is indispensable to apply, use a safe method of application, use fewer toxic pesticides, and pay special heed in applying pesticides near freshwater resources.

There is a rising concern about pesticide pollution all over the world. In the past few decades, many methods of removal of pesticides from aquatic media have been proposed. There are three basic methods to remove pesticides. These are based on the use of the characteristic of the therapeutic process used. Currently, physical, biological, and chemical remedial methods are used. Chemical and physical methods are used in coordination. The main factors on which removal of pesticides depend are the nature of pesticides, pH, temperature, cost of investment, and kind of matrix.

Physical means are used in the physical removal of pesticides (Tan, 2009). Some of the physical means of removal of pesticides are zeolites, activated carbon, clays, and polymer materials. Zeolites are a family of crystalline aluminosilicates that have incredibly narrow pore size distribution with interconnected micropores (Yang et al., 2016; Pham et al., 2016). Due to its physicochemical properties, low cost, and easy availability, it is widely being used in industry, pollution control, and agriculture mainly to remove pesticides (Huong et al., 2016). For instance, De Smedt et al. (2016) reported it as an absorbent for treating pesticide-contaminated water. The absorption rate depends on the polarity and mobility of pesticides. Zeolite is very effective against chlorinated pesticides, especially against hexachlorobenzene (Valičková et al., 2013). The effective method against carbamate pesticide in aqueous solutions is surfactant-modified sorbents. For example, bromide modified zeolites are used to tackle carbamates due to its hydrophobic property. Over 94% of carbamates were absorbed by zeolites (Arnok and Burakham, 2014). The adsorption rate is affected by two main factors polarity and mobility of pesticides. Mobile pesticides are partitioned in the water while immobile pesticides are associated with zeolites based on affinity for zeolites. This depends on polarity. Trapping rate of zeolites depends on polarity and mobility of the pesticides (De Smedt et al., 2015).

One of the most superabsorbents with a high capacity of pesticides in water is activated carbon (Bonvin et al., 2016). The properties of adsorption solution (temperature, the concentration of adsorbate, pH, etc.), structure, surface chemistry, and chemical properties of adsorbate are main factors which influence mechanism adsorption processes of activated carbon (Merczewski et al., 2016). Powder activated carbon (PAC) and granular active carbon (GAC) are the most used activated carbon. The flexibility of manipulation and cost-effectiveness makes PAC is more advantageous over GAC. Srivastava et al., 2009). The iron catalyst, when used on activated carbon, is used efficiently to eliminate atrazine from water. Biochar has also shown a high capacity to remove pesticides from aquatic media. Particle size plays a crucial role in the removal of pesticides from the system. Other properties like solid to solution ratio and pH also affect biochar's ability to remediate chemicals (Zheng et al., 2010). At pH 7, the efficiency of AC is maximum and economical due to the fact that no needs to neutralize water afterward (Gupta et al., 2011).

Cationic pesticides are readily removed by clays because clays are negatively charged and hydrophilic in nature. They are mostly used in soils to remove pesticides. However, some studies suggest their use to remove pesticides from marine systems also. El Ouardi et al. (2013) states that carbaryl can be removed from aqueous solution. Absorption increases as the concentration of pesticide increases in the system. This adsorption capacity is increased at acidic conditions.

Biological remediation is the use of organisms to convert organic compounds into less harmful and products like water and carbon dioxide. Its main significant advantage is that it is environment-friendly and cost-effective (Nwankwegu and Onwosi, 2017). Following are three main types of biological remediation:

a. Bioaugmentation;
b. Through natural attenuation; and
c. Biostimulation.

The introduction of genetically modified microorganisms in a system is called bioaugmentation. By taking advantage of the natural microbial population, remediation can be done through natural attenuation. Biostimulation is the addition of nutrients in the system (Helbing, 2015; Nwankwegu and Onswosi, 2017). Marine sediments or other contaminated sites can provide many microorganisms which can degrade organic pollutants (Ferreira et al., 2016). Bioaugmentation shows best results in wastewater removal of pesticides. There are a lot of strategies stated by Herrero and Stuckey (2015). The rice fields are successfully remediated by bioremediation (Pimmata et al., 2013). The advancement in bioremediation is the use of enzymes from microorganisms which are capable of degrading pesticides. The main pathway is bioaugmentation, which is an alternate path to revolutionize bioremediation. The enzymes are immobilized on solids, which lead to the recovery of these enzymes, making them lower costs due to reuse (Gao et al., 2014). Temperature, organic solvents, pH, and other environmental factors are those to which enzymes are susceptible and may ultimately lose the degradable ability (Xie et al., 2010).

Fungi are another tool to tackle pesticide pollution. These are effectively being used as biodegrades against pesticides like endosulfan, atrazine, dieldrin, heptachlor, etc. However, the main disadvantage of this tool is its speed of degradation depends on many factors like pH, oxygen level, temperature, nutrient availability, soil moisture content, etc. The processes like dechlorination, hydroxylation, oxidation, esterification, demethylation, and dehydrochlorination play a crucial role in fungal strains during the biological degradation of pesticides. Many pesticides are degraded by esterase, hydrolase, lactase, lignin peroxidase enzymes (Maqbool et al., 2016).

Some chemical agents are applied to extract hazardous chemicals such as pesticides to harmless or less harmful chemicals through chemical reactions (Hamby, 1996; Tyler and Spoolman, 2014). Due to high economic cost, physical, and chemical remedial processes are used in combination (Dhal et al., 2013). The advanced chemical techniques which are environment-friendly are used at ambient temperature and pressure. These are advanced oxidation processes (AOPs). Due to not transferring of contaminants from one phase to another, these are regarded as environmentally friendly. All types of organic pesticides are converted to harmless products by AOPs. Hydroxyl radicals are produced in these processes having redox potential of 2.8 V. They are as reactive as fluorine. Mineralization products like carbon dioxide and water are yielded by a series of oxidation reactions started by hydroxyl radical (Cheng et al., 2016).

Another chemical technique to tackle pesticides is the Fenton reaction. The reaction in which oxidation of organic pollutants by hydrogen peroxide and Fe (II) is known as Fenton Reaction. This is one of the best methods for the oxidation of organic pesticides (Barbusiński, 2009). Hydrogen peroxide produces hydroxyl radical (OH), which is a strong oxidizing agent. Following is its equation:

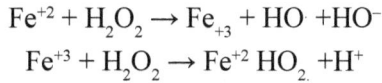

$$Fe^{+2} + H_2O_2 \rightarrow Fe_{+3} + HO \cdot + HO^-$$
$$Fe^{+3} + H_2O_2 \rightarrow Fe^{+2} HO_2 \cdot + H^+$$

From the second equation, a catalyst is generated (Oturan and Aaron, 2014). This process is improved by applying an electrical current. In electro-Fenton by in situ, reductions of oxygen in the presence of Fe^{2+} generates hydrogen peroxide. You don't have to add hydrogen peroxide continually in this reaction (Iglesisas et al., 2014). The processes of UV hydrogen peroxide and UV-ozone is also used.

6.9 BIOPESTICIDE TECHNIQUE TO REMEDIATE PESTICIDES IN POLLUTED ECOSYSTEMS

Biopesticides are those pesticides which are developed from natural materials such as bacteria, plants, animals, and certain other minerals. Following are three main types of biopesticides:

- Plant pesticides;
- Microbial pesticides; and
- Biochemical pesticides.

Plant pesticides are those which are derived from plants produced from their genetic material. For instance, Bt pesticides protein genes are incorporated in plants, and plants produce substances that destroy the pest. The active ingredient in microbial pesticides is microorganisms including fungus, alga, bacterium, or protozoa. The most used varieties of microbial pesticides are from the bacterium Bacillus thuringiensis or commonly known as Bt. these are mostly used in potatoes, cabbage, and other vegetables. Biochemical pesticides are prepared from naturally occurring compounds that control pests by nontoxic mechanisms. Pheromones and plant growth regulators are mainly included in biochemical pesticides. The main advantages of pesticides are they are target specific, i.e., that only affect their target organisms; they are beneficial even at low concentrations, they decompose quickly, environment-friendly, far less harmful as compared to conventional pesticides, and there are no residue problems. They often compete with synthetic pesticides when applied in fields. However, their efficiency is less as compared to other types of pesticides but not very less that could make them unable to use (Joshi, 2006).

6.10   CONCLUSION

The sullying of the amphibian environment from pesticides, creature squanders, and bug sprays, manures, and so on is the most widely recognized issue of water these days. In the amphibian condition, the pesticide can adsorb or desorb on suspended solids and further settle down in the base silt. Pesticides achieve oceanic frameworks fundamentally through two procedures, to be specific overflow, and filtering. These two fundamental procedures are straightforwardly associated with the hydrological cycle. Because of fast urbanization and industrialization, the utilization of dangerous synthetic compounds in pesticides expanded. Pesticides bring down the dimensions of oxygen, particulate natural carbon, and disintegrated natural carbon (DOC) from diminished essential generation and expanded breath in amphibian plants. At the point when atrazine is presented to an oceanic domain, it builds calcium in the water. This is because of the diminished generation of calcium carbonate ($CaCO_3$) precipitation because of the restraint of precious stone development or the reduction in pH. OCPs, atrazine, and so on build ammonium (NH_4^+), nitrate (NO_3^-), sulfate (SO_4^{-2}), and nitrite in the amphibian framework. There are a few strategies to expel pesticide sullying from an oceanic condition like organic (bioaugmentation, characteristic constriction, and bioincitement), synthetic (denitrification, switch assimilation, electrodialysis, and reactant denitrification), and physical (zeolites, enacted carbon, dirts, and polymer materials).

KEYWORDS

- aquatic microorganisms
- biopesticide technique
- organochlorine pesticides
- pesticide sources
- pesticides
- water pollutants

REFERENCES

(2003). *UNEP Global Report on Regionally Based Assessment of Persistent Toxic Substances.* Geneva, Switzerland. 2003: UNEP.

Ahlgren, G., Lundstedt, L., Brett, M., & Forsberg, C., (1990). Lipid composition and food quality of some freshwater phytoplankton for *Cladoceran* zooplankters. *J. Phytoplankton Res., 12,* 809–818.

Akhtar, M., Iqbal, S., Bhanger, M. I., Zia-Ul-haq, M., & Moazzam, M., (2009). Sorption of organophosphorus pesticides onto Chickpea husk from aqueous solutions. *Colloids and Surfaces. B, Biointerfaces, 69,* 63–70.

Alexander, A. C., Heard, K. S., & Culp, J. M., (2008). Emergent body size of mayfly survivors. *Freshw. Biol., 53,* 171–180.

APHA, (1975). *Standard Method for Examination of Water and Wastewater* (14th edn.). Washington; AWWA/WPCE.

Arnnok, P., & Burakham, R., (2014). Retention of carbamate pesticides by different surfactant-modified sorbents: A comparative study. *J. Braz. Chem. Soc., 25,* 1720–1729.

Azab, M. M., Darwish, A. A., Mahmoud, A. H., & Sdeek, A. F., (2012). Study on the pesticides pollution in Manzala Lake, Egypt. *Journal of Applied Sciences, 27*(8), 105–122.

Barakat, A. O., Mostafa, A., Wade, T. L., Sweet, S. T., & El Sayed, N. B., (2012). Assessment of persistent organochlorine pollutants in sediments from Lake Manzala, Egypt. *Mar. Pollut. Bull., 64*(8), 1713–1720.

Barbusiński, K., (2009). Fenton reaction-controversy concerning the chemistry. *Ecol. Chem. Eng., S 16,* 347–358.

Bonvin, F., Jost, L., Randin, L., Bonvin, E., & Kohn, T., (2016). Super-fine powdered activated carbon (SPAC) for efficient removal of micropollutants from wastewater treatment plant effluent. *Water Res., 90,* 90–99.

Castillo, L., & Wesseling, C., (1992). *Monitoring PIC: Setting the Pesticides Agenda: Update for Costa Rica.* Programa de Plaguicidas, Universidad Nacional, Heredia, Costa Rica.

Cheng, M., Zeng, G., Huang, D., Lai, C., Xu, P., Zhang, C., & Liu, Y., (2016). Hydroxyl radicals based advanced oxidation processes (AOPs) for remediation of soils contaminated with organic compounds a review. *Chem. Eng., J., 284,* 582–598.

Chou, J. P., Chen, H. C., Chen, C. S., Chang, P. C., & Chou, S. S., (2003). Survey on the organochlorine pesticide residues in fish and shellfish in Taiwan. *Ann. Rept. BFDA Taiwan., 21,* 266–291.

Corbett, J. R., Wright, K., & Baillie, A. C., (1984). *The Biochemical Mode of Action of Pesticides.* Academic, London, UK.

Cragg, B. A., & Fry, J. C., (1984). The use of microcosms to simulate filed experiments to determine the effects of herbicides on aquatic bacteria. *J. Gen. Microbiol., 130,* 2309–2316.

Davies, P. E., Cook, L. S. J., & Goenarso, D., (1994). Sublethal responses to pesticides of several species of Australian freshwater fish and crustaceans and rainbow trout. *Environ. Toxicol. Chem., 13*(8), 1341–54.

De Smedt, C., Ferrer, F., Leus, K., & Spanoghe, P., (2015). Removal of pesticides from aqueous solutions by adsorption on zeolites as solid adsorbents. *Adsorpt. Sci. Technol., 33*(5), 457–485.

DeNoyelles, F., & Kettle, W. D., (1983). *Site Studies to Determine the Extent and Potential of Herbicide Contamination in Kansas Waters* (p. 239). Kansas: Kansas Water Resources Research Institute, University of Kansas.

Detenbeck, N. E., Hermanutz, R., Allen, K., & Swift, M. C., (1996). Fate and effects of the herbicide atrazine in flow-through wetland mesocosms. *Environ. Toxicol. Chem., 15*(6), 937–946.

Dewey, S. L., (1986). Effects of the herbicide atrazine on aquatic insect community structure and emergence. *Ecology, 67*(1), 148–62.

Dhal, B., Thatoi, H. N., Das, N. N., & Pandey, B. D., (2013). Chemical and microbial remediation of hexavalent chromium from contaminated soil and mining/metallurgical solid waste: A review. *J. Hazard. Mater., 250, 251,* 272–291.

El Ouardi, M., Alahiane, S., Qourzal, S., Abaamrane, A., & Assabbane, A., (2013). Removal of carbaryl pesticide from aqueous solution by adsorption on local clay in Agadir. *Am J. Anal. Chem., 4*(07), 72–79.

FAO/WHO, (1978). *Joint FAO/WHO Food Standards Program Codex Alimentarius Commission Twelfth Session* (pp. 1–74).

Fedtke, C., (1982). *Biochemistry and Physiology of Herbicide Action.* Springer-Verlag, New York, NY, USA.

Ferreira, L., Rosales, E., Danko, A. S., Sanromán, M. A., & Pazos, M. M., (2016). Bacillus thuringiensis a promising bacterium for degrading emerging pollutants. *Process Saf. Environ., 101,* 19–26.

Gao, Y., Truong, Y. B., Cacioli, P., Butler, P., & Kyratzis, I. L., (2014). Bioremediation of pesticide contaminated water using an organo-phosphate degrading enzyme immobilized on nonwoven polyester textiles. *Enzym. Microb. Technol., 54,* 38–44.

Gibbons, D., Morrissey, C., & Mineau, P., (2014). A review of the direct and indirect effects of Neonicotinoids and fipronil on vertebrate wildlife. *Environ. Sci. Pollut. Res.,* 1–16.

Goldsborough, L. G., & Brown, D. J., (1986). *Effects of Aerial Spraying of Forestry Herbicides on Aquatic Ecosystems* (pp. 86–11). III. Baseline physical and chemical characterization of experimental study ponds. Manitoba Dept. Environment and Workplace Safety and Health, Water Standards and Studies Report.

Gruessner, B., & Watzin, M. C., (1996). Response of aquatic communities from a Vermont stream to environmentally realistic atrazine exposure in laboratory microcosms. *Environ. Toxicol. Chem., 4,* 410–419.

Gupta, V. K., Gupta, B., Rastogi, A., Agarwal, S., & Nayak, A., (2011). Pesticides removal from wastewater by activated carbon prepared from waste rubber tire. *Water Res., 45*(13), 4047–4055.

Hamby, D. M., (1996). Site remediation techniques supporting environmental restoration activities: A review. *Sci. Total Environ., 191*(3), 203–224.

Hamilton, P. B., Lean, D. R. S., Jackson, G. S., Kaushik, N. K., & Solomon, K. R., (1989). The effect of two applications of atrazine on the water quality of freshwater enclosures. *Environ. Pollut., 60,* 291–304.

Heimbach, F., (1992). *Ecotoxicology of Earthworms.* Intercept, Andover.

Helbling, D. E., (2015). Bioremediation of pesticide-contaminated water resources: The challenge of low concentrations. *Curr. Opin. Biotechnol., 33,* 142–148.

Herrero, M., & Stuckey, D. C., (2015). Bioaugmentation and its application in wastewater treatment: A review. *Chemosphere, 140,* 119–128.

Holm, L. G., Plucknett, D. L., Pancho, J. V., & Herberger, J. P., (1977). *The World's Worst Weeds: Distribution and Biology* (p. 609). Honolulu: University Press of Hawaii.

Hudson, R. H., Richard, K. T., & Haegele, M. A., (1984). *Handbook of Toxicity of Pesticides to Wildlife.* USDI Fish and Wildlife Service Resource Publication Number: 153. Washington, D.C.

Huong, P. T., Lee, B. K., Kim, J., & Lee, C. H., (2016). Nitrophenols removal from aqueous medium using Fe-nano mesoporous zeolite. *Mater Des., 101*, 210–217.

Iglesias, A. O., Gomez, O., Pazos, M., & Sanroman, M. A., (2014). Electro-Fenton oxidation of imidacloprid by Fe alignate gel beads. *Appl. Catal. B environment., 144*, 416–424.

Jampani, C. S. R., & Kumari, D. S., (1988). Toxicity of pesticides to *Scenedesmus incrassatulus. Indian J. Bot., 11*, 44–47.

Jemec, A., Tisler, T., Drobne, D., Sepcic, K., Fournier, D., & Trebse, P., (2007). Comparative toxicity of imidacloprid of its commercial liquid formulation and of diazinon to a non-target arthropod, the microcrustacean daphnia magna. *Chemosphere, 68*, 1408–1418.

Joshi, S. R., (2006). *Biopesticides* (1st edn.), *01*, 1–134.

Karl, H., Bladt, A., Rottler, H., Ludwigs, R., & Mathar, W., (2010). Temporal trends of PCDD, PCDF, and PCB levels in muscle meat of herring from different fishing grounds of the Baltic Sea and actual data of different fish species from the Western Baltic Sea. *Chemosphere, 78*(2), 106–112.

Kelderman, P., Xuedong, Y., & Drossaert, W. M. E., (2005). *Sediment Pollution with Respect to Heavy Metals and Organic Micropollutants in the City Canals of Delft (The Netherlands)—Assessment of a Data Base of 188 Sediment Stations.* Hennef: European Water Association (EWA).

Kent, R. A., & Currie, D., (1995). Predicting algal sensitivity to a pesticide stress. *Environ. Toxicol. Chem., 14*, 983–991.

Khan, Q., (1997). *Pesticides in Environment* (p. 127). New York: Plenum.

Lakshminarayana, J. S. S., O'Neill, H. J., Jonnavithula, S. D., Leger, D. A., & Milburn, P. H., (1992). Impact of atrazine-bearing agricultural tile drainage discharge on planktonic drift of a natural stream. *Environ Pollut., 76*, 201–210.

Lal, R., & Lal, S., (1988). *Pesticides and Nitrogen Cycle* (Vol. 3). CRC, Boca Raton, FL, USA.

Lee, S. S., Fang, S. C., & Freed, V. H., (1976). Effect of DDT on photosynthesis of *Selenastrum capricornutum. Pestic. Biochem. Physiol., 6*, 46–51.

Leyva-Cardoso, D., Ponce-Velez, G., Botello, A., & Diaz-Gonzalez, D., (2003). Persistent organochlorine pesticides in coastal sediments from petacalco bay, Guerrero, Mexico. *Bulletin of Environmental Contamination and Toxicology, 71*, 1244–1251.

Lopez, L., Pozo, C., Gomez, M. A., Calvo, C., & Gonzalez-Lopez, J., (2002). Studies on the effects of the insecticide aldrin on aquatic microbial populations. *International Biodeterioration and Biodegradation, 50*, 83–87.

Macek, K. J., Buxton, R. S., Sauter, S., Ginilka, S., & Dean, J. W., (1976). *Chronic Toxicity of Atrazine to Selected Aquatic Invertebrates and Fishes.* Washington (DC), Environmental Protection Agency, (Ecological Research Series).

Mageed, A. A. A., (2007). Distribution and long-term historical changes of zooplankton assemblages in Manzala Lake (South Mediterranean Sea, Egypt). *Egyptian Journal of Aquatic Research, 33*(1), 183–192.

Malhat, F. M. M., (2010). *Organochlorines and Organophosphorus Pesticides, Petroleum Hydrocarbons, Polychlorinated Biphenyls, and Trace Metal Monitoring of Nile River in Egypt.* PhD Thesis, Chemistry Department, Faculty of Science, Menofiya University.

Mansour, S. A., Mahram, M. R., & Sidky, M. S., (2001). Ecotoxicological studies. Monitoring of pesticide residues in the major components Lake Qarun, Egypt. *Journal of Egyptian Academic Society for Environmental Development*, 83–116.

Maqbool, Z., Hussain, S., Imran, M., Mahmood, F., Shahzad, T., Ahmed, Z., Azeem, F., & Muzammil, S., (2016). Perspectives of using fungi as bioresource for bioremediation of pesticides in the environment: A critical review. *Environ. Sci. Pollut. Res., 23*, 1–22.

Marczewski, A. W., Seczkowska, M., Deryło-Marczewska, A., & Blachnio, M., (2016). Adsorption equilibrium and kinetics of selected phenoxyacid pesticides on activated carbon: Effect of temperature. *Adsorption, 22*(4–6), 777–790.

Mayer, F. L., & Mark, R. E., (1986). *Manual of Acute Toxicity: Interpretation and Data Base for 410 Chemicals and 66 Species of Freshwater Animals* (p. 160). U.S. Fish and Wildlife Service. Washington, DC, Publication.

Megharaj, M., Pearson, H. W., & Venkateswarlu, K., (1991). Toxicity of phenol and three nitrophenols toward growth and metabolic activities of *Nostoc linckia* isolated from soil. *Arch Environ. Contam. Toxicol., 21,* 578–584.

Melendez, A. L., Kepner, R. L., Balczon, J. M., & Pratt, J. R., (1993). Effect of diquant on microbial communities. *Arch. Environ. Contam. Toxicol., 25,* 95–101.

Netrawali, M. S., Gandhi, S. R., & Pednekar, M. D., (1986). Effect of endosulfan, malathion and permethrin on sexual life cycle of *Chlamydomonas reinhardtii. Bull. Environ. Contam. Toxicol., 36,* 412–420.

Nwankwegu, A. S., & Onwosi, C. O., (2017). Bioremediation of gasoline contaminated agricultural soil by bioaugmentation. *Environ. Technol. Innov., 7,* 1–11.

Outorun, M. A., & Aaron, J. J., (2014). Advanced oxidation processes in water and wastewater treatment: Principle and applications: A review. *Crit. Rev. Environ. Sci. Technol., 44,* 2577–2641.

Pérez, G. L., Vera, M. S., & Miranda, L., (2011). In: Andreas, K., (ed.), *Effects of Herbicide Glyphosate and Glyphosate-Based Formulations on Aquatic Ecosystems, Herbicides and Environment.* ISBN: 978-953-307-476-4, InTech.

Pesce, S., Fajon, C., Bardot, C., Bonnemoy, F., Portelli, C., & Bohatier, J., (2006). Effects on the phenylurea herbicide diuron on natural rivierine microbial communities in an experimental study. *Aquat. Toxicol., 78,* 303–314.

Pham, T. H., Lee, B. K., Kim, J., & Lee, C. H., (2016). Enhancement of CO_2 capture by using synthesized nano-zeolite. *J. Taiwan Inst. Chem. Eng., 64,* 220–226.

Pimmata, P., Reungsang, A., & Plangklang, P., (2013). Comparative bioremediation of carbofuran contaminated soil by natural attenuation, bioaugmentation and biostimulation. *Int. Biodeterior. Biodegrad., 85,* 196–204.

Radosevich, S. R., Holt, J. S., & Ghersa, C. M., (2007). *Ecology of Weeds and Invasive Plants: Relationship to Agriculture and Natural Resource Management* (3rd edn.). Wiley-Inter-science, Hoboken, USA.

Raveton, M., Ravanel, P., Kaouadji, M., Bastide, J., & Tissut, M., (1997). The chemical transformation of atrazine in corn seedling. *Pestic. Biochem. Physiol., 58,* 199–208.

Roberts, D., Dissanayake, W., Sheriff, M. H. R., & Eddleston, M., (2004). Refractory status epilepticus following self-poisoning with the organochlorine pesticide endosulfan. *Journal of Clinical Neuroscience, 11,* 760–762.

Sanchez-Bayo, F., (2006). Comparative acute toxicity of organic pollutants and reference values for *Crustaceans.* I. *Branchiopoda, Copepoda and Ostracoda. Environ. Pollut., 139,* 385–420.

Scott, G. I., (1990). *Agricultural Insecticide Runoff Effects on Estuarine Organisms.* Correlating laboratory and field toxicity.

Sikka, H. C., & Pramer, D., (1968). Physiological effects of fluometuron on some unicellular algae. *Weed Sci., 16,* 296–299.

Singh, V. K., Singh, K. P., & Mohan, D., (2005). Status of heavy metals in waterbed sediments of River Gomti: A tributary of the Ganga River, India. *Environmental Monitoring and Assessment, 105,* 43–67.

Solomon, K. R., Baker, D. B., Richards, R. P., Dixon, K. R., Klaine, S. J., La, P. T. W., Kendall, R. J., et al., (1996). Ecological risk assessment of atrazine in North American surface waters. *Environ. Toxicol. Chem., 15*(1), 31–76.

Srivastava, B., Jhelum, V., Basu, D., & Patanjali, P., (2009). Adsorbents for pesticide uptake from contaminated water: A review. *J. Sci. Ind. Res., 68*, 839–850.

Sumpono, Pierotti, P., Belan, A., Forestier, C., Lavendrine, B., & Bohatier, J., (2003). Effect of Diuron on aquatic bacteria in laboratory-scale wastewater treatment ponds with special reference to Aeromonas species studied by colony hybridization. *Chemosphere, 50*, 445–455.

Tan, K. H., (2009). *Environmental Soil Science.* CRC Press, Boca Raton.

Tennekes, H. A., (2010). The Significance of the druckrey-kupfmuller equation for risk assessment-the toxicity of neonicotinoid insecticides to arthropods is reinforced by exposure time. *Toxicology, 276*, 1–4.

Tyler, M. G., & Spoolman, S., (2014). Solid and hazardous waste. In: *Living in the Environment* (18th edn., p. 592). Cengage Learning, Boston.

Ukeles, R., (1962). Growth of pure cultures of marine phytoplank-ton in the presence of toxicants. *Appl. Microbiol., 10*,532–537.

Valičková, M., Derco, J., & Šimovičová, K., (1995). Removal of selected. In: Pesti-Van, D. B. P. J., Van, D. E., Gylstra, R., Crum, S. J. H., & Brock, T. C. M., (eds.), *Effects of Chronic Low Concentrations of the Pesticides Chlorpyrifos and Atrazine in Indoor Freshwater Microcosms, Chemosphere* (Vol. 31, pp. 3181–3200).

Valičková, M., Derco, J., & Šimovičová, K., (2013). Removal of selected pesticides by adsorption. *Acta Chim. Slov., 6*, 25–28.

Vance, B. D., & Drummond, W., (1969). Biological concentration of pesticides by algae. *J. Phycol., 61*, 360–362.

Ware, & George, W., (1994). *The Pesticide Book.* Thompson Publications.

Wilde, F. D., (2005). *National Field Manual for the Collection of Water-Quality Data.* Chapter A1. Preparations for Water.

Winkelmann, D. A., & Klaine, S. J., (1991). Degradation and bound residue formation of four atrazine metabolites, deethylatrazine, deisopropylatrazine and hydroxyatrazine, in a western Tennessee soil. *Environ. Toxicol. Chem., 10*, 347–354.

Xie, H., Zhu, L., Ma, T., Wang, J., Wang, J., Su, J., & Shao, B., (2010). Immobilization of an enzyme from a fusarium fungus WZ-I for chlorpyrifos degradation. *J. Environ. Sci., 22*(12), 1930–1935.

Yamamoto, I., Yabuta, G., Tomizawa, M., Saito, T., Miyamoto, T., & Kagabu, S., (1995). Molecular mechanism for selective toxicity of nicotinoids and neonicotinoids. *J. Pestic. Sci., 20*, 33–40.

Yang, H., Yang, P., Liu, X., & Wang, Y., (2016). Space-confined synthesis of zeolite beta microspheres via steam-assisted crystallization. *Chem. Eng. J., 299*, 112–119.

Zhang, G., Pan, Z., Bai, A., Li, J., & Li, X., (2014). Distribution and bioaccumulation of organochlorine pesticides (OCPs) in food web of Nansi Lake, China. *Environ. Monit. Assess, 186*(4), 2039–2051.

Zheng, W., Guo, M., Chow, T., Bennett, D. N., & Rajagopalan, N., (2010). Sorption properties of green waste biochar for two triazine pesticides. *J. Hazard Mater., 181*(1–3), 121–126.

CHAPTER 7

Benthic Macroinvertebrates as Bioindicators of Water Quality in Freshwater Bodies

ZAHOOR AHMAD MIR, MOHAMMAD YASIR ARAFAT, and
YAHYA BAKHTIYAR

Fish Biology and Limnology Research Laboratory,
Department of Zoology, University of Kashmir, Hazratbal,
Srinagar–190006, Jammu and Kashmir, India,
E-mail: yahya.bakhtiyar@gmail.com (Y. Bakhtiyar)

ABSTRACT

Benthic macroinvertebrates are the potential bioindicators of water quality as they provide a clear picture of past and present health conditions of aquatic bodies. The ubiquitous distribution, life cycle of considerable duration, high accumulating capacity, and sedentary nature of benthic macroinvertebrates make them a great tool in assessing the health condition of aquatic ecosystems. The present paper summarizes the role of benthic macroinvertebrates as bioindicators and to evaluate the water quality in freshwater bodies using various bioassessment approaches which include diversity index, saprobic index, FBI index, NSFWQI score, taxa richness, and EPT ratio. The presence or absence of benthic macroinvertebrates is related to the tolerance towards pollution. The presence of pollution sensitive benthic macroinvertebrates indicates good water quality while the presence of pollution tolerant benthic macroinvertebrates indicates poor water quality. The classification of benthic macroinvertebrates with respect to tolerance values indicates the status of freshwater bodies. The paper also highlights the methods involved in biomonitoring and their need for the conservation of aquatic resources and the management programs of freshwater bodies.

7.1   INTRODUCTION

The freshwater resources such as rivers, streams, lakes, ponds, and wetlands provide multiple ecosystem services which may boost the economy of society. These freshwater resources provide a medium to many aquatic organisms such as fishes, zooplankton, benthic macroinvertebrates, macrophytes, and aquatic birds for their survival. Presently, the outbreak of anthropogenic activities such as habitat alteration, nutrient enrichment, pollutants, and sediment deposition from agricultural practices (Dudgeon et al., 2006; Feld and Hering, 2007; Luana et al., 2009; Jansson et al., 2015; Debordae et al., 2016) which have posed a serious threat on aquatic life and freshwater resources. In this alarming aquatic ecosystem degradation, there is a continuous need of implication of monitoring protocols and appropriate conservative measures (Sharma and Sharma, 2010). Traditionally, water quality was being monitored through physical and chemical measurements, which can screen the environmental conditions present at the time of sampling only and gives a 'snapshot' of water quality parameters. Presently, the use of bioindicators can provide a 'moving picture' of previous and latest environmental conditions. Bioindicators being potential in response to changes in water quality can indicate the level of pollution in water body (Benetti et al., 2012; Carter and Resh, 2006). Biomonitoring involves the use of aquatic organisms to determine the water quality of the habitat they live. It deals with the water quality assessment of rivers, lakes, streams, and wetlands by examining the resident aquatic organisms. It is the most common form of biomonitoring in which an aquatic ecosystem can be studied in the best way (Gerhardt, 1999; Oertel and Salanki, 2003). Biomonitoring is one of the vital components of environmental monitoring. The International Organization for Standardization (ISO) have stated that biomonitoring is a planned process of sampling, measuring, and analyzing different water parameters using respective standards and protocols. It may be either passive or active. Passive biomonitoring is the use of organisms or their responses, occurring spontaneously within the natural ecosystem. At the same time, active biomonitoring involves the introduction of organisms at the monitoring site under controlled conditions (Szczerbinska and Gaczynska, 2015). The aquatic biologists have a crucial role in highlighting the environmental management program. Biomonitoring involves bioassay and community assessment approaches. Bioassay is a method to determine the effect of substance on living organisms. It is often used to monitor the water quality and its impact upon the surroundings. Typical organisms used in bioassays are mostly fishes. Community assessment is the sampling of entire community of different organisms to determine taxa composition. Typical organisms are mostly benthic macroinvertebrates to monitor aquatic ecosystem (Karr, 1981). There are three steps during the biological assessment of water quality (Sharma and Sharma, 2010). The survey is the initial step that deals with the ecological aspects of a given site to be monitored. For example, in case of a river, its geographic location, catchment area, tributaries, sources of pollution as well as biotic community are taken under consideration during survey. The surveillance is the whole practice of repeated measurements of different parameters of a particular site. Monitoring is the final step concerned to pollution status of water body (Mason, 1996; Sharma and Sharma, 2010). The nature and degree of pollution of water body can be determined from the presence or absence of either pollution-sensitive or pollution tolerant organisms based on tolerance values (Carter and Resh, 2013; Chang et al., 2014).

7.2   PURPOSE OF BIOMONITORING

The aims of biomonitoring of freshwater resources are as under (Sharma and Sharma, 2010):

1.   To expose the incoming health risks or threats, in order to avoid the deleterious impacts on aquatic ecosystem.
2.   To detect the illegal disposal of pollutants or failure of previous implicated models.
3.   To detect graphical trends or cycles in parameters over time.
4.   To determine the alternative approaches.
5.   To evaluate negative impacts associated with the introduction of exotic species in aquatic ecosystems.

7.3   BIOINDICATORS

Bioindicators are an important tool to monitor the health status of the aquatic ecosystem in order to detect the changes in the environment either positive or negative and their impacts upon human society (Plafkin et al., 1989; Jain et al., 2010; Khatri and Tyagi, 2015). Bioindicators are presently utilized and promoted by various organizations (World Conservation Union or International Union for Conservation of Nature and Natural Resources), as a means to deal with biomonitoring and evaluate impacts on humans. In freshwater ecosystems, the living organisms which are commonly used as bioindicators include macrophytes, plankton, fishes, amphibians, and benthic macroinvertebrates (Li et al., 2010; Parmer, 2016). As for selection is concerned, a biological indicator should be taxonomically easy and well-known, wide range of distribution, abundant, and easy to capture, preferably big size, sedentary, and long life cycle, low genetic and ecological variability, presence of well-known ecological characteristics and possibility of used in laboratory studies (Johnson et al., 1993; Bellinger and Sigee, 2010; Holt and Miller, 2011). The advantages associated with using bioindicators are as follows (Khatri and Tyagi, 2015; Parmer et al., 2016):

1. Use of bioindicators can detect biological impacts upon the ecosystem.
2. Bioindicators can access synergetic and antagonistic impacts of various pollutants on living creatures.
3. Early-stage diagnosis on harmful effects of toxins and pollutants can be monitored.
4. They are easy to count, due to prevalence.
5. Bioindicators are cost-effective as compared with other specialized measuring tools.

7.4   BENTHIC MACROINVERTEBRATES AS BIOINDICATORS

Benthic macroinvertebrates are bottom-dwelling organisms without backbone in the aquatic ecosystem (APHA, 1985) that can be visible to the naked eye without the use of a microscope and may retain on 0.5 mm mesh size sieve (Mason, 1981). They can be found on, under, and around the bottom substrates of water bodies. Benthic community usually consist of the immature stages of mayflies, caddisflies, stoneflies, dragonflies, damselflies, dobsonflies, craneflies, midges, crustaceans, mites, aquatic worms, snails, clams, leeches, and numerous other organisms (Meritt et al., 2008; Allan, 1995) and shows a strong correlation with water quality parameters and habitat morphology (Lamouroux et al., 2004). These organisms are widely distributed. They can be found in large rivers, small creeks, small ponds, wetlands, and lakes and can live on all bottom types, such as mud, sand, gravel, pebble, boulder, or rocks (Creek, 2013). Benthic macroinvertebrates are present throughout the year with high prevalence in the summer season. During the winter season, many species hide deep in the sediment or remain inactive on rock surfaces. Due to various characteristics of benthic macroinvertebrate, they serve as potential bioindicators of water quality. Unlike fish and other vertebrates, benthic macroinvertebrates are sedentary and cannot escape the impacts of pollutants that diminish water quality. Therefore, the richness of benthic macroinvertebrate community composition can be used to estimate the health of the aquatic ecosystem (Armitage et al., 1983; Rosenberg and Resh, 1993). Benthic macroinvertebrates have a long history of use in biomonitoring programs (Barbour et al., 1999).

7.5   ADVANTAGES OF USING BENTHIC MACROINVERTEBRATES AS BIOINDICATORS

The advantages of benthic macroinvertebrates as potential bioindicators are as under (Rosenberg and Resh, 1993; Bae et al., 2005):

1. Benthic macroinvertebrates are widely distributed and occur in every aquatic habitat (Lenat et al., 1980).
2. Most of the physicochemical parameters have direct impact upon the benthic macroinvertebrates.
3. Benthic macroinvertebrates are mostly sedentary in nature and have a history of pollution impacts which makes them indicator for assessing water quality (Slack et al., 1973).
4. They are most diverse in aquatic habitats and are less or more tolerant to pollution and their occurrence indicate the water quality (Carter and Resh, 2013).
5. They can retain the toxic substances within their body and can be detected when toxin levels are undetectable in the water resource.
6. Small order streams mostly support extensive macroinvertebrate communities.
7. Macroinvertebrate sampling is easy, requires less people and simple equipment (e.g., nets, buckets, trays), and does not adversely affect other organisms.
8. Macroinvertebrate monitoring is a low-cost technique that can be completed in the field and is easy for lab analysis.

7.6   POLLUTION TOLERANCE OF BENTHIC MACROINVERTEBRATES

Benthic macroinvertebrates respond to pollution in different ways that can be either sensitive or tolerant to pollution level (Carter and Resh, 2013).

Pollution-sensitive benthic macroinvertebrates are more susceptible to the effects of physicochemical changes in the water quality and their presence indicates the absence of pollution. Pollution-tolerant benthic macroinverte-brates are less susceptible to the effects of physicochemical changes in the water quality and their presence indicates the level of pollution (Rosenberg and Resh, 1993). As the water body gets polluted, pollution-sensitive organ-isms decrease in number or disappear while pollution-tolerant organisms increase in their number and variety. As organic pollution increases, dissolved oxygen levels decreases and pollution sensitive benthic macroinvertebrates also decreases. As organic pollution continues to increases, pollution-tolerant benthic macroinvertebrates become more dominant, while other sensitive and semi-sensitive benthic macroinvertebrates will not be able to survive under increased pollution (Dudgeon et al., 2006).

Benthic macroinvertebrates are classified into four categories in response to level of pollution which are most sensitive, semi-sensitive, semi-tolerant, and very tolerant (Sallenave, 2015; Ojija and Laizer, 2016), as shown in Figure 7.1 and Table 7.1.

- **Group I: (Most Sensitive):** These are very sensitive to organic pollutants and require high dissolved oxygen and can survive in good water quality. Some examples are Stonefly nymph, Mayfly nymph, and Caddisfly larvae.
- **Group II: (Semi-Sensitive):** These are semi-sensitive to pollutants and require moderate to high dissolved oxygen levels and can survive in fair water quality. Some instances are Scud, Cranefly larvae and Clam.
- **Group III: (Semi-Tolerant):** These are semi-tolerant of pollutants and can survive with moderate oxygen levels and fair or poor water quality. Some examples are Leech, Midge larvae and Blackfly larvae.
- **Group IV: (Very Tolerant):** These are tolerant of pollutants and can survive with low dissolved oxygen and poor water quality. They mostly include Aquatic worm, Blood midge larvae and Pouch snail.

7.7 BIOASSESSMENT APPROACHES USING BENTHIC MACROINVERTEBRATES

Analysis of benthic macroinvertebrate communities depends upon the presence or absence, sensitivity, richness, abundance, and diversity of particular taxa. The information collected can be converted into numerical values, including indices and scores (Mathews et al., 1982; Metcalfe, 1989; Karr and Chu, 1999). Various approaches have been developed for bioassessment of water quality using benthic macroinvertebrates, which are as under:

FIGURE 7.1 Some common stream benthic macroinvertebrates as bioindicators: (A) Stonefly nymph, (B) Mayfly nymph and (C) Caddisfly larva (most sensitive); (D) Scud, (E) Cranefly larva and (F) Clam (semi-sensitive); (G) Leech, (H) Midge larva, (I) Blackfly larva (semi-tolerant); (J) Aquatic worm, (K) Blood midge larva, (L) Pouch snail (Very tolerant).

7.8   DIVERSITY APPROACH

The diversity indices relate the number of observed species (richness) to the number of individuals (abundance). The aim of diversity is to evaluate the

TABLE 7.1   Pollution Tolerance of Benthic Macroinvertebrates

Pollution Tolerance Index (PTI) (Mitchell and Stapp, 1996)			
I. Most Sensitive	**II. Semi-Sensitive**	**III. Semi-Tolerant**	**IV. Very Tolerant**
Stonefly nymph	Damselfly nymph	Leech	Aquatic worms
Mayfly nymph	Dragonfly nymph	Midge Larvae	Blood midge larvae
Caddisfly larvae	Scud	Planaria/Flatworm	Rat-tailed Maggot
Riffle Beetle	Sowbug	Blackfly larva	Pouch snail
Dobsonfly larvae	Cranefly larva	Water mite	
Gilled snail	Clam/Mussel		
Water penny	Crayfish		

benthic community structure with respect to occurrence of species. Diversity indices are easy to use and calculate, applicable to communities, and are best used for comparative study (Washington, 1984). These indices include:

1. **Simpson's Index 'D' (Simpson, 1949):** This index measures the probability of randomly selected two individuals from a sample belonging to the same species. So it is used to find out the dominant species.

$$D = \frac{\sum n(n-1)}{N(N-1)}$$

where, n = total no. of organisms of a particular species; N = total no. of organisms of all species. Value of D ranges from 0–1 and is inversely proportional to the diversity.

2. **Shannon-Wiener Index 'H' (Shannon and Wiener, 1949):** It is commonly used to characterize species diversity in a community.

$$H = -\sum_{i=1}^{s} Pi \ln Pi$$

where, S = total no. of species in the sample; Pi = n/N; ln = natural logarithm. n= no. of individuals of one particular species; N= total no. of individuals found. The higher value of 'H' indicates the increase in the number and distribution of benthic taxa.

3. **Margalef Diversity Index 'MD' (Margalef, 1958):** This index is used to measure the richness of species.

$$MD = \frac{(S-1)}{\ln N}$$

where, S = total no. of species; N = total no. of individuals in the sample; ln = natural logarithm.

4. **Pielou Evenness Index 'J' (Pielou, 1966):** It is used to calculate the evenness of species.

$$J = \frac{H'}{\ln S}$$

where, H' = Observed value of Shannon index; S = total no. of species in the sample.

J value ranges from 0–1 and is inversely proportional to the variation between the species.

7.9 SAPROBIC APPROACH

It represents the water quality classification based on the presence of pollution tolerant indicator species. Every species has tolerance to organic substances and requirement for dissolved oxygen. This tolerance is expressed in a saprobic indicator value. The approach provides the classification of the investigated benthic macroinvertebrate community universal scale (Table 7.2) (Kolkwotz and Marsson, 1902).

1. **Saprobic Index (Pantle and Buck, 1955):**

$$S = \frac{\sum s \times h}{\sum h}$$

where, S = Saprobic index, s = indicator value of each species, h = frequency of each species.

TABLE 7.2 Biological Classification of Water Quality Using Saprobic Index

Saprobic Index	Degree	Water Quality
0.5–1.5	Oligosaprobic	Clear water with high dissolved oxygen
1.5–2.5	β-mesosaprobic	Moderately polluted with still high dissolved oxygen
2.5–3.5	α-mesosaprobic	Polluted water with low dissolved oxygen
3.5–4.5	Polysaprobic	Strongly polluted water with no dissolved oxygen

7.10 BIOTIC APPROACH

Various Biotic indices have been used to get numerical scores of indicators at specific taxonomic level derived from summation of individual scores.

The presence or absence of macroinvertebrates, abundance, morphology, physiology, or behavior may indicate the physicochemical conditions of aquatic ecosystems (Rosenberg and Resh, 1993). The number of taxonomic groups reduces with the increase in organic pollution (Mackenthum, 1969; Woodiwiss, 1980). The presence of highly tolerant species indicates poor water quality (Hynes, 1972). Biotic indices used in the protocols for benthic macroinvertebrates requires 'tolerance scores' derived from databases of all major groups of benthic macroinvertebrate taxa (Mackie, 2001). The following indices are applied to calculate numerical values in order to classify water quality of aquatic bodies.

1. **Modified Family Biotic Index (Hilsenhoff, 1988):** It is based upon the species level index and is used to detect organic pollution in an aquatic body by using formula:

$$FBI = \frac{\sum x_i t_i}{N}$$

where, x_i = number of individuals within a taxon; t_i = tolerance value of a taxon, N = total number of organisms in the sample. This index provides a single tolerance value from the average values of all species within the benthic community (Table 7.3). The tolerance values ranges from 0 (very intolerant) to 10 (highly tolerant) based on their tolerance to organic pollution.

TABLE 7.3 Pollution Tolerance Values 'PTV' for Benthic Macroinvertebrates at the Family Level

Taxonomic Name	Common Names	PTV
Elmidae	Riffle Beetle	4
Psephenidae	Waterpenny Beetle	4
Gammaridae	Scud	4
Helobdella	Leech	9
Sphaeriidae	Fingernail Clam	7
Corixidae	Water Boatman	5
Chironomidae	Midge	6
Ceratopogonidae	Biting Midge	6
Ephemeroptera	Mayfly	4
Oligochaeta	Aquatic Worm	8
Nematoda	Thread Worm	8
Adelaide	Sowbug	8

TABLE 7.3 *(Continued)*

Taxonomic Name	Common Names	PTV
Prosobranch	Gilled Snail	6
Pulmonate	Lunged Snail	8
Decapoda	Crayfish	6
Chloroperlidae	Stone Fly	1
Sialidae	Alder Fly	4
Hydropsychidae	Caddisfly	4
Tipulidae	Crane Fly	3
Simuliidae	Black fly	6
Planaria	Flatworm	4
Anisoptera	Dragonfly	3
Arachnida	Water Mites	6
Tabanidae	Horsefly/Deerfly	6
Corydalidae	Dobsonfly	4
Zygoptera	Damselfly	5
Stratiomyidae	Soldierfly	7
Hydrometridae	Water Measurer	5
Gerridae	Water Strider	5
Lepidoptera	Pyralid	5
Hydrophilidae	Water Scavenger	4
Ephydridae	Brine fly	6
Helophorus	Beetle	5
Haliplidae	Crawling Beetle	5
Dytiscidae	Predaceous Diving Beetle	5
Haliplidae	Predaceous Water Beetle	5
Dytiscidae	Diving Beetle	5
Empididae	Dancefly	6
Hydrophilidae	Water Scavenger Beetle	5
Athericidae	Snipefly	4

Source: Mandaville (2002); Bode et al. (1991, 1996, 2002); Hauer and Lamberti (1996); Hilsenhoff (1988); Plafkin et al. (1989); Creek (2013).

2. **National Sanitation Foundation Water Quality Index 'NSFWQI' (Alexakis et al., 2016):** It is used to assess the water quality classes for public health, safety, and protection of the environment. Its value

ranges from 0 to 100. Water quality has been divided into five classes (Table 7.4).

TABLE 7.4   NSFWQI Rating for Different Water Quality Classes

NSFWQI Score	Water Quality	WQ Class
91–100	Excellent water quality	I
71–90	Good water quality	II
51–70	Medium or average water quality	III
26–50	Poor water quality	IV
0–25	Bad water quality	V

Source: Brown et al. (1970); Alexakis et al. (2016).

7.11   MULTIVARIATE APPROACH

Several multivariate approaches have been used to assess water quality through benthic macroinvertebrates (Norris and Georges, 1993). These approaches include:

1. **Jaccard Coefficient of Community Similarity (Jaccard, 1908):** It measures the degree of similarity in taxonomic composition between two sites by either presence or absence of taxa. Its values range from 0 to 10.

   Formula:

   $$\text{Jaccard Coefficient} = \frac{a}{a+b+c}$$

   where, a = number of taxa common to both samples; b = number of taxa present in sample B but not in sample A; c = number of taxa present in sample A but not in sample B.

2. **Sorenson's Index 'S' (Sorenson, 1948):** This index is used for comparing the similarity between two samples taken at different stations. Its value ranges from 0–1, closer the value to 1, the more the communities have in common.

   $$S = \frac{2C}{S1+S2}$$

   where, S1 = no. of species present in first community; S2 = no. of species present in second community.
   C = no. of species common to both of the communities.

3. **Community Loss Index (Plafkin et al., 1989):** It measures the loss of benthic taxa between two sampling sites. The formula for the Community Loss Index is:

$$\text{Community Loss} = \frac{d-a}{e}$$

where, a = number of taxa common to both samples; d = total number of taxa present in sample A; e = total number of taxa present in sample B. Its value ranges from 0 to 'infinity.'

7.12   ECOLOGICAL QUALITY RATIO APPROACHES

The value obtained with any index system is compared with a reference status, by calculating the proportion between both values. Some of these ratios are as under:

1. **Taxa Richness 'TR' (Plafkin et al., 1989):** Taxa richness indicates the health of the community using diversity and increases with increase in habitat diversity, suitability, and water quality. The total number of taxa within the sample represents the taxa richness and the greater value indicates the health condition within the community.
2. **EPT Index (Plafkin et al., 1989):** The EPT Index represents three orders of aquatic insects that are common in the stream benthic macroinvertebrate community: Ephemeroptera (mayflies), Plecoptera (stoneflies), and Trichoptera (caddisflies). The EPT Index is based on high water quality streams with greatest species richness. The Ephemeroptera, Plecoptera, and Trichoptera index represents the taxa richness within the insect groups which are sensitive to pollution and increases with water quality. It is equal to the number of families present in these three orders in the sample (Table 7.5).
3. **Ratio of EPT and Chironomidae 'EPT/C' (Plafkin et al., 1989):** The abundance of EPT and Chironomidae indicates stability of the benthic community, because EPT are more sensitive and Chironomidae are less sensitive to environmental stress. A community with even distribution of these four groups displays good biotic condition, while high numbers of Chironomidae indicates environmental stress. The index is calculated by dividing the total number of individuals classified as Ephemeroptera, Plecoptera, and Trichoptera by the total number of individuals classified as Chironomidae (Table 7.6).

TABLE 7.5 Benthic Macroinvertebrates as Bioindicators of Indian Freshwater Bodies

Water Bodies	Dominant Taxa	Water Quality	References
Behta River, Uttaranchal	Ephemeroptera, Trichoptera, Coleoptera, Plecoptera, Diptera, Mollusca	S-1: Class II (Good) S-2: Class III (moderately polluted)	Sharma et al. (2006)
Ninglad stream, Uttarakhand	Ephemeroptera, Coleoptera, Trichoptera, Odonata, Lepidoptera, Diptera	S-1: Moderate pollution S-2: Good water quality	Sharma et al. (2008)
Tawi River, Jammu and Kashmir	*Chironomus, Tubifx*	High nutrient content, Low Dissolved Oxygen, poor water quality	Sharma and Choudhary (2011)
Koratty Chal stream, Kerala	Ephemeroptera, Trichoptera, Plecoptera, Coleoptera, Odonata, Hemiptera, Diptera	Upstream: Fair Mid-stream: Marginal Downstream: Poor	Kripa et al. (2013)
Bakuamari stream, Assam	*Gerridae, Veliidae*	Moderate water quality	Barman and Gupta (2015)
River Ujh, J, and K	*Baetidae, Hydropsychidae, Philopotamidae*	Slightly polluted to moderatively polluted	Singh et al. (2017)

TABLE 7.6 Benthic Macroinvertebrates as Bioindicators of Freshwater Bodies Outside India

Water Bodies	Dominant Taxa	Water Quality	References
Lake Nemi (Central Italy)	Oligochaeta, Diptera: Chironomidae, Crustacea Nematoda	Oligo-mesotrophic	Mastrantuono (1985)
Lake Etela-Saimaa, Southern Finland	*Chaoborus flavicans* and *Chironomus plumosus*	Moderately oligotrophic	Kansanen et al. (1990)
Langat River, Malaysia	Upstream: Ephemeroptera, chironomid dipterans, Downstream: Oligochaeta: *Limnodrilus sp. Branchiodrilus sp.*, Hirudinea	Upstream-good water quality Downstream-poor water quality	Azrina et al. (2006)
Moiben River, Kenya	Headwaters: Ephemeroptera, Plecoptera, Trichoptera Downstream: Coleoptera, Oligochaeta, and Chironomidae	Upstream: Good water quality Downstream: Poor water quality	Masese et al. (2009)
Rampur Ghol wetland, Nepal	Dry season: Thiaridae Rainy season: Dytiscidae	Dry season: moderate to heavily polluted Rainy season: moderate to critically polluted	Gautam et al. (2014)
Mkondoa River, Tanzania	Ephemeroptera, Plecoptera, and Trichoptera	Upstream: Good water quality	Shimba and Jonah (2016)
Afenourir lake, Morocco	Chironomidae and Annelids	Poor water quality	Griba et al. (2017)
Cirhanyobowa River, DR Congo	Lepidomastidae, Petaluridae, Coenagrionidae, Hydropsychidae	Group 1: good Group 2: moderate Group 3: poor	Bagalwa et al. (2019)

7.13 CONCLUSION

Water quality has impacts upon the distribution and abundance of aquatic living organisms with focus upon the benthic macroinvertebrates which have a long history of use as bioindicators in both lentic and lotic water bodies. The methods employed in the benthic macroinvertebrate protocols and manuals for monitoring water quality are based upon the family or genus level identifications which provide tolerance values or scores. The tolerance values of benthic macroinvertebrates towards the pollution classifies them into four categories which are most sensitive, semi sensitive, semi tolerant and very tolerant. The scores given through various biotic approaches classify the freshwater bodies into good, fair, and poor water quality. Pollution sensitive benthic macroinvertebrates require high dissolved oxygen while the pollution tolerant requires low dissolved oxygen. An increase in the level of organic pollutants lowers the dissolved oxygen and subsequently the pollution sensitive benthic macroinvertebrates will not be able to survive. The occurrence of sensitive species indicates the health status of water bodies and the richness of sensitive species indicates the stability of freshwater ecosystems. The high prevalence of pollution tolerant benthic macroinvertebrates in water resources reveals the ecological risk assessment.

KEYWORDS

- **benthic macroinvertebrates**
- **bioindicators**
- **biomonitoring**
- **freshwater bodies**
- **pollution tolerance**
- **water quality**

REFERENCES

Alexakis, D., Tsihrintzis, V. A., Tsakiris, G., & Gikas, G. D., (2016). Suitability of water quality indices for application in lakes in the Mediterranean. *Water Resour. Manage*, *30*, 1621–1633.

Allan, J. D., (1995). *Stream Ecology: Structure and Function of Running Waters* (p. 388). Chapman & Hall. Boston, Massachusetts, USA.

APHA, (1985). *Standard Methods of the Examination of Waste and Wastewater* (16th edn.) American Public Health Association, Washington, D. C.

Armitage, P. D., Moss, D., Wright, J. F., & Furse, M. T., (1983). The performance of a new biological water quality score system based on macroinvertebrates over a wide range of unpolluted running-water sites. *Water Res., 17*, 333–347.

Azrina, M. Z., Yap, C. K., Ismail, A. R., Ismail, A., & Tan, S. G., (2006). Anthropogenic impacts on the distribution and biodiversity of benthic macroinvertebrates and water quality of the Langat River, Peninsular Malaysia. *Ecotoxicology and Environmental Safety, 64*(3), 337–347.

Bae, Y. J., Kil, H. K., & Bae, K. S., (2005). *Benthic Macroinvertebrates for Uses in Stream Biomonitoring and Restoration, 9*(1), 55–63.

Bagalwa, M., Mukumba, I., Ndahama, N., Zirirane, N., & Kalala, A. O., (2019). Assessment of river water quality using macroinvertebrate organisms as pollution indicators of Cirhanyobowa River, Lake Kivu, DR Congo. *Int. J. Curr. Microbiol. App. Sci., 8*(4), 2668–2680.

Barbour, M. T., Gerritsen, J., Snyder, B. D., & Stribling, J. B., (1999). *Rapid Bioassessment Protocols for use in Wadeable Streams and Rivers: Periphyton, Benthic Invertebrates, and Fish* (2nd edn.). Report EPA 841-B-99–002. US Environmental Protection Agency, Office of Water, Washington DC.

Barman, B., & Gupta, S., (2015). Aquatic insects as bio-indicator of water quality: A study on bakuamari stream, chakras hila wildlife sanctuary, Assam, North East India. *Journal of Entomology and Zoology Studies, 3*(3), 178–186.

Bellinger, E. G., & Sigee, D. C., (2010). *Freshwater Algae: Identification and Use as Bioindicators*. UK: Wiley-Blackwell.

Benetti, C. J., Bilbao, A. P., & Garrido, J., (2012). Macroinvertebrates as indicators of water quality in running waters: 10 Years of research in rivers with different degrees of anthropogenic impacts. In: Voudouris, (ed.), *Ecological Water Quality-Water Treatment and Reuse.* ISBN: 978-953-51-0508-4, In Tech, Available from: http://www.intechopen.com/books/ecologicalwater-quality-water-treatment-andreuse/macroinvertebrates-as-indicators-of-water-quality-inrunning-waters-10-years-of-research-in-rivers-w (accessed on 26 October 2020).

Bode, R. W., Novak, M. A., & Abele, L. E., (1991). *Methods for Rapid Biological Assessment of Streams* (p. 57). NYS Department of Environmental Conservation, Albany, NY.

Bode, R. W., Novak, M. A., & Abele, L. E., (1996). *Quality Assurance Work Plan for Biological Stream Monitoring in New York State* (p. 89). NYS Department of Environmental Conservation, Albany, NY.

Bode, R. W., Novak, M. A., & Abele, L. E., (2002). *Quality Assurance Work Plan for Biological Stream Monitoring in New York State.* NYS Department of Environmental Conservation, Albany, NY.

Brown, R. M., McClelland, N. I., Deininger, R. A., & Tozer, R. G., (1970). A water quality index: Do we dare? *Water and Sewage Works,* 339–343.

Carter, J. L., & Resh, V. H., (2013). *Analytical Approaches Used in Stream Benthic Macroinvertebrate Biomonitoring Programs of State Agencies in the USA* (p. 1129). Menlo Park: United States Geological Survey.

Carter, J. L., Resh, V. H., Hannaford, M. J., & Myers, M. J., (2006). *Macroinvertebrates as Biotic Indicators of Environmental Quality* (pp. 805–830).

Chang, F. H., Lawrence, J. E., Touma, B. R., & Resh, V. H., (2014). Tolerance values of benthic macroinvertebrates for stream biomonitoring: Assessment of assumptions underlying scoring systems worldwide. *Environ. Monit. Assess, 186*, 2135–2149.

Creek, D., (2013). *Benthic Macroinvertebrate Study.* Environmental and Engineering Services Department Wastewater Treatment Operations, London.

Deborde, D. D. D., Hernandez, M. B. M., & Magbanua, F. S., (2016). Benthic macroinvertebrate community as an indicator of stream health: The effects of land use on stream benthic macroinvertebrates. *Science Diliman., 28,* 25–26.

Dudgeon, D., Arthington, A. H., Gessner, M. O., Kawabata, Z. I., Knowler, D. J., Leveque, C., Naiman, R. J., et al., (2006). Freshwater biodiversity: Importance, threats, status and conservation challenges. *Biological Reviews, 81,* 163–182.

Feld, C. K., & Hering, D., (2007). Community structure or function: Effects of environmental stress on benthic macroinvertebrates at different spatial scales. *Freshwater Biology, 52,* 1380–1399.

Gautam, B., Byanju, R. M., Sapkota, R. P., & Dangol, D. R., (2014). Aquatic macro-invertebrates as bio-indicators: An approach for wetland water quality assessment of Rampur Ghol, Chitwan, Nepal. *Journal of Institute of Science and Technology, 19*(2), 58–64.

Gerhardt, A., (1999). *Biomonitoring of Polluted Water-Reviews on Actual Topics* (Vol. 9, pp. 1–13). Environmental Research Forum, Trans Tech Publications-Sci. Tech. Publications.

Griba, J., Laadel, N., Rhafouri, H. E., Serghini, A., & Fekhaoui, M., (2017). Inventory of benthic macroinvertebrates as bio-indicators of afenourirlake (Morocco). *Journal of Materials and Environmental Sciences, 8*(11), 3986–3992.

Hauer, F. R., & Lamberti, G. A., (1996). *Methods in Stream Ecology* (p. 696). Academic Press. ISBN: 0-12-332906-X.

Hilsenhoff, W. L., (1988). Rapid field assessment of organic pollution with a family-level biotic index. *J. North Am. Benthol. Soc., 7,* 65–68.

Holt, E. A., & Miller, S. W., (2011). Bioindicators: Using organisms to measure environmental impacts. *Nature Education Knowledge, 2*(2), 8.

Hynes, H. B. B., (1972). *The Ecology of Running Water* (p. 555). University of Toronto Press.

Jaccard, P., (1908). Nouvelles researches Sur la distribution florale. *Bull. Soc. Vaud. Sci. Nat., XLIV,* 223–269.

Jain, A., Singh, B. N., Singh, S. P., Singh, H. B., & Singh, S., (2010). Exploring biodiversity as bioindicators for water pollution. *National Conference on Biodiversity, Development, and Poverty Alleviation.* Uttar Pradesh State Biodiversity Board, India.

Jansson, A. B. G., Arhonditis, G. B., Beuson, A., & Bolding, K., (2015). Exploiting and exploring diversity of aquatic ecosystem models: A community perspective. *Aquatic Ecology, 49,* 513–548.

Johnson, M. K., Hashtroudi, S., & Lindsay, D. S., (1993). Source monitoring. *Psychological Bulletin, 114*(1), 3.

Kansanen, P. H., Lauri, P., & Tarja, V., (1990). Ordination analysis and bio-indices based on zoobenthos communities used to assess pollution of a lake in southern Finland. *Hydrobiologia, 202*(3), 153–170.

Karr, J. R., & Chu, E. W., (1999). *Restoring Life in Running Waters, Better Biological Monitoring.* Island Press: Washington, DC, USA.

Karr, J. R., (1981). Assessment of biotic integrity using fish communities. *Fisheries, 6*(6), 21–27.

Khatri, N., & Tyagi, S., (2015). Influences of natural and anthropogenic factors on surface and groundwater quality in rural and urban areas. *Frontiers in Life Science, 8*(1), 23–39.

Kolkwitz, R., & Marsson, M., (1902). Grundsatze fur die biologische Beurteiung des wassers nach zeiner Flora und Dauna. *Prufungsanst. Wasserversorg. Abwasserrein, 1,* 33–72.

Kripa, P. K., Prasanth, K. M., Sreejesh, K. K., & Thomas, T. P., (2013). Aquatic macroin-vertebrates as bioindicators of stream water quality-a case study in Koratty, Kerala, India. *Research Journal of Recent Sciences, 2,* 217–222.

Lamouroux, N., Doledec, S., & Gayraud, S., (2004). Biological traits of stream macroinvertebrate communities: Effects of microhabitat, reach, and basin filter. *Journal of the North American Benthological Society, 2,* 449–466.

Lenat, D. R., Smock, L. A., & Penrose, D. L., (1980). Use of benthic macroinvertebrates as indicators of environmental quality. In: *Biological Monitoring for Environmental Effects,* 97–112.

Li, L., Zheng, B., & Liu, L., (2010). Biomonitoring and bioindicators used for river ecosystems: definitions, approaches, and trends. *Procedia Environmental Sciences, 2,* 1510–1524.

Luana, G., De Filippis, A., Mezzotero, A., Voelz, N. J., & Lucadamo, L., (2009). Assessment of the effect of hydrological variations on macrobenthic communities in pools and riffles of a Mediterranean stream. *Environ Monit. Assess, 101,* 187–190.

Mackenthun, K. M., (1969). *The Practice of Water Pollution Biology.* FWPCA: Washington, DC, USA.

Mackie, G. L., (2001). Applied Aquatic Ecosystem Concepts (p. 744). Kendall/Hunt Publishing Company.

Mandaville, S. M., (2002). *Benthic Macroinvertebrates in Freshwater-Taxa Tolerance Values, Metrics, and Protocols.* Project H-1. Nova Scotia: Soil and Water Conservation Society of Metro Halifax.

Margalef, R., (1958). Temporal succession and spatial heterogeneity in phytoplankton. In: Buzzati-Traverso, (ed.), *Perspectives in Marine Biology* (pp. 323–347). Univ. Calif. Press, Berkeley.

Masese, F. O., Muchiri, M., & Raburu, P. O., (2009). Macroinvertebrate assemblages as biological indicators of water quality in the Moiben River, Kenya. *African Journal of Aquatic Science, 34*(1), 15–26.

Mason, C. F., (1981). *Biology of Freshwater Pollution* (p. 23). Longman International, New York.

Mason, C. F., (1996). Water pollution biology. *Pollution: Causes, Effects and Control,* 82–112.

Mastrantuono, L., (1985). Littoral sand zoobenthos and its relation to organic pollution in Lake Nemi (Central Italy). *Hydrobiological Bulletin., 19*(2), 171–178.

Matthews, R. A., Buikema, A. L., Cairns, J., & Rogers, J. H., (1982). Biological monitoring Part IIA-Receiving system functional methods, relationship, and indices. *Water Res., 6,* 129–139.

Merritt, R. W., Cummins, K. W., & Berg, M. B., (2008). *An Introduction to the Aquatic Insects of North America* (4th edn., p. 1158). Kendall Hunt Publishing. Dubuque, Iowa, U.S.A.

Metcalfe, J. L., (1989). Biological water quality assessment of running water based on macroinvertebrate communities: History and present status in Europe. *Environt. Pollut., 60,* 101–139.

Mitchell, M. K., & Stapp, W. B., (1996). *Field Manual for Water Quality Monitoring* (12th edn.).

Norris, R. H., & Georges, A., (1993). Analysis and interpretation of benthic macroinvertebrate surveys. In: Rosenberg, D. M., & Resh, V. H., (eds.), *Freshwater Biomonitoring and Benthic Macroinvertebrates* (pp. 243–286). Chapman and Hall: New York, NY, USA.

Oertel, N., & Salanki, J., (2003). Biomonitoring and bioindicators in aquatic ecosystems. In: Ambasht, R. S., & Ambasht, N. K., (eds.), *Modern Trends in Applied Aquatic Ecology* (pp. 219–246). Kluwer Academic/Plenum Publishers, New York.

Ojija, F., & Laizer, H., (2016). Macroinvertebrates as bioindicators of water quality in Nzovwe stream in Myeba, Tanzania. *International Journal of Scientific and Technology Research, 5*(6), 211–222.

Pantle, R., & Buck, H., (1955). Die biologische uberwachung der gewasser und die darstellung der ergebnisse. *Gas-u Wasserfach, 96,* 604.

Parmar, T. K., Rawtani, D., & Agrawal, Y. K., (2016). Bioindicators: The natural indicator of environmental pollution. *Frontiers in Life Science, 9*(2), 110–118.

Pielou, E. C., (1966). The measurement of diversity in different types of biological collections. *Journal of Theoretical Biology*, *13*, 131–144.

Plafkin, J. L., Barbour, M. T., Porter, K. D., Gross, S. K., & Hughes, R. M., (1989). Rapid bioassessment protocols for use in streams and rivers: Benthic macroinvertebrates and fish. Washington (DC), EPA. In: Rosenberg, D. M., & Resh, V. H., (eds.), *Freshwater Biomonitoring and Benthic Macroinvertebrates*. New York (NY), Chapman & Hall.

Rosenberg, D. M., & Resh, V. H., (1993). *Freshwater Biomonitoring and Benthic Macroinvertebrates*. Chapman and Hall, New York, NY, USA.

Sallenave, R., (2015). Stream biomonitoring using benthic macroinvertebrates. *Circular*, *677*, 1–12.

Shannon, C. E., & Wiener, W., (1949). *The Mathematical Theory of Communication* (p. 177). Urbana, University of Illinois Press.

Sharma, K. K., & Chowdhary, S., (2011). Macroinvertebrate assemblages as biological indicators of pollution in a Central Himalayan River, Tawi (JK). *International Journal of Biodiversity and Conservation*, *3*(5), 167–174.

Sharma, M. P., Sharma, S., Goel, V., Sharma, P., & Kumar, A., (2006). Water quality assessment of Behta river using benthic macroinvertebrates. *Life Science Journal*, *3*(4), 68–74.

Sharma, M. P., Sharma, S., Goel, V., Sharma, P., & Kumar, A., (2008). Water quality assessment of Ninglad stream using benthic macroinvertebrates. *Life Science Journal*, *5*(3), 67–72.

Sharma, S., & Sharma, P., (2010). Biomonitoring of aquatic ecosystem with concept and procedures particular reference to aquatic macroinvertebrates. *Journal of American Science*, *6*(12), 1246–1255.

Shimba, M. J., & Jonah, F. E., (2016). Macroinvertebrates as bioindicators of water quality in the Mkondoa River, Tanzania, in an agricultural area. *African Journal of Aquatic Science*, *41*(4), 453–461.

Simpson, E. H., (1949). Measurement of diversity. *Nature*, *163*, 688.

Singh, V., Sharma, M. P., Sharma, S., & Mishra, S., (2017). Bio-assessment of River Ujh using benthic macro-invertebrates as bioindicators, India. *Intl J. River Basin Management*, https://doi.org/10.1080/15715124.2017.1394318.

Slack, K. V., Averett, R. C., Greeson, P. E., & Lipscomb, R. G., (1973). Methods for collection and analysis of aquatic biological and microbiological samples. In: *Techniques of Water-Resources Investigations of the United States Geological Survey*, *4*, 1–65.

Sorensen, T., (1948). A method of establishing groups of equal amplitude in plant sociology based on similarity of species content and its application to analysis of the vegetation on Danish commons. *Biologiske Skifter. Det Konglige Danske Videnskabernes Selskab*, *5*, 1–34.

Szczerbinska, N., & Gaczynska, M., (2015). Biological methods used to assess surface water quality. *Arch Pol. Fish*, *23*, 185–196.

Washington, H. G., (1984). Diversity, biotic and similarity indices: A review with special relevance to aquatic ecosystems. *Water Research*, *18*(6), 653–694.

Woodiwiss, F. S., (1980). *Biological Monitoring of Surface Water Quality*. Summary Report, ENV/787/80-EN. Commission of the European Communities: Brussels Belgium.

CHAPTER 8

Biomonitoring and Bioindicators

AARIF ALI,[1] SHAHZADA MUDASIR RASHID,[2] SHOWKAT AHMAD GANIE,[1]
MANZOOR UR RAHMAN MIR,[2] SHEIKH BILAL,[2] HENNA HAMADANI,[3]
FOZIA SHAH,[4] MASRAT RASHID,[5] MUNEEB U. RAHMAN,[6,7] and
SABHIYA MAJID[6]

[1]Department of Clinical Biochemistry, University of Kashmir,
Jammu and Kashmir, India

[2]Division of Veterinary Biochemistry, FVSc & AH, SKUAST-K, Shuhama,
Jammu and Kashmir, India

[3]Mountain Livestock Research Institute, Manasbal SKUAST-K,
Jammu and Kashmir, India

[4]Division of Veterinary Physiology, FVSc & AH, SKUAST-K, Shuhama,
Jammu and Kashmir, India

[5]Food and Drug Control Organization, Government of Jammu and Kashmir,
India

[6]Department of Biochemistry, Government Medical College, Srinagar,
Jammu and Kashmir, India

[7]Department of Clinical Pharmacy, College of Pharmacy,
King Saud University, Kingdom of Saudi Arabia

ABSTRACT

The health and well-being of species, populations, ecosystems can be assessed by a broad vary of scientists, ecologists, environmentalists, governmental agencies, NGO's and, most importantly by the common masses. Due to the immense growth in the area of research, a large number of abiotic will be used to screen, assess, and monitor the health and biogeographic changes occurring in the environment. To determine the overall health of ecosystems and their components, number of tools and biomonitoring plans are used. As an ecosystem consists of thousands of living species, it is very much difficult to assess the status and monitor changes that occur within the ecosystem. An organism that provides information both qualitative as well as quantitative are called biological monitors or bio-monitors and the process of measuring toxic substances in the body is known as biomonitoring. It is very important to develop suitable markers which can be used to assess status and monitor the changes. All of this can be made possible by identifying key living organisms which could predict variability in environment in good or bad way and act as bioindicators. Various abiotic like plants, planktons, animals, and microbes will be used to find environmental health and therefore, will be considered as bioindicators.

8.1 INTRODUCTION

Due to the rapid growth in population and developmental patterns, a tremendous change has occurred in the environment, thereby deteriorating the environmental performance. Most of these changes that have occurred in the surroundings have a negative effect on the air, aquatic, land, etc., and thus demanding an immediate needs to curb or measure these changes. All of this can be achieved by using suitable and effective entities/organisms which can monitor possible negative changes occurring in the surroundings. So living organism that gives us bulk information about the surrounding it are known as biological monitors or biomonitors and the process of measuring toxic chemical compounds, elements or their metabolites in the body is referred to as biomonitoring. In order to assess the environmental contamination such as in air, water, etc., the process of biomonitoring involves the use of living organisms to measure changes either qualitative or quantitative. On the basis of these observations or studies, pollution may be suspected or inferred by measuring effects of the environment on its resident species, communities, or organisms. The presence, amount of pollutant and the intensity of exposure can easily be provided by a reliable and good biomonitor. A valuable method to protect and preserve the biological integrity of a natural ecosystem can be done through biological monitoring. All around the globe biomonitors are freely available reflecting natural impact over creatures and require minimum preparation and training (Fadila et al., 2009: Kumari et al., 2007). Living organisms are considered as biomonitors when water pollution is considered to other species of the same class (Singh et al., 2013).

Living entities consisting of abiotic organisms such as plants, animals, microbes, planktons, and lichens, etc., are known as bioindicators. In other words, bioindicators are usually the living organisms, species, or group of species, communities that help us to reveal the health, status, and changes in the environment. Various living organisms are reactive to pollutants in their environment and may change its morphology physiology or behavior, or it could even die.

Most of the changes that occur in the environment are because of anthropogenic bedevilment (e.g., pollution, land-use changes) or natural stressors (e.g., drought, late spring freeze); however anthropogenic stressors are key focus of bioindicator research (Hosmani, 2013). Since 1960s a major focus has been laid on the development of bioindicators and their practical utility to study different types of epitomes including water and terrestrial by employing each vital taxonomic group. All biological living entities cannot serve as successful bioindicators as their presence is governed by various factors in the environment like passing on of light, water, temperature, substrate, competition, and suspended solids. Besides, certain level degree of contamination can be determined by the use of bioindicators (Khatri and Tyagi, 2015). Bioindicators are found very sensitive to any changes in the environment and tend to produce molecular signals in case of environmental disturbance (Posudin, 2014). The ways in which a bioindicator delivers information regarding the deterioration in environment or the sum of pollutants involves physiological, chemical, or behavioral changes. Moreover, this information can be acquired through the study of the following things as shown in Figure 8.1.

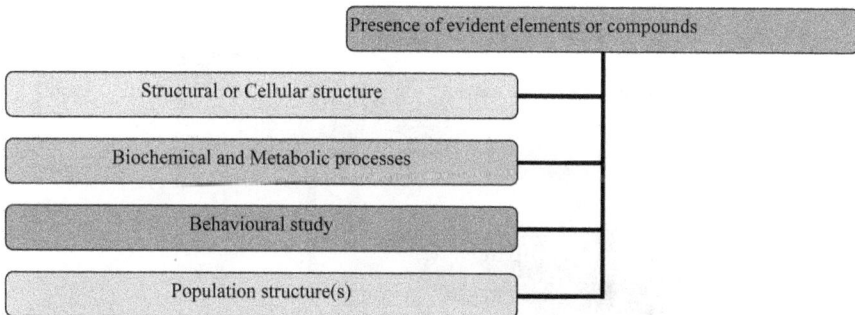

FIGURE 8.1 Different ways in which a bioindicator delivers information about changes.

8.2 BIOLOGICAL MONITORING

Biological monitoring or biomonitors involves the operation of certain properties of an organism to gather information on few ingredients of the biosphere. Various methods are being utilized to monitor air pollutants in the atmosphere and these mainly involve active and passive ways. Active method involves the presence of air pollutants in the study area is determined by placing test of known response and genotype while as in passive method naturally growing plants within the area of interest are being observed.

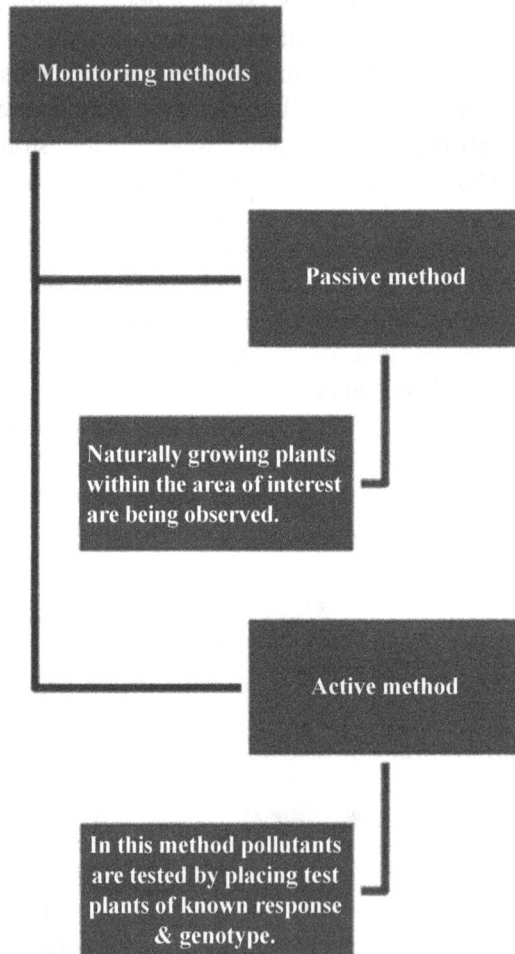

8.3   ADVANTAGES OF BIOINDICATORS

The advantages that are associated with the use of bioindicators are shown in graphical Figure 8.2.

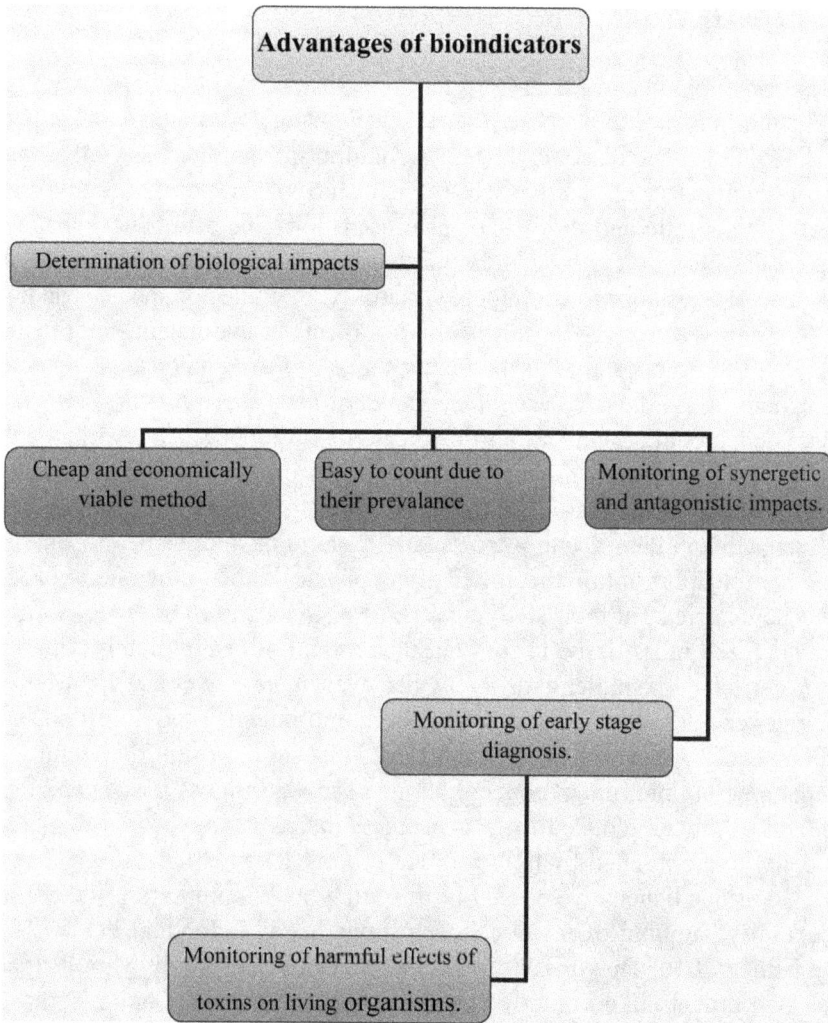

FIGURE 8.2   Flow chart representing the advantages of bioindicators.

In three situations, bioindicators are found to be useful. Firstly, in such situation where it is not possible to measure the indicated environmental factor for, e.g., studying climatic change palaeo-biomonitoring. Secondly, in such conditions where it is burdensome to compute the indicated factor, e.g., pesticides, and their residues and thirdly in such situations where it is easy to measure the environmental factor but troublesome to interpret, e.g., whether the perceived change have significance in terms of ecology.

8.4   UTILIZATION OF BIOINDICATORS

The term bioindicators usually refer to both biotic and abiotic reactions occurring in response because of ecological changes. Now taxa are utilized to show impact and discover changes in natural surroundings as well as their impact in negative or positive sense. The occurrence of pollutants adversely affects the abiotic and biotic species present in it and all of these changes can be detected by the indicators (Gerhardt, 2002; Holt and Miller, 2010). Because of resistance to ecological variability, bioindicator species can help us to monitor the condition of the environment. In the distant tundra environment of northwestern Alaska the presence of heavy metals was detected by using a moss, i.e., *Hylocomium splendens* species as a natural indicator (Hasselbach et al., 2005). From this place, i.e., Red Dog Mine which is the world's largest creator of zinc (Zn), the ore of mineral is mined and carried from singular street (75 km long) to storage spaces on the Chukchi Sea. Hasselbach and their group were eager to determine and wanted to find out whether transport of thin metal will affect physical biota or had no impact at all. They carried out their study at various distances from the area of street and analyzed moss tissue to find out the amount of heavy metals within it. They carried their studies at different distances in order to determine whether the overland transport was affecting the environment. Hasselbach and her group reported that the areas which were adjacent or within the vicinity of haul street had prominent concentrations of metal in moss tissues and their amount of metals reduced from a distance, therefore supporting the theory as being overburdened by transport can modify the encompassing environment. Further lichens were used in this study as biomonitors to determine quantitative amount of metal concentrations inside individual lichen (Holt and Miller, 2010; Thakur et al., 2013). Organisms that occupy various types of environments like natural, biological, and biodiversity markers can be found. For instance, lichens which have a symbiotic association between cyanobacteria and an algae/fungi and bryophytes particularly liverworts, are very useful to monitor air contamination. Both lichens and bryophytes acquire all their supplements from the environment as they are introduced to and thus prove to be very dominant bioindicators as they have no roots, no fingernail skin, etc. Further to support the theory of their use as bioindicators is provided by their high surface region to volume ratio and their ability to capture contaminants from the surrounding air (Holt and Miller, 2010). Thakur et al. (2013) studied rapid eutrophication in water bodies such as reservoirs, lakes, etc., which occurs due to formations of bloom and can easily be detected by Cynophyta, a type of phytoplankton thus proving to be a very powerful bioindicator. An important group of aquatic ecosystems are the algae and they have been utilized as a vital component for biological monitoring programs. Due to fast rate of reproduction and shorter life cycle, algae are valuable indicators for monitoring short-term impacts. As main producer, algae are directly affected by factors like physical and chemical means. Algae are widely used to monitor water quality. In the past few years diatoms have received an increased attention as bioindicators possibly due to their easy use.

8.5 CLASSIFICATION OF BIOINDICATORS

Bioindicators are classified into different types on the basis of aim, application, etc. From the perspective view there are different types of bioindicators and w.r.t aim of bioindication, three types of bioindicators can be described as:

1. **Compliance Indicators:** These lay stress on issues such as the sustainability of the population or community as a whole.
2. **Diagnostic Indicators:** Diagnostic indicators are to be measured at the individual levels.
3. **Early Warning Indicators:** These are biomarkers measured at the sub-organismal (biomarker) level, with major focus laid on identifying rapid and sensitive responses to environmental change.

8.5.1 *CLASSIFICATION ON THE BASIS OF DIFFERENT APPLICATIONS*

Three categories of bio-indicators are identified on the basis of their different applications. These include:

1. **Environmental Indicator:** These are simples measures, parameters, or values obtained from parameters that provide insights about the environment and its effect on human health and whole ecosystems. This type of indicator consists of species various types of species that respond to the changes or disturbances in environment and aims in examining the state of the environment and its policymaking. Examples of environmental indicators are sentinels, detectors, exploiters, accumulators, bioassay organisms).
2. **Ecological Indicator:** These consists of species which are reactive to pollutants, habitat fragmentation or other stresses and are often used

to decipate information about ecosystems. The indicators response gives the overall representation for the community and its effect on ecosystems. For instance in the field a range of beetle taxa species are found that can be used as an indicator of biodiversity (Bertollo, 1998; Girardin et al., 1999; Kurtz et al., 2001).

3. **Biodiversity Indicator:** These are statistical measures that help us to know the condition of biodiversity and the various factors affecting it. Biodiversity indicators help us to:

    i.   Monitor the threats on water bodies, habitat loss, or invasive species);
    ii. Status, state of health of species and ecosystems;
    iii. Conservation and protection of species-rich areas;
    iv. Advantages to people.

This type of indicator also determines species richness of a community. Moreover, the definition has been broadened to "measurable parameters of biodiversity," including, e.g., species richness, population-specific parameters, genetic parameters, endemism, and landscape parameters, etc.

8.5.2 *CLASSIFICATION OF BIOINDICATORS (IUBS)*

According to IUBS, bioindicators are grouped into three categories:

1. Plant system;
2. Animal system; and
3. Microbial system.

8.5.2.1 *PLANT INDICATORS*

Plants are considered as the most useful and sensitive tools for determination and recognition of environmental stresses. From the past few years a number of factors that includes rapid population growth, industrialization, and urbanization have adversely degraded the quality of water, air, land thereby having serious negative impact on the health of the living organisms in the vicinity (Joanna, 2006). In an ecosystem the presence or absence of certain plants/other vegetation in the area can provide specific information about the health of an ecosystem. Marine plants, lichens, mosses, bark pockets, tree bark, leaves, and tree rings can be used as plant biomonitors. Marine plants are immobile and provide essential knowledge about the oceanic environment status (Jain et al., 2010). Valuable information about the health of environment is provided by the absence or presence of few plants or vegetation. A valid example of bioindicator is provided by lichens, which consist of an association between fungi and algae. Lichens are small plants found living on the surfaces of rocks, trees, or soil and are very sensitive to the changes that occur in the surrounding environment. Mostly lichens obtain their food from the air and as a result they are very much sensitive to the toxins present in aerosol. The presence, amount, and type of lichens in the forests give valuable information about the quality of air. However, the decrease in number of lichens in the forest area indicates environmental stresses, like high levels of sulfur dioxide, sulfur-based pollutants, and nitrogen oxides, etc. The type of lichen species present in the forests gives an idea about the level of pollution as susceptibility of various species of lichens change as per pollutants in air. Therefore, lichens are important living tools for monitoring toxic levels of pollutants in the air and thus can serve as valuable bioindicators. *Wolffia globosa* or Asian watermeal as known locally is a species of flowering plant that is native to Asia and found in parts of America. *Wolffia globosa* is a tiny, oval-shaped plant with no leaves, stems, or roots and grows on the surface of calm, freshwater bodies, such as ponds, lakes, and marshes, etc. It is a potentially important plant to determine cadmium sensitivity and contamination. Indicators of pollutants in marine ecosystems are provided by the changes in diversity of several species of phytoplankton like *Euglena clastica*, *Phacus tortus*, and *Trachelon anas* (Phillips and Rainbow, 1993; Jain et al., 2010).

In the recent years awareness programs have been conducted to impart knowledge regarding adverse effects of soil pollution by trace metals (Adriano, 2001). As a result of this it has become a major area of interest for governments and other agencies for assessing metal contamination in order to maintain public health and sustainable development policies (Fairbrother et al., 2007). For the determination of total metal content in soils a number of rapid, accurate, and reliable measures have been developed (AFNOR, 1994, 2001, 2002). Bioindicators are often used to determine the level of pollutants in the environment and their impact. Main focus of research is to measure the amount of these metals and other toxic substances in these organisms in order to evaluate actual bioavailability of contaminants. However, the manipulation of plants as index is an old concept and many species of plants are regarded as principle bioindicators for soil contamination. These plants species include *Taraxacum officinale* (Kuleff and Djingova, 1984; Simon et al., 1996), *Capsella bursa-pastoris* (Aksoy et al., 1998), or *Populus alba*

(Madejon et al., 2004). Plant indicators have a number of advantages as compared to animal bioindicators as plants are ubiquitous, sessile organisms, easy to collect and their mineral status is dependent on the soil type. *Taraxacum officinale* is a good candidate plant to biomonitor metal contamination in the environment because of its ubiquitous scattering to tolerate a wide range of environmental conditions. The effect of environmental pollution at various levels is reflected by every organism living in a polluted environment. Inorganic pollutants include elements such as cadmium, copper, and lead at low levels is harmful to both plants and humans and because of non-biodegradable they can accumulate in vital organs. This in turn can lead to progressive toxic effects in living organisms (Demirezen and Aksoy, 2006). Several species of many plants have been identified as biological monitors of environmental metal pollution as they have the ability to assimilate metals from their surroundings (Djingova and Kuleff, 1993; Merkert, 1993). Leaves of the leafy vegetables absorb large amount of heavy metals as these metals accumulates in them. At a global level a particularly attractive candidate for biological monitoring is Dandelion (*Taraxacum officinale*). This plant is a good trace metal accumulator as it produces new leaves every year and in both native and agricultural ecosystems has a wide distribution (Keane et al., 2001; Kabata-Pendias and Pendias, 2001). As a biomonitor of environmental pollution, Dandelion revealed a correlation between the pollution level of a given element in an environmental compartment (air, soil) and its concentration in the tissues of this plant (Kuleff and Djingova, 1984; Kabata-Pendias and Dudka, 1991; Królak, 2001) while other studies did not observe such relationship (Marr et al., 1999). Various studies have collected Dandelion samples at various places to determine metals levels in the surrounding areas. From their study, they reported that smelters showed most tissue metal concentrations were distance-dependent, with plants growing nearer to the pollution source had high levels as compared to those plants growing farther away (Marr et al., 1999). Similarly, Capsella bursa-pastoris or shepherd's purse is a fully disomic tetraploid species (2n = 32) with worldwide distribution and growing under a wide range of environmental conditions has been utilized for biomonitoring of heavy metals in the environment. Capsella bursa-pastoris is one of the most common weed species growing annually highly self-fertilizing was first of all reported in the Middle East (ME), then spread to Europe (EUR) before invading Eastern Asia (ASI) and more recently spread worldwide due to human migrations (Cornille et al., 2016). Aksoy et al. (1998) carried a study in the city of Bradford, UK to find out whether *Capsella bursa-pastoris* could be used as a possible biomonitor to detect heavy metals. In their study they selected 42 sites that included, urban, urban roadside, urban park, suburban, and rural areas in Bradford UK to determine the concentrations of Pb, Cd, Zn, and Cu in soils, washed, and unwashed plant leaves. It was found that the concentrations of metal pollutant varied between the washed and unwashed samples, reflecting airborne, and soil entry routes, respectively. A significant relationship was found between the concentration of heavy metal in samples of surface soil (depth 0–10 cm) and washed leaves, the concentrations being higher with progressively increased urbanization of the sites. Thus they concluded from their studies that *Capsella bursa-pastoris* is a useful biomonitor of the four heavy metals and it may be a particularly important species since it could monitor short-term changes in pollution in urban areas.

Thus it can be concluded from these studies that plants can be important tools to screen, assess, and monitor the health and biogeographic changes occurring in the environment. Further additional studies needs to be carried out in this regard.

8.5.2.2 ANIMAL INDICATORS

An indication of damage to the ecosystem may be dissipated by an increase or decrease in the animal population. Negative impacts on the ecosystem may be reflected by changes in the population density. As animal species are dependent on food sources, depletion or reduction of important food resources will lead to decrease/decline in animal population (Phillips and Rainbow, 1993; Jain et al., 2010). An important aspect of animal indicators is that they help to detect the tissues of animals imbedded with amount of toxins (Joanna, 2006; Khatri and Tyagi, 2015). The early onset of stress in animal species are expressed and observed at population level which is manifested in physiology, morphology, and behavior of animal species long before the early onset of stress (Beaulieu et al., 2014). These sub-lethal responses can be very significant to determine how the animal populations will respond further and such responses could be utilized as early warning signals. Some of the animal species which have the potential of being used as indicators of environmental changes are discussed in subsections.

8.5.2.2.1 *Frogs and Toads as Bioindicators*

Amphibians, particularly frogs and toads, are mostly used in pollution studies as bioindicators to monitor the quality and changes in the environment

(Simon et al., 2010). In freshwater and terrestrial habitats, frogs are usually interfered by the surrounding changes that occur in their vicinity, thereby making them prime bioindicators to monitor ecological quality and changes. These organisms absorb toxic chemical substances permeable through their skin and larval gill membranes as these are sensitive to changes that occur in their surrounding environment (Lambert, 1997). Toxic substances like pesticides enter the bodies of these organisms when they eat contaminated food that is either absorbed, inhaled or ingested, and these organisms have poor ability to detoxify these toxic chemicals (Lambert, 1997). As a result of this, residual substances, particularly containing organochlorine pesticides (OCPs), accumulate in their systems (Lambert, 1997). Due to their permeable skin that helps them to absorb toxic chemicals, they can be easily used as model organisms to assess the environmental changes caused by various factors that may contribute to the decrease/decline in the amphibian population (Lambert, 1997).

8.5.2.2.2 *Fish as Bioindicators*

The overgrowing human populations and industrialization have led to significant changes in aquatic ecosystems, thereby deteriorating the quality of water by causing an increase in the level of pollutants in it, thus affecting the aquatic environment. Therefore, indicator organisms are required to find the effect of human activities on the aquatic ecosystems and fish being an important component of the aquatic environment. Fish is a vital component of the freshwater ecosystems because of their differential sensitivity to pollution. They have widely been utilized as indicators to monitor river ecosystem health/environmental water quality (Plafkin et al., 1989; Simon, 1991). Fish are the important components of the food web and are consumed by human beings as food, thus making them significant to assess contamination (Barbour et al., 1999). Fish can serve as good indicators to monitor long-term changes as they have a relatively long life cycle and are mobile (Barbour et al., 1999). Anthropogenic disturbances like eutrophication, acidification, chemical pollution, flow regulation, physical habitat alteration and fragmentation by human exploitation to physical habitat and introduction of species, etc., the communities of fish respond significantly to all the changes and are predictable to all kinds of changes (Karr et al., 1986; Hartman, 1973; Ormerod, 2003). The differential sensitivity of fishes is prime for using the basis of using them to monitor environmental degradation (Fausch et al., 1990). A case study was carried out by Naigaga et al. (2011) on a total of 29 fish species to determine fish as bioindicators in aquatic environmental pollution and reported that the distribution and abundance of the fish species were strongly influenced by the quality of water. From his studies, he suggested that the diversity of fish may be a useful biological indicator to monitor the quality of water. Further studies regarding the use of fish as a bioindicator for biomonitoring purposes need to be validated.

8.5.2.2.3 *Zooplanktons as Bioindicators*

Zooplanktons are heterotrophic animal planktons living on the surface of sea, river, or any other water bodies. Their size varies from microscopic organisms to large species. In aquatic ecosystems of ponds, lakes, rivers, estuaries, and marine environments, zooplanktons are the secondary producers. Their population number most often depends on the phytoplankton population. They feed on phytoplanktons, bacterioplankton, or detritus (i.e., marine snow) and rely on tides and currents for transport system as they are poor swimmers. They help to evaluate the level of water pollution therefore acting as bioindicators. Zooplanktons play a crucial role in determining water quality, eutrophication, and production of a freshwater body. The capability of zooplankton as a bioindicator species is based on the fact that their advancement and conveyance are liable to some abiotic (e.g., temperature, saltiness, stratification, and pollutants) and biotic parameters (e.g., limitation of food, predation, and competition) (Ramchandra et al., 2006). Zooplanktons such as *Alona guttata, Moscyclopes edex, Cyclips,* and *Aheyella* are zone dependent indicators of pollution. Zooplanktons are important components of aquatic food webs and respond strongly to the changes in environment. They are also helpful to assess conditions in aquatic ecosystems (Brito et al., 2011; Primo et al., 2015). Zooplanktons constitute a vital source of food for small fish and predatory invertebrate diets (90% calories) (Hairston et al., 1993). Some species of zooplanktons are highly sensitive to the alteration in nutrient cycling, temperature, and variable environmental conditions (Hanazato et al., 2001; Primo et al., 2015). Some research studies have postulated that in systems with increased phosphorus levels, zooplankton richness decreases and that certain species of cladocerans are highly sensitive to increased amount of phosphorus (Jeppesen et al., 2011). To measure the impact of disturbance in aquatic ecosystems, zooplankton is effective bioindicators due to their trophic significance, as well as their unique responses to certain environmental dynamics. Thus zooplanktons can also serve as important indicators to monitor changes in aquatic ecosystems.

8.5.3   OTHER ANIMAL INDICATORS

Other species of animals which can serve as useful indicators to monitor alterations environment include:

1. **Leeches:** These are the segmented parasitic worms belonging to the phylum annelida and there are about 650 species of leeches. Some certain characteristics of leeches make them ideal potential bioindicators to monitor water pollution (Koperski, 2005). Leeches are the most common animals found in both standing and running freshwaters. There are numerous advantages of using leeches as indicator organisms as they are abundant and easily accessible. Leeches are important living entities that are used as sensor bioindicators to monitor river contamination by polychlorinated biphenyls (PCBs).

2. **Earthworms:** These are the components of invertebrate populations and have widely been utilized as bioindicators for monitoring purposes. Earthworms are important components of the soil system and are used to monitor soil quality. Earthworms are continuously exposed to contaminants through their alimentary surfaces as they ingest large amounts of soil specific fractions and organic matter. The most commonly studied earthworm species are *Eisenia fetida, Eisenia andrei, Lumbricus terrestris,* and *Lumbricus rubellus.* According to the studies carried out by Sanchez-Hernandez et al. (2006), skin of earthworm is an excellent route for contaminant uptake. Earthworm species such as *Eisenia fetida* are given representation of soil fauna. Under experimental conditions, other species such as *E. andrei* and *Lumbricus rubellus* are widely used for the study of soil contamination (Spurgeon et al., 2000). *E. fetida* is observed to be less sensitive to pesticides.

3. **Oyster:** These are the large flat sea creatures and include different families of salt-water bivalve mollusks that live in marine or brackish habitats. Oyster such as (*Crassostrea gigas*), crabs (*Geotica depressa*) can detect trace metal indication particularly lead contamination (Uttah et al., 2008).

4. **Honey Bee:** These react quickly to several external factors and are therefore it is considered to be an efficient bioindicator to determine the environmental quality (Gallina et al., 2006; Spodniewska and Romaniuk, 2006). Determination of the traces in plant, animal origins along with humans and honeybees helps to monitor the

existing environmental problem (Berberova et al., 2008; Petkov et al., 2010). Honeybee has also been used in nuclear testing to monitor radionuclide strontium 90 in the environment (Svoboda, 1962).

8.5.4 MICROBIAL INDICATORS

The third group of indicators is the microbial indicators that include small microorganisms that are usually used to assess health of aquatic and terrestrial ecosystems. Some microorganisms when exposed to the contaminants of cadmium and benzene synthesize new proteins known as stress proteins which can be used as early warning signs (Khatri and Tyagi, 2015). Microorganisms have rapid growth rates and are reactive to low levels of contaminants and other physicochemical and biological changes. They give important insights about the changes in the environment as per the research view of perspective (Pradhan et al., 2008; Zannatul and Muktadir, 2009; Nkwoji et al., 2010; Hosmani, 2014). Microbial indicators are widely utilized in a number of ways to detect environmental pollutants in water. The toxins that are present in waters can quite easily be monitored by the alterations that occur in the digestive system of microbes. These toxins hinder or disturb the digestive system and results in changes in the amount of light emitted by the bacteria (Butterworth et al., 2001). These tests are very quick to monitor changes as compared to other available traditional tests, but due to the presence of toxins they become limited as they indicate only the changes in the organisms (Malik and Bharti, 2012; Khatri and Tyagi, 2015). One such example is that of *Vogesella indigofera*, a bacterium that reacts quantitatively to heavy metals only. This bacterium under no metal pollution produces blue pigmentation which can be utilized as an important marker of morphological change that has taken place and can be visually observed. The production of this pigment is blocked under the vicinity of hexavalent chromium. The production of the pigment can be linked in between chromium concentration and blue pigmentation by the bacterium (Oberholster et al., 2009; Aslam et al., 2012; Malik and Bharti, 2012). Biological monitoring involves the use of specific biological response for the determination of environmental changes. The use of bacteria as bioindicators to monitor changes in surroundings needs a better understanding in terms of its application. Further, the use of bacteria as indicators has got numerous advantages as compared with other available or studied bioindicators.

8.6   CONCLUSION

Bioindicators are very useful tools to monitor the health of the environment and assess the changes that occur in the environment. Bioindicators give an index to measure or characterize an ecosystem or one of its critical components. Bioindicators help to detect changes, either negative or positive, and their consequent effects on living beings. To assess the changes that take place in a specific biological community, bioindicators can be very useful and could be utilized at various scales, i.e., cell level to the environmental level. The overall conclusion about bioindication and biomonitoring is that they have become important tools to study the impacts of various factors on an ecosystem and its role in differentiating polluted and unpolluted areas.

KEYWORDS

- **animal indicators**
- **bioindicators**
- **biomonitors**
- **environmental pollution**
- **plant indicators**
- **zooplanktons**

REFERENCES

Adriano, D. C., (2001). *Trace Elements in Terrestrial Environments: Biogeochemistry, Bioavailability and Risks of Metals* (p. 884). Springer-Verlag, New York, Berlin, Heidelberg.

AFNOR, (1994). Soil Quality Extraction elements en traces solubles dans l'eau régale. Norme française NF ISO 11466.

AFNOR, (2001). *Soil Quality e Dissolution for the Determination of Total Element Contents e Part 1: Dissolution with Hydrofluoric and Perchloric Acids.* French Standard NF ISO 14869-1.

Aksoy, A., Hale, W. H. G., & Dixon, J. M., (1999). *Capsella bursa-pastoris* (L.) medic. As a biomonitor of heavy metals. *The Science of the Total Environment, 226,* 177–186.

Aslam, M., Verma, D. K., Dhakerya, R., Rais, S., Alam, M., & Ansari, F. A., (2012). Bioindicator: A comparative study on uptake and accumulation of heavy metals in some plants leaves of M.G. Road, Agra City, India. *Res. J. Environ. Earth Sci., 4*(12), 1060–1070.

Barbour, M. T., Gerritsen, J., Snyder, B. D., & Stribling, J. B., (1999). *Rapid Bioassessment Protocols for Use in Streams and Wadeable Rivers: Periphyton, Benthic Macroinvertebrates and Fish* (2ⁿᵈ edn.). EPA 841-B-99-002., U.S. Environmental Protection Agency; Office of Water; Washington, D.C.

Beaulieu, M., & Costantini, D., (2014). *Biomarkers of Oxidative Status: Missing Tools in Conservation Physiology.*

Berberova, R., Bliznakov, A., Baykov, B., & Gyurov, R., (2008). Prinos za otshenka na okolnata sreda v oblast Stara Zagora. In: *Proceeding of 7ᵗʰ International Conference "Ecology-Stable Development"*. Vratsa, Bulgaria.

Bertollo, P., (1998). Assessing ecosystem health in governed landscapes: A framework for developing core indicators. *Ecosystem Health, 4*, 33–51.

Brito, S. L., Maia-Barbosa, P. M., & Pinto-Coelho, R. M., (2011). Zooplankton as an indicator of trophic conditions in two large reservoirs in Brazil: Zooplankton indicator of trophic conditions. *Lakes and Reservoirs: Research and Management, 16*(4), 253–264.

Butterworth, F. M., Gunatilaka, A., & Gonsebatt, M. E., (2001). *Biomonitors and Biomarkers as Indicators of Environmental Change* (Vol. 2). Boston (MA), Springer Science & Business Media.

Cornille, A., et al., (2016). Genomic signature of successful colonization of Eurasia by the allopolyploid shepherd's purse (*Capsella bursa-pastoris*). *Molecular Ecology, 25*(2), 616–629.

Demirezen, D., & Aksoy, A., (2006). "Heavy metal levels in vegetables in Turkey are within safe limits for Cu, Zn, Ni, and exceeded for Cd and Pb. *Journal of Food Quality, 29*(3), 252–265.

Djingova, R., Kuleff, I., & Andreev, N., (2007). Comparison of the ability of several vascular plants to reflect environmental pollution. *Chemosphere, 27*(8), 1385–1396, 1993.

Fadila, K., Houria, D., Rachid, R., & Mohammed, D., (2009). Reda cellular response of a pollution bioindicator model (*Ramalina farinacea*) following treatment with fertilizer (NPKs). *American-Eurasian J. Toxicol. Sci., 1*(2), 69–73.

Fairbrother, A., Wenstel, R., Sappington, K., & Wood, W., (2007). Framework for metals risk assessment. *Ecotoxicology and Environmental Safety, 68*, 145–227.

Fausch, K. D., Lyons, J., Karr, J. R., & Angermeier, P. L., (1990). Fish communities as indicators of environmental degradation. *American Fisheries Society Symposium, 8*, 123–144.

Gallina, A., Baggio, A., & Mutinelli, F., (2005). Heavy metal contamination of honey in Veneto region (Northeastern Italy), an overview of the situation 2003–2004. In: *Bioindicators. Biological Reviews, 73*, 181–201.

Gerhardt, A., (1993). *Bioindicator Species and Their Use in Biomonitoring.* Environmental monitoring I. Encyclopedia of life support systems. UNESCO ed. Oxford (UK), Eolss Publisher.

Girardin, P., Bockstaller, C., & Van, D. W. H., (1999). Indicators: Tools to evaluate the environmental impacts of farming systems. *Journal of Sustainable Agriculture, 13*(4), 6–21.

Hairston, Jr. N. G., & Hairston, Sr. N. G., (1993). Cause-effect relationships in energy flow, trophic structure, and interspecific interactions. *American Naturalist, 142*(3), 379–411.

Hanazato, T., (2001). Pesticide effects on freshwater zooplankton: An ecological perspective. *Environmental Pollution, 112*(1), 1–10.

Hartman, W. L., (1973). *Effects of Exploitation, Environmental Changes, and New Species on the Fish Habitats and Resources of Lake Eire.* Great Lakes Fishery Commission Technical Report 22, Ann Arbor, Michigan.

Hasselbach, L., et al., (2005). Spatial patterns of cadmium and lead deposition on and adjacent to National Park Service lands in the vicinity of red-dog mine, Alaska. *Science of the Total Environment, 348*, 211–230.

Holt, E. A., & Miller, S. W., (2010). Bioindicators: Using organisms to measure environmental impacts. *Nature, 3*(10), 8–13.

Hosmani, S. P., (2013). Freshwater algae as indicators of water quality. *Universal J. Environ. Res. Technol., 3*(4), 473–482.

Hosmani, S., (2014). Freshwater plankton ecology: A review. *J. Res. Manage Technol., 3*, 1–10.

Jain, A., Singh, B. N., Singh, S. P., Singh, H. B., & Singh, S., (2010). *Exploring Biodiversity as Bioindicators for Water Pollution.* National Conference on Biodiversity, Development, and Poverty Alleviation.

Jeppesen, E., Nõges, P., Davidson, T. A., Haberman, J., Nõges, T., Blank, K., & Johansson, L. S., (2011). Zooplankton as indicators in lakes: A scientific-based plea for including zooplankton in the ecological quality assessment of lakes according to the European Water Framework Directive (WFD). *Hydrobiologia, 676*(1), 279–297.

Joanna, B., (2006). Bioindicators: Types, development, and use in ecological assessment and research. *Environ. Bioind., 1*, 22–39.

Kabata-Pendias, A., & Dudka, S., (1991). Trace metal contents of *Taraxacum officinale* (dandelion) as a convenient environmental indicator. *Environmental Geochemistry and Health, 13*(2), 108–113.

Kabata-Pendias, A., & Pendias, H., (2001). *Trace Elements in Soils and Plants.* CRC Press, Boca Raton, Fla, USA.

Karr, J. R., Fausch, K. D., Angermeier, P. L., Yant, P. R., & Schlosser, I. J., (1986). *Assessing Biological Integrity in Running Waters: A Method and its Rationale.* Special Publication No. 5. Illinois Natural History Survey, Champaign, IL.

Keane, B., Collier, M. H., Shann, J. R., & Rogstad, S. H., (2001). Metal content of dandelion (*Taraxacum officinale*) leaves in relation to soil contamination and airborne particulate matter. *Science of the Total Environment, 281*(1–3), 63–78.

Khatri, N., & Tyagi, S., (2015). Influences of natural and anthropogenic factors on surface and groundwater quality in rural and urban areas. *Front Life Sci., 8*(1), 23–39.

Koperski, P., (2005). Testing the suitability of leeches (Hirudinea, Clitellata) for biological assessment of lowland streams. *Pol. J. Ecol., 53*, 65–80.

Królak, E., (2003). Accumulation of Zn, Cu, Pb, and Cd by dandelion (*Taraxacum officinale* Web.) in environments with various degrees of metallic contamination. *Polish Journal of Environmental Studies, 12*(6), 713–721.

Kuleff, I., & Djingova, R., (1984). The dandelion (*Taraxacum officinale*): A monitor for environmental pollution? *Water, Air, and Soil Pollution, 21*(1), 77–85.

Kumari, P., Dhadse, S., Chaudhari, P. R., & Wate, S. R., (2007). Bioindicators of pollution in lentic water bodies of Nagpur city. *J. Environ. Sci. Eng., 49*(4), 317–324.

Kurtz, J. C., Jackson, L. E., & Fisher, W. S., (2001). Strategies for evaluating indicators based on guidelines from the environmental protection agency's office of research and development. *Ecological Indicators, 1*, 49–60.

Lambert, M. R. K., (1997). Environmental effects of heavy spillage from a destroyed pesticide store near Hargeisa (Somaliland) assessed during the dry season, using reptiles and amphibians as bioindicators. *Archives of Environmental Contamination and Toxicology, 32*(1), 80–93.

Madejon, P., Maranon, T., Murillo, J. M., & Robinson, B., (2004). White poplar (*Populus alba*) as a biomonitor of trace elements in contaminated riparian forests. *Environmental Pollution, 132*, 145–155.

Malik, D. S., & Bharti, U., (2012). Status of plankton diversity and biological productivity of Sahastradhara stream at Uttarakhand, India. *J. Appl. Natural Sci., 4*(1), 96–103.

Marr, K., Fyles, H., & Hendershot, W., (1999). Trace metals in Montreal urban soils and the leaves of *Taraxacum officinale. Canadian Journal of Soil Science, 79*(2), 385–387.

Merkert, B., (1993). Monitoring of heavy metal pollution by *taraxacum officinale*. In: Markert, B., (ed.), *Plants as Biomonitors: Indicators for Heavy Metals in Terrestrial Environment*. VCH, New York, NY, USA.

Naigaga, et al., (2011). Fish as bioindicators of pollution, Lake Victoria. *Physics and Chemistry of the Earth, 36*, 918–928.

Nkwoji, J. A., Igbo, J. K., Adeleye, A. O., Obienu, J. A., & Tony-Obiagwu, M. J., (2010). Implications of bioindicators in ecological health: Study of a coastal lagoon, Lagos, Nigeria. *Agric. Biol. J. Noth Am., 1*(4), 683–689.

Oberholster, P. J., Botha, A., & Ashton, P. J., (2009). The influence of a toxic cyanobacterial bloom and water hydrology on algal populations and macroinvertebrate abundance in the upper littoral zone of Lake Krugersdrift, South Africa. *Ecotoxicology, 18*(1), 34–46.

Ormerod, S. J., (2003). Current issues with fish and fisheries: Editor's overview and introduction. *Journal of Applied Ecology, 40*, 204–213.

Petkov, G., Yablanski, T., Todorova, M., Pavlov, D., Kostadinova, G., & Barakova, V., (2010). Ecological assessment of soils from parks and places in Stara Zagora. *Journal of Agricultural Science and Forest Science, 9*(4), 24–31.

Phillips, D. J. H., & Rainbow, P. S., (1993). *Biomonitoring of Trace Aquatic Contaminants*. New York (NY), Elsevier Applied Science.

Plafkin, J. L., Barbour, M. T., Porter, K. D., Gross, S. K., & Hughes, R. M., (1989). *Rapid Bioassessment Protocols for Use in Streams and Rivers: Benthic Macroinvertebrates and Fish*. EPA/444/4-89-001 U.S. Environmental Protection Agency, Washington, DC.

Posudin, Y., (2005). Bioindication, in methods of measuring environmental parameters. John Wiley & Sons, Inc., Hoboken, NJ, USA. *Proceeding of XXXIX-*[th] *Apimondia International Apicultural Congress* (pp. 145–146). Dublin, Ireland.

Pradhan, A., Bhaumik, P., Das, S., Mishra, M., Khanam, S., Hoque, B. A., Mukherjee, I., Thakur, A. R., & Chaudhuri, S. R., (2008). Phytoplankton diversity as indicator of water quality for fish cultivation. *Am. J. Environ. Sci., 4*(4), 406–411.

Primo, A., Kimmel, D., Marques, S., Martinho, F., Azeiteiro, U., & Pardal, M., (2015). Zooplankton community responses to regional-scale weather variability: A synoptic climatology approach. *Climate Research, 62*(3), 189–198.

Ramchandra, T. V., Rishiram, R., & Karthik, B., (2006). *Zooplanktons as Bioindicators: Hydro Biological Investigation in Selected Bangalore Lakes*. Technical report 115.

Simon, E., Braun, M., & Tothmeresz, B., (2010). *Water Air Soil Pollut., 209*, 467.

Simon, L., Martin, H. W., & Adriano, D. C., (1996). Chicory (*Cichorium intybus L.*) and dandelion (*Taraxacum officinale* Web.) as phytoindicators of cadmium contamination. *Water, Air and Soil Pollution, 91*, 351–362.

Simon, T. P., (1991). *Development of Ecoregion Expectations for the Index of Biotic Integrity (IBI) Central Corn Belt Plain*. U.S. Environmental Protection Agency, Region V, Chicago, Illinois, EPA 905/9-91/025.

Singh, U. B., Ahluwalia, A. S., Sharma, C., Jindal, R., & Thakur, R. K., (2013). Planktonic indicators: A promising tool for monitoring water quality (early-warning signals). *Eco Environ Cons., 19*(3), 793–800.

Spodniewska, A., & Romaniuk, K., (2006). Concentration of lead and cadmium in bees and beebread. In: *Proceedings of the Second European Conference of Apidology EurBee, Praque, Czech Republic* (pp. 10–16).

Spurgeon, D. J., Svendsen, C., Rimmer, V. R., Hopkin, S. P., & Weeks, J. M., (2000). Relative sensitivity of life cycle and biomarker responses in four earthworm species exposed to zinc. *Environ. Toxicol. Chem., 19*, 1800–1808.

Svoboda, J., (1962). Teneur en strontium 90 dans les abeilles et dans leurs produits. *Bulletin Apicole, 5*, 101–103.

Thakur, R. K., Jindal, R., Singh, U. B., & Ahluwalia, A. S., (2013). Plankton diversity and water quality assessment of three freshwater lakes of Mandi (Himachal Pradesh, India) with special reference to planktonic indicators. *Environ. Monit. Assess, 185*(10), 8355–8373.

Uttah, E. C., Uttah, C., Akpan, P. A., Ikpeme, E. M., Ogbeche, J., & Usip, J. O., (2008). Bio-survey of plankton as indicators of water quality for recreational activities in Calabar River, Nigeria. *J. Appl. Sci. Environ. Manage, 12*(2), 35–42.

Zannatul, F., & Muktadir, A. K. M., (2009). A review: Potentiality of zooplankton as bioindicator. *Am. J. Appl. Sci., 6*(10), 1815–1819.

CHAPTER 9

Zooplankton Community: A Valuable Bio-Indicator Tool in Disturbed Wetlands

JAVAID AHMAD SHAH,[1,3] ASHOK K. PANDIT,[1] and G. MUSTAFA SHAH[2]

[1]Center of Research for Development (CORD), University of Kashmir, Srinagar–190006, Jammu and Kashmir, India

[2]Department of Zoology, University of Kashmir, Srinagar–190006, Jammu and Kashmir, India

[3]Government Degree College (Boys), Pulwama–192301, Jammu and Kashmir, India, E-mail: javaidshah31@gmail.com

ABSTRACT

Zooplankton is an essential constituent of freshwater ecosystems. They help in regulating phytoplankton and microbial populations via foraging, assisting in the transfer of energy to more complicated trophic ranks. They are exceptionally receptive to altering environmental situations, as the majority of species have very diminutive generation time, ensuring their abundance, diversity, or even community composition. The present chapter was aimed to assess the zooplankton community structure of Hokersar wetland, the Queen wetland of Kashmir Himalaya. The collection of zooplankton was done in Sep. 2012–Aug. 2013 by procuring the water samples in a net made up of silk with a mesh size of 75 meshes per linear cm from 6 collection points. 5% formalin was used as a preservative for the collected samples. A total of 34 taxa encompassing 14 Rotifera (wheel animalcules), 9 Cladocera (water fleas), and Copepoda each and only 2 ostracods were recorded during the entire study. During summer, higher abundance was recorded contrarily; in winter, lower diversity was recorded in the wetland. As far as the dominance pattern of the zooplankton community is concerned, there was the dominance of rotifers followed by cladocerans and copepods in the entire study. Great diversity of zooplankton was observed at the sites having ample growth of aquatic plants. From the study, it can be inferred that the species show broad ecological tolerance and stand an extensive range of environmental conditions. During the study, small body-sized plankton was abundant over large ones clearly indicate accelerated eutrophication in the water body. Furthermore, in summer season, there is profusion of small-bodied plankton to larger-bodied plankton is a noteworthy sign of increased cultural eutrophication of the wetland.

9.1   INTRODUCTION

In an aquatic ecosystem the pattern of distribution of microscopic organisms greatly varies from smaller organisms to the larger ones that can be directed by their adaptations and their mode of life to a specific system in which these organisms thrive (McManus and Woodson, 2012). The word plankton encompasses both phytoplankton and zooplankton. Etymologically the former comes from Greek word "*zoon*" means animal, and "*planktons*" meaning drafting. These planktons remain hovering, drifting with ripples of water, and having a bit of locomotory power. In the aquatic systems, zooplankton is minute organisms, varying from one-celled to many-celled, and their size varies from few microns to millimeters. Besides size disparity, some distinctions concerning their external features (in terms of morphology) and their positions in the taxonomy. These plankters play an imperative role in studying the faunal diversity in the ecosystems in which they. They nourish on small algal populations, thus change the autotrophic substances into animal tissues, which thusly encompasses the essential nourishment for upper trophic cascade encompassing fishes and the larvae therein (Pandit, 1999).

Zooplankton being an essential constituent of freshwater bodies (Gannon and Stemberger, 1978; Sladecek, 1983; Huys and Boxshall, 1991). They regulate algal and diversity of microbes by acting as grazers and assist in the transfer of energy to the next trophic levels in the systems (Dejen et al., 2004). They are said to be enormously sensitive to varying abiotic situations, and the majority of the taxa have very small reproductive cycle life history, consequentially altering their composition, evenness, diversity, or community structure as well (Pinto-Coelho et al., 2005; Sharma et al., 2008). Freshwater zooplankton encompasses two major invertebrates viz.: rotifers and crustaceans. The Rotifera consist of phylum and are freshwater dwellers (Wallace and Snell, 1991) and the remaining two groups (Copepoda and belonging to class Crustacea of Arthropoda phylum.

Rotifera, has a Latin origin (*rota* = "wheel"; *fera* = "to bear"), were deliberated by Anton Van Leuwenhoek in 1703. They are pseudocoelomatic dwelling animals, purported as one amongst the established creatures in freshwater (Sladecek, 1983). Rotifers achieved crucial significance in freshwater ecosystems as they occupy benthal, littoral, and limnetic zones of the lakes and wetlands (Wallace and Snell, 2010), and the majority of these are supposed to encompass all types of habitats (Hyman, 1951; Ricci and Melone, 2000). Wheel animalcules assume a noteworthy function in the transfer of energy and nutrient recycling (Makarevic and Likens, 1979).

The word Cladocera has a Greek origin "*klados*" which means branch, and "*keras*" meaning horn. Among the subclass Branchiopoda, cladocerans cover the taxonomically nearly all numerous biological group (Fritsch et al., 2013). Water fleas, constituting a vital portion of zooplankton community and are inhibiting all types of aquatic ecosystems like ponds, streams, lakes, and rivers, etc. (Guher, 2000) and in addition, assuming a conclusive function in reusing of various materials in the water systems (Urabe et al., 2002).

Copepoda being most copious creatures on the globe and moderate estimations uncovered that theses nearly outnumbered the diversity of insects and encompasses major biomass and productivity of freshwater systems (Hwang et al., 2010; Reid and Williamson, 2010; Ka and Hwang, 2011; Shah et al., 2013). They constitute a significant part of the trophic levels by playing an important role in both food chains and webs in the aquatic biotopes, forms the inevitable food for fingerlings and fishes, and depict broad conveyance design in a wide range of water environs (Boxshall and Defaye, 2008; Silva; 2008; Wu et al., 2010; Shah et al., 2103).

As all we know major part of the water on the earth is inhibiting salt waters in the oceans or in solid form in polar ice caps, the tiny proportion accessible as freshwater, the sole kind accessible in profusion to land dwellers are restricted in the inland water bodies like streams, wetlands, and lakes. Among the aforementioned systems, wetlands and lakes have an immense quantity of freshwater, besides enjoying an important position in the recycling of nutrients and utilization of energy by acting as biological grounds for inland ecosystems.

Around 7% of the earth's surface is engaged by wetlands (MEA, 2005) were accounts for acting as sources, sinks (Peterjohn and Correll, 1984) or even act as transformers of inorganic nutrients to organic nutrients (Kemp and Day, 1984; Elder, 1985) that mainly relies on the kind of wetland and the hydrobiological situation of the same (Bayley et al., 1991). At a global level, these wetlands are degrading acutely, more than other ecosystems, gravely disturbing their bio-diversity. Because of their speedy squeezing rate, a large number of wetlands which were permanent in nature were changed into semi-lasting or seasonal ones with the underground level decreasing down swiftly (Pal and Akoma, 2009).

Species organization or even in major cases richness, abundance, diversity of these planktons in floodplain lakes vary marvelously between or within the seasons of the year in reaction to changing ecological conditions. Data on zooplankton species architecture, diversity of floodplain wetlands of India is inadequate and dissipated besides a few publications of Khan (1987), Pandit, and Kaul (1982) and Pandit (1999, 2008) from this region, Rai, and Datta Munshi from India (1988) and a couple of a few reports.

Maintaining in scrutiny the role of these wetlands having wide scarcity of available data, the present chapter is planned to carry out to evaluate the abundance, taxonomic nature and their variations on the basis of seasons of zooplankton community architecture in Hokersar wetland being recognized as Ramsar Site of International Importance of this valley.

9.2   STUDY AREA

Vale of Kashmir being on the top of the map of the Indian subcontinent covering around 425 km from north to south and covering more than 250 km from east to west. The state of Jammu and Kashmir is separated into three parts, i.e., Jammu, Kashmir, and Ladakh. As far as the position of this state is governed, being situated at the latitude of $32°.17''–37°.06''$N and a longitude of $37°26''–80°26''$E. The Vale of Kashmir is situated amidst the Himalaya somewhere in the range $33°20'''$ and $34°54''$N and $73°55''$ and $75°35''$E and covering a long area of 15,948 km². The valley, which is also known as the "Paradise on earth," is famous for its high peaked Himalayan mountains, meadows, spouting watercourses and warm springs put forward the most delightful destination in the globe. The Vale of Kashmir gets absolutely no parallel in the Himalayas region when taking into account its water-bodies, including lakes, wetlands, rivers, and streams. This region has an immense variety of freshwater ecosystems that include wetlands, lakes, fast, and turbulent streams and pools, and among these, the first two assumes vital importance in exemplifying very high floral and faunal composition. Amongst these the former two (wetlands and lakes), Hokera wetland assume vital importance being documented as "Ramsar Site," Wetland of global value.

Hokera, the Queen of wetlands in Kashmir, situated at $34°05'$N–$34°06'$N latitude and $748–74°12'$E longitude) situated 12 kilometers away from Srinagar on the west side besides Baramullah-Srinagar Highway in the northernmost part of Doodhganga catchment (Figure 9.1). It is positioned at an altitude of 1,584 m (amsl), that was once extended more than an area of 19.5 km², and is currently condensed to 13.26 km² (Ahmad et al., 2014). This wetland, assigned as Ramsar Site of International Importance on 08-11-2005, is in managed by the State Government since 1945 (Wild-life Protection Department of Jammu and Kashmir Govt.). Hokera is the only wetland in this region which is home for about 5 lakh migratory avian species during winter, which includes ducks, geese, and rails that come from the Palaearctic area, making way from northern Europe to central Asian, finally to this wetland crossing over the greater Himalayan region (Pandit, 1999; Ahmad et al., 2014). This wetland provides habitat for about 80–90% of the migratory birds visiting the valley in winters. Around 68 species of these birds were reported from the wetland and its immediate catchment that includes the endangered species called white-eyed pochard (*Aythya nyroca*) that is listed in Red category of IUCN (Information Sheet on Ramsar Sites, 2004). Hokersar has two inlets in the form of streams, namely Sukhnag Nalla (on west side) and Doodhganga (on eastern side). The wetland records a greatest water depth of 2.5 m in blooming season, i.e., spring (Romshoo and Rashid, 2012). Contrarily, lowest water depth of 0.7 m that was observed

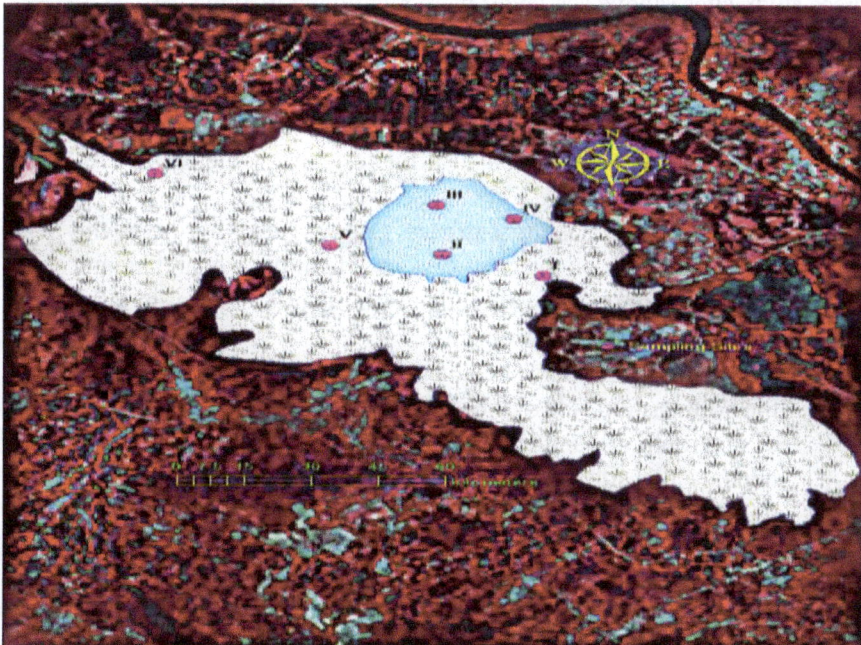

FIGURE 9.1 Hokersar wetland with six study sites.

infall (Pandit and Kumar, 2006; Ahmad et al., 2014). On the north-west side of the wetland there is a channel which acts as an outlet called the Needle Gate with a weir and lock system near the Sozieth village through which excess of water is drained out. In the wetland, the water level is upheld by this weir and lock arrangement as it helps to give a good environment for the migrant species of birds throughout the colder season.

The adjoining area of Hokersar is made up of an impressive extend of arable land having paddy farming, moreover with a little grazing land for the cattle. Anthropogenic activities in the catchment of wetland from the various rural areas like: Zainkot, Khoshipora, Hajibagh, Sharifabad, Soibogh, Dharamuna, Gotapora, Mraglar, Choripora, and Gund Khalil, Sozeth, and Lawipora have tremendously disturbed and modified the ecological and aesthetic health of the Hokersar wetland (Joshi et al., 2002).

9.2.1   STUDY SITES

Six examining sites were chosen during the investigating period varying in-depth, floral conditions, near to inflow and outflow streams, and some additional interrelated features. The physical position with coordinates and these study sites (above sea level) were mentioned in Table 9.1.

1. **Site I:** Being positioned on the western side of the Hokera near the Hajibagh area side and having slight growth of aquatic plants (macrophytes). Water depth at this site ranges as of a lowest of 1.1 meters in the fall season to a highest of 2.7 meters in the bloom season (spring).

2. **Site II:** Being close to entrance gateway neighboring to the avian watching tower on the northeastern side of the wetland having abundant macrophytic growth with a water depth of 0.7 m in fall and 1.9 m in the blooming season.

3. **Site III:** This site is located on the Chakdewan Shah village that is on the northeastern side of the wetland. This site harbors profuse growth of macrophytes, and has negligible anthropogenic effect. The water depth of this site ranges from a maximum of 2.4 m to a minimum of 0.9 m.

4. **Site IV:** Is positioned near to Zainkot village on the eastern side of the Hokera. Being close to the village has very high pollution pressures from the same village. Its water depth fluctuates from a least of 0.5 meters to an utmost of 1.6 meters.

5. **Site V:** Is positioned close to the center of the Hokersar. As this site has no emergent vegetation so considered as open water site. During spring maximum depth (2.2 m) was noticed and in autumn minimum depth (0.4 m) was recorded in this site.

6. **Site VI:** This site is on the rural area namely Gund Hassi Bhat from where a channel (an outlet) of the Hokersar is drained out on the northwestern side of the wetland. Water depth of this very site ranges from 1 meter in fall to 2.5 meters in the blooming season (spring).

TABLE 9.1 Geographical Position with Coordinates of the Study Sites in Hokera

Sites	Latitude (N)	Longitude (E)	Elevation (m)
I	34°06' 09.20"	74°42'59.32"	1580
II	34°06' 08.10"	74°42'53.26"	1582
III	34°06' 23.81"	74°42'53.41"	1585
IV	34°06' 16.26"	74°42'58.90"	1583
V	34°06' 04.88"	74°42'02.16"	1583
VI	34°06' 30.23"	74°41'23.48"	1586

9.3   HOW TO COLLECT AND PRESERVE THE ZOOPLANKTON SAMPLES

Fifty liters of subsurface water was procured from six sites for the quantitative purposes of zooplankton analysis. Standard plankton net with mesh size 75 meshes/cm was used to sieve the water. The collected content composed of various planktons was gathered in a plastic vial fixed at the inferior part of the plankton net. The content collected were shifted to bottles made up of polyethylene and were marked with respect to site collection. 4% formalin with 4–5 glycerine drops (for maintaining the flexibility of zooplankton) and 5% sucrose (to retain eggs in the broad chambers) were used as preservative.

9.3.1   FOR QUALITATIVE PURPOSES

The samples preserved were identified by the standard keys of Koste (1978), Pennak (1978), Adoni (1985), Battish (1992), Edmondson (1992), etc.

9.3.2   FOR QUANTITATIVE ESTIMATION OF ZOOPLANKTON

For the quantitative account of these plankters a specialized chamber called Sedge-wick rafter counting chamber (1 ml capacity) were used for enumeration of zooplankton organisms in each preserved sample. For the accuracy of results the sample was enumerated 10 times at least for quantitative purpose. To identify the preserved samples a microscope of binocular nature (Magnus MLX-DX, no. 4B 523861) was employed. For dissecting various important taxonomic parts such as post abdomen of water fleas 4th/5th legs of various copepods of zooplankton micro-needles of various sizes were used. For calculating the abundance of this zooplankton in the sample was calculated according to the formula devised by Welch (1952) and the results were recorded in terms of organisms per liter (org./L).

$$\text{Organism} / L = \frac{\text{Organisms per mL of concentrate} \times 1000}{\text{concentration factor(C.F.)}}$$

9.4   RESULTS

9.4.1   SPECIES COMPOSITION OF ZOOPLANKTON

During the investigation 34 taxa encompassing 14 Rotifera, 9 Cladocera, and Copepoda each and 02 Ostracoda were recorded. In general terms, wheel animalcules were the abundant phylum followed by order water fleas, order copepods and the class ostracods. Most of the species (for example *Brachionus calyciflorus*, *Filinia* sp., *Keratella* sp., *Lecane luna* and *Squatinella* sp. among rotifers, *Alona monacantha*, *Bosmina longirostris*, *Chydorus sphaericus* and *Sida crystallina* among cladocerans, *Bryocamptus minutus*, *Cyclops bicolor*, *C. scutifer* and *C. viridis* among copepods, *Cyclocypris* sp. in ostracods) stands extensive environmental conditions, hence show fair distribution in the Hokera wetland (Table 9.2).

9.4.2   SPATIAL VARIATIONS OF ZOOPLANKTON DIVERSITY

During the investigation, very clear spatial divergences were noticed at the majority of sites under observation. Among the sites, maximum zooplankton registering 24 taxa from site V (eight-wheel animalcules, seven water fleas, and nine copepods), closely followed by 23 at study site III (nine rotifers, six cladocerans, seven copepods, and one ostracod, and study site IV (thirteen rotifers, three cladocerans, five copepods, and two ostracod) at study site II, 20 (six-wheel animalcules, four water fleas, eight copepods and two ostracods) at study site VI, 18 (Five rotifers and cladocerans each, six copepods and two ostracods) and declining to a lowest of 17 (seven wheel animalcules, four cladocerans, four copepods and two ostracods) on study site I.

TABLE 9.2   Distribution Pattern of Zooplankton at Six Sites in Hokersar Wetland from Sep. 2012 to Aug. 2013

Zooplankton	Site I	Site II	Site III	Site IV	Site V	Site VI
(a) Rotifera						
Brachionus calyciflorus (Ehrenberg, 1838)	P	P	P	P	P	P
B. quadridentata (Hermann, 1783	A	A	A	P	A	A
Brachionus sp.	A	A	P	P	A	A
Cephalodella megalocephala (Glascott, 1893)	A	A	A	P	P	A
Filinia sp.	P	P	P	P	P	A
Keratella cochlearis (Gosse, 1851)	A	A	A	P	A	A
Keratella sp.	P	P	P	P	P	P
Keratella valga (Ehrenberg, 1834)	P	P	A	P	A	A
Lecane flexilis (Gosse, 1886)	A	A	A	P	A	A
Lecane luna (Müller, 1776)	P	P	P	P	P	A
Lecene sp.	P	A	P	A	P	A
Platyias patulus (Muller, 1786)	A	A	P	P	P	P
Squatinella sp.	P	P	P	P	P	P
Trichoderma sp.	A	A	P	P	A	P
Total = 14	**7**	**6**	**9**	**13**	**8**	**5**
(b) Cladocera						
Alona affinis (Leydig, 1860)	A	A	A	A	P	A
A. monacantha (Sars, 1901)	P	P	P	P	P	P
A. quadrangularis (Müller, 1776)	A	A	A	A	A	A
Bosmina longirostris (Müller, 1785)	P	P	P	P	P	P
Ceriodaphnia reticulata (Jurine, 1820)	A	P	P	A	P	P

TABLE 9.2   *(Continued)*

Zooplankton	Site I	Site II	Site III	Site IV	Site V	Site VI
Chydorus ovalis (Kurz, 1875)	A	A	A	A	A	A
C. sphaericus (Muller, 1785)	P	P	P	P	P	P
Sida crystallina (Muller, 1875)	P	A	P	A	P	A
Simocephalus vetulus (Müller, 1776)	A	A	P	A	P	P
Total = 09	**4**	**4**	**6**	**3**	**7**	**5**
(c) Copepoda						
Bryocamptus minutus (Claus, 1863)	A	P	P	P	P	A
Cyclops bicolor (Sars, 1863)	A	P	P	P	P	P
C. nanus (Sars, 1863)	A	P	P	A	P	P
C. scutifer (Sars, 1863)	P	P	A	P	P	P
C. vernalis (Fischer, 1853)	A	P	P	A	P	P
C. vicinus (Ulyanin, 1875)	A	A	A	A	P	A
C. viridis (Jurine, 1820)	P	P	P	P	P	P
Eucyclops macrurus (Sars, 1863)	P	P	P	A	P	A
Eucyclops sp.	P	P	P	P	P	P
Total = 09	**4**	**8**	**7**	**5**	**9**	**6**
(d) Ostracoda						
Cypris sp.	P	P	A	P	A	P
Cyclocypris sp.	P	P	P	P	A	P
Total = 02	**2**	**2**	**1**	**2**	**0**	**2**
Grand Total = 34	**17**	**20**	**23**	**23**	**24**	**18**

Note: P = Present and A = Absent.

9.4.3   SEASONAL VARIATION OF ZOOPLANKTON COMMUNITY

While investigating the zooplankton community on seasonal basis there were great diversity of species at majority of study sites in the wetland under observation. During the study, the following zooplankton are not present or feebly represented in winter season were *Keratella valga, Brachionus* sp., *Lecene* sp. *L. flexilis, Alona Monacantha, A. quadrangularis, Cyclops viridis, C. bicolor, C. vernalis,* and *Eucyclops macrurus.* However, *Filinia* sp., *Keratella* sp., *Squatinella* sp., *Bosmina longirostris, Sida crystallina, Chydorus sphaericus, C. scutifer, Eucyclops* sp, *Cypris*, and *Cyclocypris* sp. were present throughout the year. Moreover, *Keratella valga, Platyias patulus, Lecene luna, Brachionus calyciflorus* (Rotifera), *Alona monacantha, Bosmina longirostris, Ceriodaphnia reticulata* (Cladocera), *Eucyclops macrurus,* and *Cyclops viridis* (Copepoda) were rich in summer season displaying a peak population growth. Among the wheel animalcules taxa like *Squatinella* sp., *Filinia* sp., and *Lecene* sp. were associated with fall. For the interpretation of seasonal variation, data based on monthly variation on population density of various zooplankton groups was analyzed and calculated.

In terms of the population density of rotifers generally depicted dominant growth peak in summer at majority of study sites registering a highest of 2346 org./L at site IV in the wetland. However, intermediate population growth peaks were observed in autumn (except sites II and III) and spring. Further, the lowest population growths were noticed in the winter season during the entire study course. Interestingly, the comparable minimum organisms were noticed in the winter season (21 org./L) at site VI.

The greatest population density of water flies was also observed in summer registering 1013 org./L at site V. However, the minimum number of taxa being maintained in winter maintaining 14 org./L at site IV. Among the zooplankton copepods are characterized by varying number of taxa in various sites of the Hokersar wetland. In summer season there were abundant in number registering, the maximum density values of with 925.3 org./L at study site II. However, lowest seasonal density values were evinced in winter in entire. Among the zooplankton, ostracods were represented by least values and show meager distribution with no seasonal trend in the entire period of investigation. It is imperative to note that this group contributes very little in terms of population density during the study (Tables 9.3–9.8).

9.4.4 TOTAL POPULATION OF ZOOPLANKTON

On studying the entire zooplankton population, it was evident that lowest population density were observed during winter (393.3 org./L). Modest population density were evinced in spring, abrupt increase in density of these plankters were noticed in summer (3370.7 org./L), followed by autumn (2262.3 org./L) during the study (Tables 9.9–9.14).

TABLE 9.3 Monthly Variations in the Population Density of Zooplankton (org./L) at Site I in Hokersar Wetland from Sep. 2012 to Aug. 2013

Seasons	Autumn			Winter			Spring			Summer		
Zooplankton	Sep.	Oct.	Nov.	Dec.	Jan.	Feb.	Mar.	Apr.	May	Jun.	Jul.	Aug.
(a) Rotifera												
Brachionus calyciflorus (Ehrenberg, 1838)	78	43	0	17	21	14	12	33	44	97	120	241
Filinia sp.	143	233	16	11	31	11	44	56	67	122	136	142
Keratella sp.	267	146	65	44	51	61	76	84	98	198	241	178
Keratella valga (Ehrenberg, 1834)	145	98	22	0	0	0	45	37	43	265	566	278
Lecane luna (Müller, 1776)	340	45	33	21	11	34	24	35	55	49	190	245
Lecane sp.	244	122	141	0	0	0	98	134	234	77	102	98
Squatinella sp.	345	259	134	78	41	56	67	82	76	109	89	122
Total = 07	**1562**	**946**	**411**	**171**	**155**	**176**	**366**	**461**	**617**	**917**	**1444**	**1304**
(b) Cladocera												
Alona monacantha (Sars, 1901)	132	202	156	0	0	0	78	322	198	100	167	189
Bosmina longirostris (Müller, 1785)	109	76	82	32	45	32	87	171	172	234	347	266
Chydorus sphaericus (Muller, 1785)	121	252	156	123	156	161	78	87	44	31	111	76
Sida crystallina (Muller, 1875)	23	55	89	44	76	23	52	44	156	140	244	161
Total = 04	**385**	**585**	**483**	**199**	**277**	**216**	**295**	**624**	**570**	**505**	**869**	**692**
(c) Copepoda												
Cyclops scutifer (Sars, 1863)	134	23	51	78	45	56	23	27	35	176	134	127
C. viridis (Jurine, 1820)	121	11	22	0	0	0	23	46	22	197	144	123

TABLE 9.3 *(Continued)*

Seasons	Autumn			Winter			Spring			Summer		
Zooplankton	Sep.	Oct.	Nov.	Dec.	Jan.	Feb.	Mar.	Apr.	May	Jun.	Jul.	Aug.
Eucyclops macrurus (Sars, 1863)	100	74	44	0	0	0	78	61	65	234	322	259
Eucyclops sp.	154	123	51	187	98	89	34	52	44	145	272	280
Total = 04	**509**	**231**	**168**	**265**	**143**	**145**	**158**	**186**	**166**	**752**	**872**	**789**
(d) Ostracoda												
Cypris sp.	13	45	67	44	13	18	34	11	16	14	18	11
Cyclocypris sp.	45	59	11	16	45	43	67	44	50	12	54	14
Total = 02	**58**	**104**	**78**	**60**	**58**	**61**	**101**	**55**	**66**	**26**	**72**	**25**
Grand Total = 17	**2224**	**1929**	**1179**	**754**	**762**	**686**	**921**	**1355**	**1462**	**2147**	**3212**	**2810**

TABLE 9.4 Monthly Variations in the Population Density of Zooplankton (org./L) at Site II in Hokersar Wetland from Sep. 2012 to Aug. 2013

Seasons	Autumn				Winter			Spring			Summer	
Months	Sep.	Oct.	Nov.	Dec.	Jan.	Feb.	Mar.	Apr.	May	Jun.	Jul.	Aug.
(a) Rotifera												
Brachionus calyciflorus (Ehrenberg, 1838)	145	131	34	11	25	27	35	49	57	124	214	256
Filinia sp.	298	214	133	32	37	18	44	37	122	136	252	276
Keratella sp.	439	134	265	19	22	29	49	61	74	86	422	304
Keratella valga (Ehrenberg, 1834)	34	87	22	0	0	0	24	39	42	267	504	422
Lecane luna (Müller, 1776)	111	90	44	23	41	32	56	81	73	189	173	144
Squatinella sp.	123	'105	121	55	37	41	122	178	146	78	69	58
Total = 06	**1150**	**656**	**619**	**140**	**162**	**147**	**330**	**445**	**514**	**880**	**1634**	**1460**
(b) Cladocera												
Alona monacantha (Sars, 1901)	45	56	14	0	0	0	87	231	311	89	216	79
Bosmina longirostris (Müller, 1785)	43	32	14	11	22	0	35	46	53	134	158	178
Ceriodaphnia reticulata (Jurine, 1820)	52	40	56	0	0	0	45	56	67	89	96	74
Chydorus sphaericus (Muller, 1785)	145	121	167	141	174	100	46	78	84	71	89	122
Total = 04	**285**	**249**	**251**	**152**	**196**	**100**	**213**	**411**	**515**	**383**	**559**	**453**
(c) Copepoda												
Bryocamptus minutus (Claus, 1863)	11	33	67	98	189	178	33	43	22	67	91	82
Cyclops bicolor (Sars, 1863)	32	11	22	0	0	0	11	23	34	163	182	88
C. nanus (Sars, 1863)	14	45	67	22	68	32	0	0	0	123	172	67
C. scutifer (Sars, 1863)	11	23	52	122	87	98	11	23	31	72	81	57

TABLE 9.4 *(Continued)*

Seasons	Autumn				Winter			Spring		Summer		
Months	Sep.	Oct.	Nov.	Dec.	Jan.	Feb.	Mar.	Apr.	May	Jun.	Jul.	Aug.
C. vernalis (Fischer, 1853)	44	31	11	0	0	0	11	17	23	133	167	156
C. viridis (Jurine, 1820)	23	12	24	0	0	0	41	37	32	149	162	167
Eucyclops macrurus (Sars, 1863)	22	34	15	0	0	0	16	34	51	87	94	109
Eucyclops sp.	71	45	27	45	56	62	25	41	26	74	102	131
Total = 08	**228**	**234**	**285**	**287**	**400**	**370**	**148**	**218**	**219**	**868**	**1051**	**857**
(d) Ostracoda												
Cypris sp.	34	51	27	15	22	11	56	13	44	12	23	35
Cyclocypris sp.	19	13	18	43	19	23	31	23	31	16	18	22
Total = 02	**53**	**64**	**45**	**58**	**41**	**34**	**87**	**36**	**75**	**28**	**41**	**57**
Grand Total =20	**1727**	**1223**	**1200**	**673**	**891**	**758**	**797**	**1124**	**1356**	**2033**	**3119**	**2852**

TABLE 9.5 Monthly Variations in the Population Density of Zooplankton (org./L) at Site III in Hokersar Wetland from Sep. 2012 to Aug. 2013

Seasons	Autumn			Winter			Spring			Summer		
Months	Sep.	Oct.	Nov.	Dec.	Jan.	Feb.	Mar.	Apr.	May	Jun.	Jul.	Aug.
(a) Rotifera												
Brachionus calyciflorus (Ehrenberg, 1838)	145	83	44	12	9	19	32	41	53	91	142	189
Brachionus sp.	98	32	11	0	0	0	14	18	33	67	81	76
Filinia sp.	144	187	122	31	24	22	45	53	77	134	264	344
Keratella sp.	143	344	213	12	32	48	45	31	82	111	329	208
Lecane luna (Müller, 1776)	241	134	76	32	51	17	44	56	64	133	151	164
Lecene sp.	567	422	316	0	0	0	189	244	311	77	61	55
Platyias patulus (Muller, 1786)	231	143	33	0	0	81	74	98	122	72	189	210
Squatinella sp.	245	214	191	31	49	53	142	125	111	62	45	35
Trichocerca sp.	351	145	111	45	44	13	134	152	167	155	478	262
Total = 09	**2165**	**1704**	**1117**	**163**	**209**	**253**	**719**	**818**	**1020**	**902**	**1740**	**1543**
(b) Cladocera												
Alona monacantha (Sars, 1901)	33	41	32	0	0	0	98	145	249	94	109	81
Bosmina longirostris (Müller, 1785)	44	28	37	22	15	33	45	62	79	246	376	209
Ceriodaphnia reticulata (Jurine, 1820)	41	35	23	18	14	21	44	111	142	131	214	157
Chydorus sphaericus (Muller, 1785)	187	155	133	86	70	85	44	87	79	76	49	64
Sida crystallina (Muller, 1875)	46	37	41	37	24	46	82	155	109	133	151	176
Simocephalus vetulus (Müller, 1776)	22	45	41	0	0	0	67	73	61	158	191	218
Total = 06	**373**	**341**	**307**	**163**	**123**	**185**	**380**	**633**	**719**	**838**	**1090**	**905**

TABLE 9.5 *(Continued)*

Seasons	Autumn			Winter			Spring			Summer		
Months	Sep.	Oct.	Nov.	Dec.	Jan.	Feb.	Mar.	Apr.	May	Jun.	Jul.	Aug.
(c) Copepoda												
Bryocamptus minutus (Claus, 1863)	9	14	54	98	104	91	14	11	22	74	85	11
Cyclops bicolor (Sars, 1863)	78	22	18	0	0	0	32	14	36	198	216	189
C. nanus (Sars, 1863)	65	22	35	78	84	111	0	0	0	67	85	79
C. vernalis (Fischer, 1853)	54	67	22	0	0	0	36	42	46	183	215	224
C. viridis (Jurine, 1820)	0	22	39	0	0	0	11	0	27	55	111	96
Eucyclops macrurus (Sars, 1863)	46	35	22	0	0	0	12	11	52	87	132	128
Eucyclops sp.	0	0	27	66	96	89	38	42	64	72	64	52
Total = 07	252	182	217	242	284	291	143	120	247	736	908	779
(d) Ostracoda												
Cyclocypris sp.	43	51	35	41	56	14	56	14	34	61	19	22
Total = 01	**43**	**51**	**35**	**41**	**56**	**14**	**56**	**14**	**34**	**61**	**19**	**22**
Grand Total = 23	**2833**	**2278**	**1676**	**781**	**844**	**908**	**1298**	**1604**	**2044**	**2408**	**3585**	**3149**

TABLE 9.6   Monthly Variations in the Population Density of Zooplankton (org./L) at Site IV in Hokersar Wetland from Sep. 2012 to Aug. 2013

Seasons	Autumn			Winter			Spring			Summer		
Months	Sep.	Oct.	Nov.	Dec.	Jan.	Feb.	Mar.	Apr.	May	Jun.	Jul.	Aug.
(a) Rotifera												
Brachionus calyciflorus (Ehrenberg, 1838)	240	115	56	11	19	26	45	54	70	82	142	171
B. quadridentatus (Hermann, 1783)	241	132	111	34	54	32	44	63	81	132	174	139
Brachionus sp.	315	252	145	0	0	0	167	182	177	214	345	219
Cephalodella megalocephala (Glascott, 1893)	212	152	76	22	32	41	56	82	144	312	422	189
Filinia sp.	344	356	232	67	82	34	145	122	156	341	386	461
Keratella cochlearis (Gosse, 1851)	135	100	87	22	35	11	67	74	83	91	122	141
Keratella sp.	102	89	72	12	0	0	0	0	12	55	71	82
Keratella valga (Ehrenberg, 1834)	61	13	52	0	0	0	41	33	61	134	411	189
Lecane flexilis (Gosse, 1886)	147	86	12	0	0	0	23	45	67	123	241	251
Lecane luna (Müller, 1776)	35	19	12	0	0	0	0	0	0	31	55	19
Platyias patulus (Muller, 1786)	231	132	71	0	0	49	45	51	111	189	456	312
Squatinella sp.	112	178	131	24	19	36	89	142	175	42	51	53
Trichocerca sp.	11	13	17	19	33	22	19	44	63	69	57	64
Total = 13	**2186**	**1637**	**1074**	**211**	**274**	**251**	**741**	**892**	**1200**	**1815**	**2933**	**2290**
(b) Cladocera												
Alona monacantha (Sars, 1901)	67	81	12	0	0	0	123	176	193	78	143	198
Alona quadrangularis (Müller, 1776)	123	111	19	0	0	0	78	84	103	124	182	134
Bosmina longirostris (Müller, 1785)	32	42	15	0	0	0	34	42	56	144	176	186
Chydorus sphaericus (Muller, 1785)	22	33	24	17	12	13	39	43	22	51	44	62
Total = 04	**244**	**267**	**70**	**17**	**12**	**13**	**274**	**345**	**374**	**397**	**545**	**580**

TABLE 9.6 *(Continued)*

Seasons	Autumn			Winter			Spring			Summer		
Months	Sep.	Oct.	Nov.	Dec.	Jan.	Feb.	Mar.	Apr.	May	Jun.	Jul.	Aug.
(c) Copepoda												
Bryocamptus minutus (Claus, 1863)	0	0	11	67	82	74	13	21	11	34	51	22
Cyclops bicolor (Sars, 1863)	37	21	11	0	0	0	14	19	45	187	157	184
C. scutifer (Sars, 1863)	18	0	23	98	65	43	44	37	32	11	34	41
C. viridis (Jurine, 1820)	34	48	69	0	0	0	22	34	41	113	132	94
Eucyclops sp.	42	58	69	84	100	107	24	46	57	67	131	79
Total = 05	**131**	**127**	**183**	**249**	**247**	**224**	**117**	**157**	**186**	**412**	**505**	**420**
(d) Ostracoda												
Cypris sp.	34	67	44	52	16	31	45	56	51	42	37	47
Cyclocypris sp.	44	52	41	23	43	25	65	33	16	21	31	14
Total = 02	**78**	**119**	**85**	**75**	**59**	**56**	**110**	**89**	**67**	**63**	**68**	**61**
Grand Total = 24	**2639**	**2150**	**1412**	**727**	**782**	**754**	**1253**	**1483**	**1827**	**2538**	**3970**	**3211**

TABLE 9.7 Monthly Variations in the Population Density of Zooplankton (org./L) at Site V in Hokersar Wetland from Sep. 2012 to Aug. 2014

Seasons	Autumn				Winter			Spring			Summer	
Months	Sep.	Oct.	Nov.	Dec.	Jan.	Feb.	Mar.	Apr.	May	Jun.	Jul.	Aug.
(a) Rotifera												
Brachionus calyciflorus (Ehrenberg, 1838)	98	72	56	22	11	9	22	36	41	105	125	168
Cephalodella megalocephala (Glascott, 1893)	123	100	68	14	18	22	34	54	78	145	178	166
Filinia sp.	144	122	131	22	31	44	49	55	77	212	262	278
Keratella sp.	603	544	294	84	62	77	55	44	31	361	450	379
Lecane luna (Müller, 1776)	341	134	43	13	36	22	71	69	54	180	249	207
Lecene sp.	229	198	122	0	0	0	367	489	268	78	64	82
Platyias patulus (Muller, 1786)	244	113	56	0	0	0	61	74	45	139	216	251
Squatinella sp.	239	217	156	67	45	22	63	121	143	74	52	64
Total = 08	**2021**	**1500**	**926**	**222**	**203**	**196**	**722**	**942**	**737**	**1294**	**1596**	**1595**
(b) Cladocera												
Alona affinis (Leydig, 1860)	51	44	49	0	0	0	51	62	79	132	151	173
A. monacantha (Sars, 1901)	44	31	26	0	0	0	89	109	144	64	145	143
Bosmina longirostris (Müller, 1785)	22	37	11	0	0	0	38	41	69	89	132	170
Ceriodaphnia reticulata (Jurine, 1820)	18	43	22	11	21	24	41	52	62	122	146	182
Chydorus sphaericus (Muller, 1785)	167	141	98	145	122	100	64	55	74	146	123	72
Sida crystallina (Muller, 1875)	56	35	42	14	33	15	33	51	156	143	214	239
Simocephalus vetulus (Müller, 1776)	67	54	31	0	0	0	44	122	102	143	178	134
Total = 07	**425**	**385**	**279**	**170**	**176**	**139**	**360**	**492**	**686**	**839**	**1089**	**1113**

TABLE 9.7 *(Continued)*

Seasons	Autumn			Winter			Spring			Summer		
Months	Sep.	Oct.	Nov.	Dec.	Jan.	Feb.	Mar.	Apr.	May	Jun.	Jul.	Aug.
(c) Copepoda												
Bryocamptus minutus (Claus, 1863)	11	18	32	78	89	94	35	13	17	14	38	45
Cyclops bicolor (Sars, 1863)	47	44	22	0	0	0	12	32	41	174	168	141
C. nanus (Sars, 1863)	29	56	61	134	152	145	0	0	0	63	74	86
C. scutifer (Sars, 1863)	0	0	11	45	57	63	63	44	51	56	61	12
C. verralis (Fischer, 1853)	55	42	28	0	0	0	24	51	62	149	174	182
C. vicinus (Ulyanin, 1875)	55	44	68	89	122	120	53	22	46	19	41	56
C. viridis (Jurine, 1820)	19	56	74	67	154	111	32	41	37	156	163	144
Eucyclops macrurus (Sars, 1863)	67	16	9	0	0	0	11	16	44	85	123	145
Eucyclops sp.	43	47	56	63	74	84	21	27	37	62	71	84
Total = 09	**326**	**323**	**361**	**476**	**648**	**617**	**251**	**246**	**335**	**778**	**913**	**895**
Grand Total = 24	**2772**	**2208**	**1566**	**868**	**1084**	**1012**	**1595**	**1695**	**1790**	**2809**	**3598**	**3603**

TABLE 9.8 Monthly Variations in the Population Density of Zooplankton (org./L) at Site VI in Hokersar Wetland from Sep. 2012 to Aug. 2014

Seasons	Autumn			Winter			Spring			Summer		
Months	Sep.	Oct.	Nov.	Dec.	Jan.	Feb.	Mar.	Apr.	May	Jun.	Jul.	Aug.
(a) Rotifera												
Brachionus calyciflorus (Ehrenberg, 1838)	19	23	31	0	0	0	14	17	15	34	41	29
Keratella sp.	10	22	31	0	0	0	0	0	0	32	39	44
Platyias patulus (Muller, 1786)	44	13	23	0	0	0	61	74	45	61	74	51
Squatinella sp.	19	12	14	9	11	8	22	16	32	12	19	11
Trichocerca sp.	19	14	18	8	16	11	34	33	41	53	49	43
Total = 05	**111**	**84**	**117**	**17**	**27**	**19**	**131**	**140**	**133**	**192**	**222**	**178**
(b) Cladocera												
Alona monacantha (Sars, 1901)	14	13	0	0	0	0	14	52	36	12	19	13
Bosmina longirostris (Müller, 1785)	12	23	11	0	0	0	0	14	18	23	31	36
Ceriodaphnia reticulata (Jurine, 1820)	42	14	9	0	0	0	11	13	72	55	89	44
Chydorus sphaericus (Muller, 1785)	41	33	44	37	41	33	19	22	31	24	36	25
Simocephalus vetulus (Müller, 1776)	35	38	23	15	18	14	21	37	22	67	73	81
Total = 05	**144**	**121**	**87**	**52**	**59**	**47**	**65**	**138**	**179**	**181**	**248**	**199**
(c) Copepoda												
Cyclops bicolor (Sars, 1863)	51	34	19	0	0	0	19	23	41	126	163	144
C. nanus (Sars, 1863)	33	44	49	89	121	136	0	0	0	134	156	161
C. scutifer (Sars, 1863)	0	9	32	55	89	72	24	11	25	37	62	64
C. vernalis (Fischer, 1853)	44	22	11	0	0	0	35	46	76	122	146	157

TABLE 9.8 *(Continued)*

Seasons	Autumn			Winter			Spring			Summer		
Months	Sep.	Oct.	Nov.	Dec.	Jan.	Feb.	Mar.	Apr.	May	Jun.	Jul.	Aug.
C. viridis (Jurine, 1820)	67	45	53	0	0	0	27	11	31	93	81	72
Eucyclops sp.	34	31	73	45	61	53	35	19	45	67	84	78
Total = 06	229	185	237	189	271	261	140	110	218	579	692	676
(d) Ostracoda												
Cypris sp.	46	71	44	34	22	11	34	52	59	64	14	43
Cyclocypris sp.	51	43	62	57	82	32	54	44	49	51	59	41
Total = 02	97	114	106	91	104	43	88	96	108	115	73	84
Grand Total = 18	581	504	547	438	585	508	455	496	694	1067	1235	1137

TABLE 9.9   Seasonal Variations in the Population Density of Zooplankton (org./L) at Site I in Hokersar Wetland from Sep. 2012 to Aug. 2013

Zooplankton	Autumn	Winter	Spring	Summer
Rotifera	973.0	167.3	481.3	1221.7
Cladocera	484.3	230.7	496.3	688.7
Copepoda	302.7	184.3	170.0	804.3
Ostracoda	80.0	59.7	74.0	41.0
Grand Total	**1840.0**	**642.0**	**1221.7**	2755.7

TABLE 9.10   Seasonal Variations in the Population Density of Zooplankton (org./L) at Site II in Hokersar Wetland from Sep. 2012 to Aug. 2013

Zooplankton	Autumn	Winter	Spring	Summer
Rotifera	849.0	149.7	429.7	1324.7
Cladocera	261.7	149.3	379.7	465.0
Copepoda	249.0	352.3	195.0	925.3
Ostracoda	54.0	44.3	66.0	42.0
Grand Total	**1413.7**	**695.7**	**1070.3**	**2757.0**

TABLE 9.11   Seasonal Variations in the Population Density of Zooplankton (org./L) at Site III in Hokersar Wetland from Sep. 2012 to Aug. 2013

Zooplankton	Autumn	Winter	Spring	Summer
Rotifera	1662.0	208.3	852.3	1395.0
Cladocera	340.3	157.0	577.3	944.3
Copepoda	217.0	272.3	170.0	807.7
Ostracoda	43.0	37.0	34.7	34.0
Grand Total	2262.3	674.7	1634.3	3181.0

TABLE 9.12   Seasonal Variations in the Population Density of Zooplankton (org./L) at Site IV in Hokersar Wetland from Sep. 2012 to Aug. 2013

Zooplankton	Autumn	Winter	Spring	Summer
Rotifera	1632.3	245.3	944.3	2346.0
Cladocera	193.7	14.0	331.0	507.3
Copepoda	147.0	240.0	153.3	445.7
Ostracoda	94.0	63.3	88.7	64.0
Grand Total	2067.0	562.7	1517.3	3363.0

TABLE 9.13 Seasonal Variations in the Population Density of Zooplankton (org./L) at Site V in Hokersar Wetland from Sep. 2012 to Aug. 2013

Zooplankton	Autumn	Winter	Spring	Summer
Rotifera	1482.3	207.0	800.3	1495.0
Cladocera	363.0	161.7	512.7	1013.7
Copepoda	336.7	469.7	277.3	862.0
Grand Total	2182.0	838.3	1590.3	3370.7

TABLE 9.14 Seasonal Variations in the Population Density of Zooplankton (org./L) at Site VI in Hokersar Wetland from Sep. 2012 to Aug. 2013

Zooplankton	Autumn	Winter	Spring	Summer
Rotifera	104.0	21.0	134.7	197.3
Cladocera	117.3	52.7	127.3	209.3
Copepoda	217.0	240.3	156.0	649.0
Ostracoda	105.7	79.3	97.3	90.7
Grand Total	544.0	393.3	515.3	1146.3

9.4.5 SEASONAL SUCCESSION OF ZOOPLANKTON (RELATED TO ZOOPLANKTON DENSITY)

Seasonal succession pattern of plankton community provides us and design regarding to the ascendancy pattern of various groups in various seasons of the year based on percentage density values). During the whole investigation time frame no particular seasonal succession trend was observed in the wetland.

During the examination period (Sep. 2012 to Aug. 2013), the harvest time of these plankters in the Hokersar being ruled by wheal animalcules contributing 58.74%. Genuinely equivalent extents of 17.91% and 17.68% separately contributed by water fleas and Copepoda. The extent of Ostracoda only contributes 5.67% of the whole zooplankton diversity (Figure 9.2). Accordingly, the pattern of dominance in zooplankton community rotifers> cladocerans ≥ copepods>ostracods.

In the colder months (winter season) that was typify with Copepoda contributing 46.58%, trail in declining number by Rotifera, Cladoceran, and Ostracoda contributing 25.35%, 19.30%, and 8.76%, respectively (Figure 9.3). The order of domination in the winter season was: copepods> rotifers>cladocerans>ostracods.

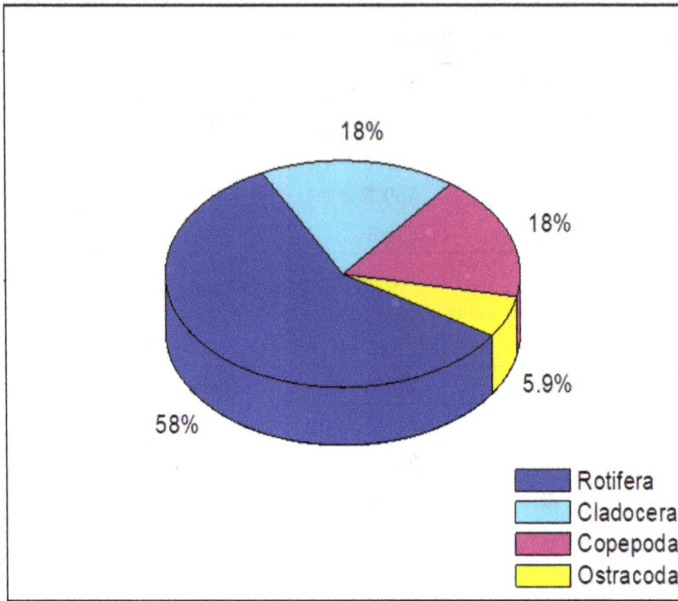

FIGURE 9.2   Relative density of zooplankton in autumn during Sep. 2012 to Aug. 2013 in Hokersar wetland.

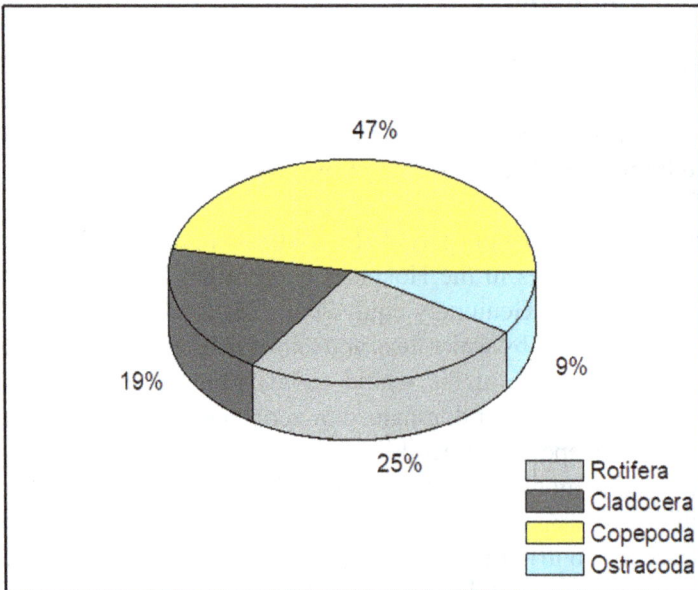

FIGURE 9.3   Relative density of zooplankton in winter during Sep. 2012 to Aug. 2013 in Hokersar wetland.

An increasing population trend was noticed for Rotifera and Cladocera from that of winter, making 45% and 31.% correspondingly, which show the way to the significant lessening the contribution of Copepoda (16.73%) and somewhat that of Ostracoda (6.59%). Along the succession, the occurrence during spring was: rotifers>cladocerans >copepods>ostracods (Figure 9.4).

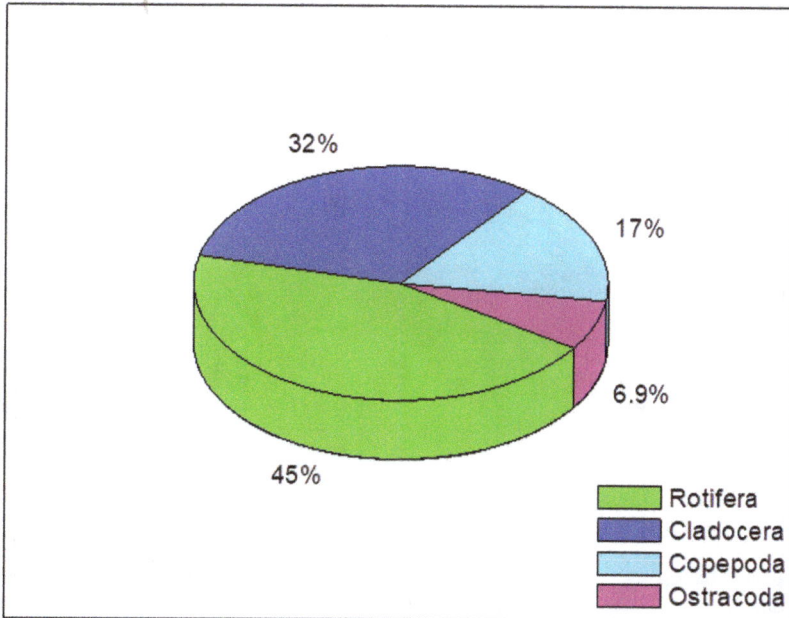

FIGURE 9.4 Relative density of zooplankton in spring during Sep. 2012 to Aug. 2013 in Hokersar wetland.

In the summer season, t rotifers were most abundant in the plankton community, followed by copepods respectively contributing (44.59%) and (30.60%), which as a result, lead to the declining population of water fleas with 22.49% and ostracods with 2.32%. So the dominance order during this summer was rotifers (Figure 9.5).

9.5 DISCUSSION

In aquatic environments, there are dynamic cycles and procedures working on temporal as well as on spatial scales. The dynamic nature with regular disorderness, operate to assign the biotic communities into different spatial and temporal mosaics (Ruokolainen et al., 2009; Shah and Pandit, 2013a). In addition to these, resource accessibility in the encompassing medium challengingly impacts the zooplankton community and are emphatically related (Beisner, 2001). Further, zooplankton diversity and composition give us an idea regarding water quality and the nature of the water body (Okayi et al., 2001).

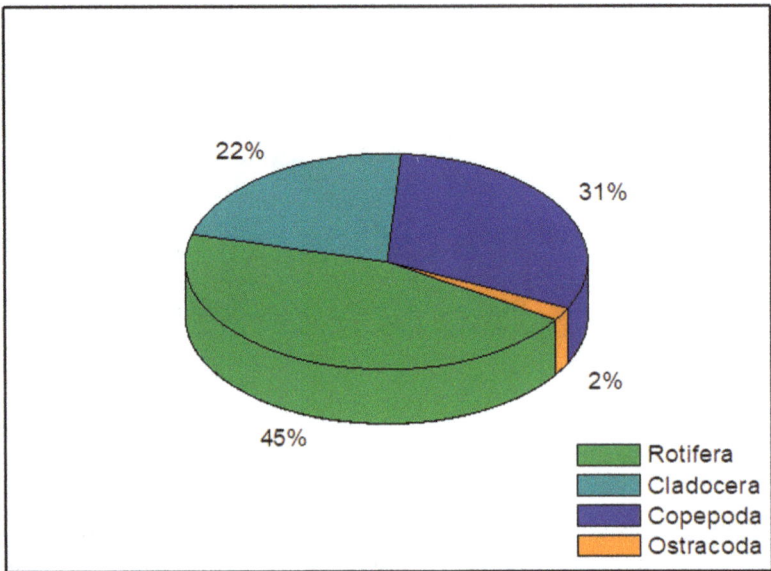

FIGURE 9.5   Relative density of zooplankton in summer during Sep. 2012 to Aug. 2013 in Hokersar wetland.

9.5.1   SPECIES COMPOSITION OF ZOOPLANKTON

During the investigation period, a sum of 34 taxa, including 14 wheel animalcules, 9 water fleas, and copepods each, and just 02 Ostracod taxa were reported (Table 9.2). By and large, rotifers were the most abundant phylum pursued by the cladocerans, copepods, and ostracods. In the entire course of the investigation, the abundance of Rotifera and Cladocera in these aquatic ascribed by numerous researchers to the eutrophication of these environs (Guevare et al., 2009; Tasevska et al., 2010). Even more, during our investigation, there is an abundance of Rotifera and Cladocera at those study points s which get more pollution/contamination stacks sewage, obviously showing that rotifers and cladocerans are highly tolerable to pollution stress

(Gannon and Stemberger, 1978). As per the reports of Balkhi et al. (1987), the abundance of these wheel animalcules is principally the consequence of their very smaller size, a good way to stay away from fish predation and predation by other rapacious water animals. In any case, Pandit (1980) connected the higher abundance of these wheel animalcules to food availability in the form of protozoans in the wetlands and lakes of Kashmir Himalaya. The higher amount of protozoan food in these systems was, thus, identified with the large percentage of organic matter (Pandit, 1980; Pandit and Kaul, 1982).

9.5.2 *SEASONAL VARIATIONS IN ZOOPLANKTON ARCHITECTURE*

The abundance and density of planktons are restricted through a variety of features (including both physical and biological), and the majority of them are changeable, and few of these features are reliant on one another in the normal circumstances of aquatic ecosystems. Temperature one of the abiotic factor has a leading factor in shaping the plankton abundance and diversity (Pandit, 1999). Among the biotic components available food (Pandit and Kaul, 1982; Hairston, 1987) predation by the other trophic levels (Pandit, 1980; Pandit and Kaul, 1982; Gulati, 1990; Santer and Lampert, 1995), intra, and inter specific contest (Lampert and Rothhaupt, 1991; Dohet and Hoffmann, 1995) and the type of vegetation (Moya and Duggen, 2011) are some of the important variables that affect the community architecture, abundance, and population density of various plankters in shallow lakes of freshwater type (Dumont, 1977).

At the time of investigation, there were more or less evident seasonal changes in the plankton community in the Hokersar wetland. It is noteworthy to note that some of the recorded taxa did not display any seasonal variation, so these species can be categorized to exhibit long variation of ecological/ ecological situations (Pandit; 1980; Berzins and Pejiler, 1987, 1989).

On analyzing the data regarding the population density of these plankters, their exhibits varied growth pattern in the study. Rotifers generally were abundant at maximum exiting sites throughout the investigation period. The dominance pattern of wheel animalcules can be assigned to a number of reasons: (i) have a high power of reproducing their young ones/larvae and have a comparably very short life cycle (Lansac-Toha et al., 1997; Estves, 1998), (ii) have in between trophic levels in aquatic food chains and webs (Pandit, 1980; Sanders and Wicham, 1993), (iii) have the very wide thriving ability in nutrient-rich waters so can be perceived as pollution indicators (Pandit, 1980, 2002), and (iv) have the dominating ability in the plankton community in summer season (Pandit, 1980; Hansson et al., 1998). It is because of these facts wheel animalcules were characterized by a different number of taxa at various sites and demonstrate a high summer growth peak in the wetland under study.

As reported by Rutner-Kolisko (1974), the abundance of rotifers in the zooplankton community doesn't rely upon the living food availability in these aquatic ecosystems but rather attributed it to the detritus food availability, a fact also opined by Pandit (1980).

Extensive growth and development of wheel animalcules were reported in the study area, which can be assigned to (i) availability of food sources, (ii) low population of fishes leading to less predation prey interaction (iii) very high reproduction rate of these plankters (Gilbert, 1988). Herzing (1987) is of the view that the temperature of the water is one of the plausible reasons for the constrained occurrence of wheel animalcules in temperate regions of the globe. Further, as indicated by Berzens and Pejler (1989), these creatures stand broad changes in temperature, and along these lines show peak growth in terms of abundance in the summer season (Ji-Gaohua et al., 2013).

The distinguishing aspect of shallow lakes and wetlands is the plenitude of aquatic plants (macrophytes) and having no thermal stratification (Chen et al., 2010). Besides above, the high population growth of zooplankton can also be ascribed to the lush development of macrophytes in summer act as microhabitats for the zooplankton, protect them from prey predation interaction in the Hokersar wetland (Sommer et al., 1986; Hann and Zrum, 1997; Moya and Duggen, 2011).

Further, the plenitude of small-sized planktons in summer season over hefty sized planktons as was reported from the wetland is a clear warning sign of nutrient enrichment (eutrophication) in Hokersar, as opined by the early workers of Jarnrnefelt (1956), Pandit (1980, 1999, 2002) and Shah and Pandit (2014). Furthermore, Moore et al. (1996) are of the view that large-sized taxa exhibit constricted tolerance for temperature as compared to the small-bodied plankton, so easily substituted by smaller sized taxa at maximum temperature and vice-versa at minimum temperature (Feniova et al., 2014). This whole supports our observation, there is significant confirmation that increased temperature supports the abundance and density of smaller planktons and restrains the development of larger species of zooplankton (Feniova et al., 2013, 2014).

During winter lower density in terms of the population were evinced were in accordance to the precious works of Pennak (1978), Pandit (1980),

Yousuf et al. (1986), Hann and Zrum (1997), Pandit (1998), and Shah and Pandit (2013a–c). In addition to the aforesaid researchers, our observation got additional hold from the annotations of by Gyllström and Hansson (2004) who was of the viewpoint that the lower water temperature and decrease in photoperiod results in the diapause period during the harsh environmental conditions (Hann and Zrum, 1997). Martinez et al. (2013) further observed that most of the plankton undergoes diapause or resting stage eggs that link the time of unfavorable ecological circumstances. Moreover, during colder periods, there is meager or even no growth of aquatic plants in the wetland under examination that also result in diminishing the zooplankton growth as the apt biological niches are on the whole absent (Tessier et al., 2004; Thomaz et al., 2008; Moya and Duggen, 2011).

During the entire course of study, ostracods were the least predominant group, characterized by just two taxa in the wetland. The infrequent occurrence of these taxa might be because of their benthic dwelling nature (Shah and Pandit, 2013a).

9.6 CONCLUSIONS

1. During the investigation period zooplankton abundance, diversity, and density varied between the sites and seasons of the wetland. Among the observed taxa, some of the species were present throughout the year, reflecting that these plankters have a very wide tolerance for the changing abiotic as well as biotic conditions.

2. It was observed that wheel animalcules were dominant over water flea and copepods, indicating that Hokersar wetland understudy is moving rapidly towards cultural eutrophication. Still further, the former two groups (rotifers and cladocerans) show dominant nature at those study sites, which were more prone to the pollution, reflecting that rotifers and cladocerans show more tolerance towards the pollution load. Even more, it was also noticed that calanoids were made a small contribution in the zooplankton community in comparison to copepods indicate that the Hokersar is under remarkable stress due to various anthropogenic actions prevailing in the immediate catchment.

3. During the investigation, it is believed that the diapause in Copepoda isn't confined to a specific period of the year. Further, it is opined that some biological factors and some abiotic clues might be conceivable

explanations behind this unpredictable behavior in the lakes and wetlands of the Kashmir region.

4. The dominant nature of small-bodied plankton in the summer season over the larger planktons is a reasonable sign of cultural eutrophication of the wetland.

KEYWORDS

- **eutrophication**
- **Hokersar wetland**
- **seasonal variations**
- **zooplankton**
- **zooplankton architecture**
- **zooplankton community**

REFERENCES

Adoni, A. D., (1985). *Work Book of Limnology*. Pritibha Publication, Sagar, M.P.; India.

Ahmad, S. S., Reshi, Z. A., Shah, M. A., Rashid, I., Ara, R., & Andrabi, S. M. A., (2014). Phytoremediation potential of *Phragmites australis* in Hokersar Wetland: A Ramsar Site of Kashmir Himalaya. *International Journal of Phytoremediation, 16*(12), 1183–1191.

Balkhi, M. H., Yousuf, A. R., & Qadri, M. Y., (1987). Hydrobiology of Anchar Lake. *J. Comp. Phys. Ecol., 12*, 131–139.

Battish, S. K., (1992). *Freshwater Zooplankton of India*. Oxford and IBH Publishing Co.; New Delhi.

Bayley, S. E., Zoltek, Jr. J., Hermann, A. J., Dolan, T. J., & Tortora, L., (1991). Experimental manipulation of nutrients and water in a freshwater marsh: Effects on biomass, decomposition, and nutrient accumulation. *Limnol. and Oceanogr., 30*, 500–512.

Beisner, B. E., (2001). Plankton community structure in fluctuating environments and the role of productivity. *Oikos, 95*, 496–510.

Berzins, B., & Pejler, B., (1987). Rotifer occurrence in relation to pH. *Hydrobiologia, 147*, 107–116.

Berzins, B., & Pejler, B., (1989). Rotifer occurrence and trophic degree. *Hydrobiologia, 182*, 171–180.

Boxshall, G. A., & Defaye, D., (2008). Global diversity of copepods (Crustacea: Copepoda) in freshwater. *Hydrobiologia, 595*, 195–207.

Chen, M. R., Kâ, S., & Hwang, J. S., (2010). Diet of the copepod *Calanus sinicus* Brodsky, 1962 (Copepoda, Calanoida, Calanidae) in northern coastal waters of Taiwan during the northeast monsoon period. *Crustaceana, 83*, 851–864.

Dejen, E., Vijverberg, J., Nagelkerke, L., & Sibbing, F., (2004). Temporal and spatial distribution microcrustacean zooplankton in relation to turbidity and other environmental factors in large tropical lake (Lake Tana, Ethiopia). *Hydrobiologia, 513*, 39–49.

Dohet, A., & Hoffmann, L., (1995). Seasonal succession and spatial distribution of the zooplankton community in the reservoir Esch-sur-Sure (Luxembourg). *Belg. J. Zool., 125*, 109–123.

Dumont, H. J., (1977). Biotic factors in the population of dynamics of rotifers. *Arch. Hydrobiol. Ergenbn., 8*, 98–122.

Edmondson, W. T., (1992). *Freshwater Biology* (2nd edn., p. 1248). John Wiley and Sons Inc.

Elder, J. F., (1985). Nitrogen and phosphorus speciation and flux in a large Florida river wetland system. *Water Resources Research, 21*(5), 724–732.

Esteves, F. A., (1998). *Fundamentos De Limnologia* (2nd ed.). Interciencia, Rio de Janeiro, Brazil.

Feniova, I. U., Razlutskij, V. I., Palash, A. L., Tunowsky, J., Sysova, E. A., & Dzialowski, A. R., (2014). Cladoceran community structure in three meso-eutrophic polish lakes with varying thermal regimes. *Limnetica, 33*(1), 13–30.

Feniova, I., Palash, A., Razlutskij, V., & Dzialowski, A., (2013). Effects of temperature and resource abundance on small- and large-bodied cladocerans: Community stability and species replacement. *Open Journal of Ecology, 3*, 164–171.

Fritsch, M., Emonds, O. B., & Richter, S., (2013). Unraveling the origin of Cladocera by identifying heterochrony in the developmental sequences of Branchiopoda. *Frontiers in Zoology*, 10–35. doi: 10.1186/1742-9994-10-35.

Gannon, J. E., & Stemberger, R. S., (1978). Zooplankton (especially crustaceans and rotifers) as indicators of water quality. *Trans. Amer. Micros. Soc., 97*, 16–35.

Gilbert, J. J., (1988). Suppression of rotifer populations by *Daphnia*: A review of the evidence, the mechanisms, and the effects of zooplankton community structure. *Limnol. Oceanogr., 33*, 1286–1303.

Guevara, G., Lozano, P., Reinoso, G., & Villa, F., (2009). Horizontal and seasonal patterns of tropical zooplankton from the eutrophic Prado Reservoir (Colombia). *Limnologica, 39*, 128–139.

Güher, H., (2000). A faunistic study on the freshwater Cladocera (Crustacea) species in Turkish Thrace (Edirne, Tekirdağ, K klareli). *Turkish Journal of Zoology, 24*, 237–243.

Gulati, R. D., (1990). Structural and grazing responses of zooplankton community to bioma-nipulation of some Dutch water bodies. *Hydrobiologia, 200, 201*, 99–118. (http://dx.doi.org/10.1007/BF02530332).

Gyllstrom, M., & Hansson, L. A., (2004). Dormancy in freshwater zooplankton: Induction, termination and importance of benthic-pelagic coupling. *Aquat. Sci., 66*(1), 274–295.

Hairston, N. G., (1987). Diapause as a predator-avoidance adaptation. In: Kerfoot, W. C., & Sih, A., (eds.), *Predation: Direct and Indirect Impacts on Aquatic Communities* (pp. 281–290). University Press of New England, London.

Hann, B. J., & Zrum, L., (1997). Littoral micro-crustaceans in a prairie coastal wetland: Seasonal abundance and community structure. *Hydrobiologia, 357*, 37–52.

Hansson, L. A., Annadotter, H., Bergman, E., Hamrin, S. F., Jeppesen, E., Kairesalo, T., Luokkanen, E., Nilsson, P. A., Søndergaard, M., & Strand, J., (1998). Biomanipulation as an application of food-chain theory: Constraints, synthesis, and recommendations for temperate lakes. *Ecosystems, 1*, 558–557.

Herzing, A., (1987). The analysis of planktonic rotifer population: A plea for long-term investigation. *Hydrobiologia, 147*, 163–180.

Huys, R., & Boxshall, G. A., (1991). *Copepoda Evolution* (p. 486). Guildford: Unwin Brothers Ltd.

Hwang, J. S., Kumar, R., Dahms, H. U., Tseng, L. C., & Chen, Q. C., (2010). Inter-annual, seasonal, and diurnal variation in vertical and horizontal distribution patterns of six *Oithona* sp. (Copepoda: Cyclopoida) in the South China Sea. *Zool. Stud., 49,* 220–229.

Hyman, L. H., (1951). *The Invertebrates: Acanthocephala, Aschelminthes and Entoprocta.* McGraw-Hill, New York.

Information Sheet on Ramsar Sites, (2004). http://archive.ramsar.org/cda/en/ramsar-news-archives-2004-ramsar-bulletin-22702/main/ramsar/1-26-45-54%5E22702_4000_0__2004 (accessed on 26 October 2020).

Järnefelt, H., (1956). Zur limnologie einiger gewässer finnlands XVI. Mit besonderer beriicksichtigung des planktons. *Ann. Zool. Soc. Zool. Bot. Fennicae, 17*(1), 1–201.

Ji-Gaohua, W. X., & Wang, L., (2013). Planktonic rotifers in a subtropical shallow lake: Succession, relationship to environmental factors, and use as bioindicators. *The Scientific World Journal.* Article ID: 702942, (http://dx.doi.org/10.1155/2013/702942).

Joshi, P. K., Humayun, R., & Roy, P. S., (2002). Landscape dynamics of hokersar wetland, Jammu and Kashmir-An application of geospatial approach. *J. of the Indian Society of Remote Sensing, 30*(1/2). 2–5.

Kâ, S., & Hwang, J. S., (2011). Mesozooplankton distribution and composition on the northeastern coast of Taiwan during autumn: Effects of the Kuroshio Current and hydrothermal events. *Zool. Stud., 50,* 155–163.

Kemp, G. P., & Day, J. W., (1984). In: Ewel, K. C., &. Odums, H. T., (eds.), *Nutrient Dynamics in a Louisiana Swamp Receiving Agricultural Runoff, Cypress Swamps.* Gainesville, University of Florida Press.

Khan, M. A., (1987). Observations on zooplankton composition, abundance, and periodicity in two floodplain lakes of Kashmir Himalayan valley. *Acta Hydrochimica et Hydrobiologica, 15,* 174–176.

Koste, W., (1978). *Rotatoria* (p. 673) Borntraeger, Berlin.

Lampert, W., & Rothhaupt, K. O., (1991). Alternating dynamics of rotifers and *Daphnia magna* in a shallow lake. *Arch. Hydrobiol., 120,* 447–456.

Lansac-Tôha, F. A., Bonecker, C. C., Velho, L. F. M., & Lima, A. F., (1997). Comunidade zooplanctônica. In: Vazzoler, A. M., Agostinho, A. A., & Hahn, N. S., (eds.), *A Planície de Inundação do Alto rio Paraná: Aspectos Físicos, Químicos, Biológicos e Socioeconômicos* (pp. 117–155). Editora da Universidade Estadual de Maringá, Maringá.

Makarewicz, J. C., & Likens, G. E., (1979). Structure and function of the zooplankton community of Mirror Lake, New Hampshire. *Ecol. Monogr., 49,* 109–127.

Martínez, M., Espinosa, N., & Calliari, D., (2013). Incidence of dead copepods and factors associated with non-predatory mortality in the Río de la Plata estuary. *Journal of Plankton Research, 36*(1), 265–270.

McManus, M. A., & Woodson, C. B., (2012). Plankton distributions and ocean dispersal. *J. Exp. Biol., 215,* 1008–1016. (doi: 10.1242/jeb.059014).

Millennium Ecosystem Assessment (MEA), (2005). *Ecosystems and Human Wellbeing: Wetlands and Water Synthesis* (p. 80). World Resources Institute, Washington.

Moore, M. V., Folt, C. F., & Stemberger, R. S., (1996). Consequences of elevated temperatures for zooplankton assemblages in temperate lakes. *Arch. Hydrobiol., 135,* 289–319.

Moya-Paloma, L., & Duggen, I. C., (2011). Macrophyte architecture affects the abundance and diversity of littoral microfauna. *Aquatic Ecology, 45*(2), 279–287.

Okayi, R. G., Jeje, C. Y., & Fagade, F. O., (2001). Seasonal patterns in the zooplankton community of River Benue (Makurdi), Nigeria. *African Journal of Environmental Studies*, *2*(1), 9–19.

Pal, S., & Akoma, O. C., (2009). Water scarcity in wetland area within Kandi Block of West Bengal: A hydro-ecological assessment. *Ethiop. J. Environ. Stud. Manag.*, *2*(3), 1–12.

Pandit, A. K., & Kaul, V., (1982). Trophic structure of some typical wetlands. In: Gopal, B., Turner, R. E., Wetzel, R. G., & Whigham, D. F., (eds.), *Wetlands-Ecology and Management*, *Part II.* (pp. 55–82). Nat. Inst. Ecol. & Int. Sci. Publi, Jaipur, India.

Pandit, A. K., & Kumar, R., (2006). Comparative studies on ecology of Hokersar wetland, Kashmir: Present and past. *Journal of Himalayan Ecology and Sustainable Development*, *1*, 73–81.

Pandit, A. K., (1980). *Biotic Factor and Food Chain Structure in Some Typical Wetlands of Kashmir.* PhD thesis, University of Kashmir, Srinagar-190006, J and K, India.

Pandit, A. K., (1998). Plankton dynamics in freshwater wetlands of Kashmir. In: Mishra, K. D., (ed.), *Ecology of Polluted Waters and Toxicology* (pp. 22–68). Technoscience Publications, Jaipur, India.

Pandit, A. K., (1999). *Freshwater Ecosystems of the Himalayas.* Parthenon Publishing, New York, London.

Pandit, A. K., (2002). Plankton as indicators of trophic status of wetlands. In: Arvind, K., (ed.), *Ecology and Ethology of Aquatic Biota* (pp. 341–360). Daya Publishing House, New Delhi-110002.

Pandit, A. K., (2008). Biodiversity of wetlands of Kashmir Himalaya. *Proc. Nat. Acad. Sci.; India B (Pt. Spl. Issue)*, *78*, 29–51.

Pennak, R. W., (1978). *Freshwater Invertebrates of the United States of America.* Wiley Inter-science Pub.; N. Y.

Peterjohn, W. T., & Correll, D. L., (1984). Nutrient dynamics in an agricultural watershed: Observations on the role of a riparian forest. *Ecology*, *65*(5), 1466–1475.

Pinto-Coelho, R., Pinel-Alloul, B., Méthot, G., & Havens, K. E., (2005). Crustacean zooplankton in lakes and reservoirs of temperate and tropical regions: Variation with trophic status. *Can. J. Fish. Aquat. Sci.*, *62*, 348–361.

Rai, D. N., & Munshi, J. D., (1980). Ecological characteristics of Chaurs of North Bihar. *Wetlands Ecology Management*, *2*, 88–95.

Rashid, I., & Romshoo, S. A., (2012). Impact of anthropogenic activities on water quality of Lidder River in Kashmir Himalayas. *Environmental Monitoring and Assessment.* doi: 10.1007/s10661-012-2898-0.

Reid, J. W., & Williamson, C. E., (2010). Copepoda. In: Thorpe, J. H., & Covich A. P., (eds.), *Ecology and Classification of North American Freshwater Invertebrates* (pp. 829–899). Academic Press, New York.

Ricci, C., & Melone, G., (2000). Key to the identification of the genera of bdelloid rotifers. *Hydrobiologia*, *418*, 73–80.

Ruokolainen, L., Lindén, A., Kaitala, V., & Fowler, M. S., (2009). Ecological and evolutionary dynamics under colored environmental variation. *Trends in Ecology and Evolution*, *24*, 555–563.

Ruttner-Kolisko, A., (1974). *Planktonic Rotifers: Biology and Taxonomy* (Vol. 26, pp. 1–146). Die Binnengewässer (Suppl.).

Sanders, R. W., & Wickham, S. A., (1993). *Planktonic protists* and metazoa: Predation, food quality and population control. *Marine Microbial Food Webs*, *7*, 197–223.

Santer, B., & Lampert, W., (1995). Summer diapause in cyclopoid copepods: Adaptive response to a food bottleneck? *Journal of Animal Ecol., 64*, 600–613.

Shah, J. A., & Pandit, A. K., (2013a). Relation between physico-chemical limnology and crustacean community in Wular Lake of Kashmir Himalaya. *Pakistan Journal of Biological Science, 16*(19), 976–983.

Shah, J. A., & Pandit, A. K., (2013b). Diversity and abundance of cladocerans zooplankton in Wular Lake, Kashmir Himalaya. *Research Journal of Environmental and Earth Sciences, 5*(7), 410–417.

Shah, J. A., & Pandit, A. K., (2013c). Some crustacean zooplankton of Wular Lake in Kashmir Himalaya. *African Journal of Environmental Science and Technology, 7*(5), 329–335.

Shah, J. A., & Pandit, A. K., (2014). Taxonomic survey of crustacean zooplankton in Wular Lake of Kashmir Himalaya. *Journal of Evolutionary Biology Research, 6*(1), 1–4.

Shah, J. A., Pandit, A. K., & Shah, G. M., (2013). Distribution, diversity and abundance of copepod zooplankton of Wular Lake, Kashmir Himalaya. *Journal of Ecology and Natural Environment, 5*(2), 24.

Silva, W. M., (2008). Diversity and distribution of the free-living freshwater cyclopoida (Copepoda: Crustacea) in the neotropics. *Braz. J. Biol., 68*(4), 1099–1106.

Sládecek, V., (1983). Rotifers as indicators of water quality. *Hydrobiologia, 100*, 169–201.

Sommer, U., Gliwicz, Z. M., Lampert, W., & Duncan, A., (1986). The PEG-model of seasonal succession of planktonic events in fresh waters. *Archiv. Für. Hydrobiologie, 106*, 433–471.

Tasevska, O., Kostoski, G., & Guseska, D., (2010). Rotifers based assessment of the lake Dojran water quality. *Ohrid. The Republic of Macedonia, 25*, 1–8.

Tessier, C., Cattaneo, A., Pinel-Alloul, B., Galanti, G., & Morabito, G., (2004). Biomass, composition, and size structure of invertebrate communities associated to different types of aquatic vegetation during summer in Lago di Candia (Italy). *J. Limnol., 63*, 190–198. (doi: 10.4081/jlimnol. 2004.190).

Thomaz, S. M., Dibble, E., Evangelista, L. R., Higuti, J., & Bini, L. M., (2008). Influence of aquatic macrophyte habitat complexity on invertebrate abundance and richness in tropical lagoons. *Freshwater Biol., 53*, 358–367.

Urabe, J., Kyle, M., Makino, W., et al., (2002). Reduced light increases herbivore production due to stoichiometric effects of light: Nutrient balance. *Ecology, 83*, 619–627.

Wallace, R. L., & Snell, T. W., (1991). *Ecology and Classification of North American Freshwater Invertebrates* (pp. 187–248). *Rotifera*. Invertebrates. Academic Press, San Diego.

Wallace, R. L., & Snell, T. W., (2010). *Rotifera*. In: Thorp, J. H., & Covich, A. P., (eds.), *Ecology and Classification of Freshwater Invertebrates* (pp. 173–235). Elsevier, Oxford.

Welch, P. S., (1952). *Limnology* (2nd edn., p. 538). Mc Graw Hill Book Co. New York.

Wu, C. H., Dahms, H. U., Buskey, E. J., Strickler, J. R., & Hwang, J. S., (2010). Behavioral interactions of *Temora turbinata* with potential ciliate prey. *Zool. Stud., 49*, 157–168.

Yousuf, A. R., Balkhi, M. H., & Qadri, M. Y., (1986). Limnological features of a forest lake of Kashmir. *J. Zool. Soc. India, 38*, 29–42.

CHAPTER 10

Mycoremediation of Pollutants in Aquatic Environs

REZWANA ASSAD, IFLAH RAFIQ, IQRA BASHIR, IRSHAD AHMAD SOFI,
ZAFAR AHMAD RESHI, and IRFAN RASHID

*Department of Botany, University of Kashmir, Srinagar–190006,
Jammu and Kashmir, India, E-mail: rezumir@gmail.com (R. Assad)*

ABSTRACT

Aquatic pollution is one of the grim ecological threats that the world faces, with the maintenance of appropriate water quality being a major challenge nowadays. Pollution of aquatic ecosystems has adverse impacts on the environment, public health, and economy. In view of that, there is an imperative obligation to safeguard the health of our aquatic ecosystems, and restoration of polluted aquatic environs has attracted worldwide attention. This review provides an overview of different types of aquatic pollutants, their consequences, and mitigation of these pollutants through mycoremediation. Emphasis is laid on different fungal species with their role in mycoremediation of chemical, suspended matter, microbiological, nutrient, groundwater, and oxygen depleting aquatic pollutants into environmentally less detrimental products via diverse mechanisms. Fungi execute a key function in environment cleanup through remediation of aquatic pollutants owing to their diverse metabolic capability comprising well-known fungal enzymes viz., catalases (CAT), cellulases, chitinases, cytochrome P450 monooxygenases, laccases, ligninase, lignocellulases, pectinases, peroxidases, oxidases, and xylanases. Further, this chapter provides comprehensive information about the mycoremediation potential of vast fungal diversity and can serve as a baseline for the selection and use of fungi either independently or in a consortium for forthcoming mycoremediation projects of aquatic ecosystems.

10.1   INTRODUCTION

In the present industrialized world, aquatic pollution is one of the grim ecological threats that the world face nowadays. It creates severe challenges globally vis-a-vis sustainability of our blue planet 'Earth.' Aquatic pollution is caused by several anthropogenic activities like ever-growing population pressure, rapid industrialization, increasing urbanization, modern unsafe agricultural practices, eutrophication, atmospheric deposition, underground storage leakages, oil pollution, nuclear waste, industrial wastewater, sewage pollution and marine dumping in aquatic environs (Owa, 2014; Water Pollution Guide, 2019) and it adversely affects aquatic flora and fauna. There is an imperative obligation to safeguard the health of our aquatic ecosystems, and restoration of polluted aquatic environs has attracted worldwide attention.

10.2   TYPES, CONSEQUENCES, AND PREVENTION OF AQUATIC POLLUTION

Following are the types of water pollution based on the type of pollutant: (a) Chemical water pollution is caused by runoff of different chemicals used in agriculture and industrial sectors like pesticides, heavy metals, dyes, oil spills, petroleum products and solvents. These chemical pollutants are poisonous to aquatic organisms, and subsequently reach other living organisms, including humans, through the food chain and result in biomagnification. Across the world, chemical pollution of aquatic ecosystems has become a major public apprehension. (b) Suspended matter water pollution is caused by particulate matter (pollutants with large size which do not dissolve in water). These suspended particles settle and form a thick slit at the bottom of water body, which is detrimental for organisms that reside on the floor of water bodies. (c) A microbiological water pollution is a natural form of water pollution caused by microorganisms such as bacteria, protozoa, and viruses. These microbes cause many serious diseases like cholera, dysentery, fever, and typhoid. (d) Nutrient pollution is caused by eutrophication (excessive release of nutrients in aquatic ecosystems which induces excessive growth of weeds and algal bloom), which results in oxygen depletion/ hypoxia in aquatic environs. (e) Groundwater pollution is most often caused by the runoff of excessive pesticides from soil. Humans use groundwater for obtaining drinking water, and its pollution causes enormous problems. (f) Oxygen depleting water pollution is caused by microorganisms that feed on biodegradable material and create oxygen depletion. After oxygen depletion, harmless aerobic microbes die and anaerobic harmful microbes commence to thrive (Water Pollution Guide, 2019).

Pollution of aquatic environs has adverse impacts on environment, public health, and economy. All kinds of aquatic pollution are detrimental to the health of living organisms including humans. Different pollutants influence environmental and public health in different ways and accordingly lead to problems like, bioaccumulation, biomagnification, and other ill-effects on aquatic flora and fauna. Simultaneously the same parameters also influence the transparency of water bodies, abnormal growth of photosynthetic plants and acute poisoning of marine life. Furthermore, the same parameters may also be responsible for the problems like, immune suppression, cholera, typhoid, cardiovascular disorders, renal failure, neurological disorders, reproductive failure, and birth defects, etc., and ultimately disturb the default structure of aquatic environs (Schwarzenbach et al., 2010). Prevention of aquatic pollution is economical as compared to cleaning up pollution that has previously taken place. Aquatic pollution is expensive to control and can prove damaging to the economy.

We can prevent aquatic pollution by following some simple guidelines viz., reduction in usage of pesticides and fertilizers, proper disposal of industrial wastes, planting more plants to prevent fertilizers and other chemicals from running off into aquatic bodies, clean up litter and avoid throwing waste material into water bodies, etc. Till date, many laws and conventions have been passed for the protection of water bodies across the world. Various organizations and agencies like Advisory Committee on Protection of the Sea (ACOPS), Clean Ocean Action, and Friends of the Earth, Greenpeace, International Water Association, Marine Conservation Society, Save Our Seas, Wetlands International, WWF, and others work to give relevant data to thwart aquatic pollution (Water Pollution Guide, 2019).

10.3 WATER QUALITY INDICATORS

Numerous parameters viz., Taste, and odor, pH, color, turbidity, temperature, hardness, alkalinity, total dissolved solids, nitrate, and nitrite, sulfate, chlorides, fluoride, arsenic, lead, phosphorus, iron, dissolved oxygen, biological oxygen demand (BOD) and total coliform are tested and regarded as indicators of water quality (Thatai et al., 2019).

Several studies have reported that fungi can act as bioindicators/ biomarkers of pollution in aquatic environs (Gadd, 2016; Bai et al., 2018). Several macrofungi act as bioindicators of metal pollution (Gadd, 2016).

Recently, Bai et al. (2018) reported that the relative abundance of fungi *Schizosaccharomyces* improved with an increase in the concentrations of nitrogen and organic carbon; thus, it can act as a prospective biomarker to reveal nutrient discharge in aquatic ecosystems.

10.4   MYCOREMEDIATION OF AQUATIC POLLUTANTS

In nature, aquatic ecosystems are polluted with many persistent pollutants simultaneously, which further complicate their cleanup process. Thus, a key concern for researchers was to find a holistic remediation technique for the restoration of degraded aquatic environs, which after proper testing in laboratories, can be commercialized in the future. Although several conventional physicochemical pollutant cleanup approaches are effective but, they are not feasible for large-scale application under natural conditions (Akcil et al., 2015). In contrast to this, bioremediation has emerged as the most desirable, efficient, safe, economical, practicable, sustainable, and environment-friendly mode of degradation or transformation and immobilization of recalcitrant environmental pollutants into non-harmful or less-harmful form, primarily by means of selected microorganisms (Tegli et al., 2014; Mishra and Sarma, 2017; Varjani and Patel, 2017; Pandey et al., 2018). Bioremediation process cleans the environmental pollutants through many phenomena like natural bioattenuation (natural degradation occur), biostimulation (applying nutrients, water, or aeration as an intentional stimulant), bioaugmentation (introduction of exogenous microorganisms), biosorption/mycosorption (surface binding and complexation with fungal cell surface), bioaccumulation (accumulation of pollutants by living cells), biovolatilization (enzymatic conversion of compounds into volatile derivatives), bioconversion (conversion into less/non-harmful state), biodegradation (ultimate degradation of complex molecules into simpler ones), and mycofiltration (filtration through fungal mycelia) or a combination of the above described processes (Zeng et al., 2010; Bhattacharya et al., 2011; Ulčnik et al., 2013; Kulshreshtha et al., 2014; Rhodes, 2015; Deshmukh et al., 2016; Gadd, 2016; Kumar, 2017).

Microorganisms like microalgae, bacteria, and fungi have the capability to degrade environmental pollutants that are harmful to human health and the environment. Numerous benefits of using microbial technology for cleanup of aquatic environs are simple methodological procedure, proficient degradation capability, low energy utilization, enduring practicability, no secondary pollution, without any need of supplementary constructions (Gao et al., 2018). Based on the type of bioremediator employed, bioremediation has several types.

Mycoremediation (Greek word, Myco: means 'Fungus' and Remediation: means 'to clean' or 'to solve a problem') also known as mycodegradation or mycodeterioration, is a form of bioremediation (Varjani and Patel, 2017). Mycoremediation is a fungal-based technology which involves degradation and removal of pollutants by means of fungi (Jain et al., 2017; Kumar, 2017). It is one of the most efficient, economic, and environment-friendly techniques and recently it has attracted substantial attraction (Yakop et al., 2019).

Mycoremediation is a fascinating area of research wherein fungi use pollutants as carbon source and degrade them into non-harmful/less-harmful substances like carbon dioxide, water, and cell biomass (Kumar, 2017; Hasan and AI-Jawhari, 2018). Application of fungi for environmental cleanup purposes emerged in the 1980's (Rodríguez-Rodríguez et al., 2013) with the discovery of mycoremediation potential of *Phanerochaete chrysosporium* (white rot fungus). Earlier mycoremediation received diminutive attention since bacterial bioremediation was the center of research. However, from past one decade, several workers reported huge potential of fungi in improving environmental quality (Pinedo-Rivilla et al., 2009; Spina et al., 2018). Till date fungi belonging to some major genera viz., *Acremonium, Agrocybe, Alternaria, Aspergillus, Bjerkandera, Boletus, Candida, Cephalosporium, Cladosporium, Cordyceps, Doratomyces, Fusarium, Ganoderma, Geotrichum, Gliocladium, Hypholoma, Lecanicillium, Lentinus, Mortierella, Mucor, Paecilomyces, Penicillium, Phanerochaete*, Phlebia, *Phoma, Pleurotus, Polyporus, Rhizopus, Rhodotolura, Saccharomyces, Streptomyces, Talaromyces, Torulopsis, Trametes*, and *Trichoderma* have been tested and employed for mycoremediation and *Phanerochaete chrysosporium* has been reported as the most capable model mycoremediator (Rodríguez-Rodríguez et al., 2013; Rhodes, 2015; Kabenge et al., 2017; Mishra and Sarma, 2017). To facilitate productive use of fungi for mycoremediation, proper knowledge of fungal biochemistry, ecology, enzymology, genetics, physiology, taxonomy, and other similar subjects is requisite (Rhodes, 2015).

In contaminated aquatic ecosystems, fungi personify an effectual toolbox for sustainable mycoremediation of pollutants that is yielding great results worldwide (Bhandari, 2017). A significant number of fungal species have capability to degrade chemical, suspended matter, microbiological, nutrient, groundwater, and oxygen depleting aquatic pollutants into environmentally less detrimental products via diverse mechanisms (Figure 10.1). The rate of degradation of any aquatic pollutant depends on its chemical nature and is further influenced by its accessibility as a carbon source to fungi and several physical factors also regulate this process (Singh and Gauba, 2014). Fungi, through the process of mycoremediation safeguard the environment as well as living organisms, including humans from the ill effects of pollution. Any mycoremediation program can be implemented by a four-phase approach which engrosses: (a) bench-scale treatability (b) on-site pilot testing (c) inoculum production and (d) full-scale application (Lamar and White, 2001; Varjani and Patel, 2017).

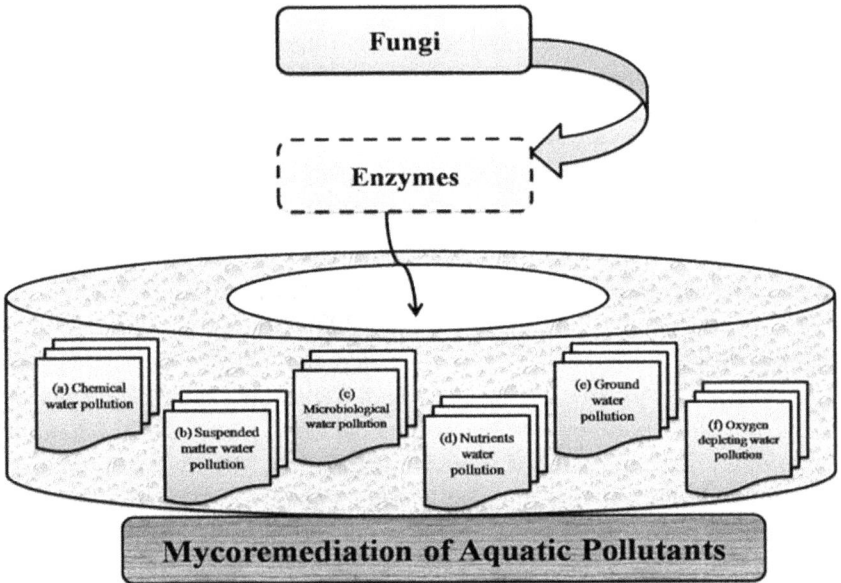

FIGURE 10.1   Mycoremediation of various groups of aquatic pollutants.

Fungi play a very crucial role in cleanup of contaminated sites through bioremediation process. These fungi by means of their unique features like diverse metabolic capability, sound protein cogs, varied morphology, rapid growth, and development under extreme conditions, low cultivation cost, potent absorption capability, and well-known nonspecific intracellular and extracellular fungal enzyme systems viz., catalases (CAT), cellulases, chitinases, cytochrome P450 monooxygenases, hemicellulase, laccases, ligninase, lignocellulases, pectinases, peroxidases, oxidases, and xylanases, act efficiently and synergistically as decontaminating agents either by chemical conversion or by affecting chemical bio-accessibility (Junghanns et al., 2005; Ulčnik et al., 2013; Singh and Gauba, 2014; Ikehata, 2015;

Marco-Urrea et al., 2015; Deshmukh et al., 2016; Jain et al., 2017; Kumar, 2017; Prakash, 2017; Ceci et al., 2018; Mohapatra et al., 2018; Rodríguez-Rodríguez et al., 2018; Spina et al., 2018). These extracellular enzymes have low substrate-specificity and can act upon variety of molecules with similar structures which makes fungi excellent tools for degradation of broad range of pollutants. Furthermore, fungi has the ability to develop extensive mycelial networks, which increase the fungal surface area for using range of pollutants as growth substrates, thereby making fungi suitable for practice of bioremediation. Around aquatic ecosystems, inoculation of fungal mycelium can even act as a myco-filter system for the runoff approaching from industrial and agricultural areas (Chiu et al., 2000; Varjani and Patel, 2017). Additionally, stress response proteins like ABC transporters confer fungi with the tolerance to many pollutants (Deshmukh et al., 2016) and the role of genes encoding such proteins must be thoroughly studied.

The fungal inoculum is either placed over the water surface in the form of mycelial mats or mixed-up into the contaminated water. Ultimately, applying living fungal cells for mycoremediation can lead to microbiological pollution of aquatic ecosystems. In order to circumvent this, desiccated or chemically pre-treated fungal biomass is preferred substitute to the treatment of live cells for mycoremediation of polluted water (Gazem and Nazareth, 2013). This will not only avert the problem of microbiological pollution, but will also prevent hindrance of microbial growth by pollutants.

It has been reported that naturally occurring microbiome acts in synergy with the fungi to degrade pollutants but, there are several reports that occasionally native microbes give a harsh competition to the mycoremediators (Rhodes, 2015). Applying a particular bioremediator can even augment the growth of local microbes, which may overshadow the remediator and lead to microbiological pollution. Thus, such studies must be validated prior to large-scale field application (Deshmukh et al., 2016).

Several fungi can degrade range of pollutants; however some are selective (Singh and Gauba, 2014). Janicki et al. (2018) reported that a single fungus can remove multiple pollutants simultaneously from co-contaminated sites. However, application of diverse fungi together in consortia can prove to be more effective in remediating complex concoction of pollutants simultaneously (Mishra and Malik, 2014). On the other hand, fungal hyphae act as vectors for effective dispersal of pollutant-degrading bacteria (Kohlmeier et al., 2005; Spina et al., 2018), thereby increasing the efficiency of bioremediation process. Hence, selection of appropriate fungal species for the remediation of particular group of pollutants is critical. In this context, this chapter is of great substance.

10.5   OVERVIEW OF MYCOREMEDIATION POTENTIAL OF FUNGI

Various fungi that have been employed for mycoremediation of different groups of aquatic pollutants. This data demonstrate enormous potential of fungi as mycoremediation agents for range of pollutants and it can serve as baseline for selection and use of fungi either independently or in consortium for forthcoming mycoremediation projects of aquatic ecosystems (Tables 10.1).

TABLE 10.1   Overview of mycoremediation potential of fungi in aquatic environs.

(I) Polyaromatic hydrocarbon (PAH)

S. No.	Type of Hydrocarbon	Fungi	References
1.	1,1'-binaphthalene	*Allescheriella* sp. *Phlebia* sp. *Stachybotrys* sp.	D'Annibale et al., (2006b)
2.	1,2,3,4, 9-phenanthrols	*Cunninghamella elegans*	Lisowska et al., (2006)
3.	1,2,3,4-tetrahydronaphthalene (THN)	*Hypoxylono ceanicum*	Li et al., (2005a)
4.	Acenaphtene	*Cunninghamella elegans*	Pothuluri et al., (1995)
5.	Anthracene	*Bjerkandera* sp. *Cunninghamella elegans* *Naematoloma frowardii* *Phanerochaete chrysosporium* *Phanerochaete laevis* *Pleurotus ostreatus* *Pleurotus sajor-caju* *Ramaria* sp. *Rhizoctonia solani* *Trametes versicolor*	Cerniglia and Yang, (1984) Hammel et al., (1992) Bogan and Lamar, (1995) Bezalel et al., (1996a, b, c) Johannes and Majcherczyk, (2000)
		Pleurotus ostreatus	Pozdnyakova et al., (2006)
		Tetrahymena pyriformis	Guiraud et al., (2008)
		Armillaria sp.	Hadibarata et al., (2013)
6.	Benzo(α)anthracene	*Candida krusei* *Cunninghamella elegans* *Phanerochaete chrysosporium* *Phanerochaete laevis* *Pleurotus ostreatus* *Rhodotorula minuta* *Syncephalastrum racemosum* *Trametes versicolor*	Cerniglia and Yang, (1984)
		Fusarium flocciferum *Pleurotus ostreatus* *Trametes versicolor* *Trichoderma* sp.	Atagana et al., (2006)

TABLE 10.1 (Continued)

S. No.	Type of Hydrocarbon	Fungi	References
7.	**Benzo(α)fluoranthene**	*Fusarium flocciferum* *Pleurotus ostreatus* *Trametes versicolor* *Trichoderma* sp.	Atagana et al., (2006)
8.	**Benzo(α)pyrene**	*Aspergillus ochraceus* *Bjerkandera adusta* *Bjerkandera* sp. *Candida maltosa* *Candida maltose* *Candida tropicalis* *Chrysosporium pannorum* *Cunninghamella elegans* *Mortierella verrucosa* *Naematoloma frowardii* *Neurospora crassa* *Penicillium janczewskii* *Penicillium janthinellum* *Phanerochaete chrysosporium* *Phanerochaete laevis* *Pleurotus ostreatus* *Ramaria* sp. *Saccharomyces cerevisiae* *Syncephalastrum racemosum* *Trametes versicolor* *Trichoderma* sp. *Trichoderma viride*	Bumpus et al., (1985) Haemmerli et al., (1986) Bezalel et al., (1996a, b, c) Bogan and Lamar, (1996)
		Trametes versicolor	Collins et al., (1996)
		Fusarium flocciferum *Pleurotus ostreatus* *Trametes versicolor* *Trichoderma* sp.	Atagana et al., (2006)
		Aspergillus ochraceus	Passarini et al., (2011)
9.	**Bisphenol A (5)**	*Ganoderma lucidum* *Irpex lacteus* *Pleurotus eryngii* *Polyporellus brumalis* *Schizophyllum commune* *Trametes versicolor*	Shin et al., (2007)
10.	**Chlorobenzoic Acids**	*Mortierella* sp. *Phanerochaete chrysosporium* *Phlebia* sp.	Bath and Arora, (2006)
11.	**Crude Oil**	*Aspergillus niger* *Penicillium documbens*	Gesinde et al., (2008)
		Pleurotus tuber-regium	Ogbo et al., (2008)

TABLE 10.1 (Continued)

S. No.	Type of Hydrocarbon	Fungi	References
		Aspergillus niger *Candida* sp. *Mucor* sp. *Penicillium* sp. *Rhizopus* sp.	Damisa et al., (2013)
		Aspergillus sp.	Zhang et al., (2016)
		Articulosporain flata *Aspergillus flavus* *Aspergillus fumigatus* *Aspergillus niger* *Bdellospora helicoides* *Botrytis cinerea* *Candida albicans* *Gonadobotricuma piculata* *Helminthosporium velutinum* *Neurospora crassa* *Pleurothecium recurvatum* *Streptothrix atra* *Thysarophora longispora* *Trichoderma viridae* *Varicosporiume lodeae* *Zoophagenito spora*	Olukunle and Oyegok, (2016)
		Penicillium sp.	Al-Hawash et al., (2018a)
12.	**Chrysene**	*Cunninghamella elegans* *Penicillum janthinellum* *Syncephalastrum racemosum*	Pothuluri et al., (1995)
		Fusarium flocciferum *Pleurotus ostreatus* *Trametes versicolor* *Trichoderma species*	Atagana et al., (2006)
13.	**Dibenzo-p-dioxins and Dibenzofurans**	*Phlebia radiata*	Kamei et al., (2005) Xu et al., (2006)
		Phlebia acerina *Phlebia brevispora* *Phlebia lindtneri*	Kamei et al., (2005)
		Fusarium sp. *Irpex* sp. *Phanerochaete chrysosporium* *Trametes* sp.	Nam et al., (2008)
		Pleurotus pulmonarius	Yamaguchi et al., (2007)
14.	**Dichloroaniline Isomers**	*Allescheriella* sp. *Phlebia* sp. *Stachybotrys* sp.	D'Annibale et al., (2006b)
15.	**Diesel**	*Cladosporium* sp.	Li et al., (2008)
16.	**Different pHs**	*Aspergillus* sp.	Al-Hawash et al., (2018b)
		Strophariarugoso annulata	Steffen et al., (2007)

TABLE 10.1 (Continued)

S. No.	Type of Hydrocarbon	Fungi	References
17.	Diphenyl Ether	*Pleurotus ostreatus* *Trametes versicolor*	Rosales et al., (2013) Wu et al., (2013)
18.	Ferulic	*Pleurotus involutus*	Zeng et al., (2006)
19.	Fluorene	*Pleurotus ostreatus*	Pozdnyakova et al., (2006)
		Cunninghamella elegans *Laetiporus sulphurous* *Phanerochaete* *chrysosporium* *Pleurotus ostreatus* *Trametes versicolor*	Bezalel et al., (1996a, b, c) Bogan and Lamar, (1996)
20.	Fluoranthene	*Cunninghamella elegans* *Laetiporus sulphurous* *Naematoloma frowardii* *Penicillium* sp. *Pleurotus ostreatus*	Sack and Günther, (1993)
		Pleurotus ostreatus	Pozdnyakova et al., (2006)
21.	Fuel oil	*Fusarium solani*	Yoshioka et al., (2006)
22.	Gasoline	*Exophiala xenobiotica*	Isola et al., (2013)
23.	Indole	*Sporotrichum thermophile* *Pleurotus ostreatus*	Ren et al., (2006) Katapodis et al., (2007)
24.	n-hexadecane	*Aspergillus niger*	Volke-Sepúlveda et al., (2003)
		Penicillium sp.	Pointing, (2001)
25.	Naphthalene	*Trichoderma harzianum*	Mollea et al., (2005)
		Aspergillus fumigatus	Ye et al., (2011)
		Pleurotus eryngii	Hadibarata et al., (2013)
		Allescheriella sp. *Phlebia* sp. *Stachybotrys* sp.	D'Annibale et al., (2006b)
26.	Naphthalan Petroleum	*Cephalosporium* sp. *Fusarium* sp. *Mucor* sp. *Penicillium* sp.	Kasumova et al., (2006, 2007)
27.	Nonylphenol (4)	*Bjerkandera* sp. *Phanerochaete* *chrysosporium Pleurotus* *ostreatus* *Trametes versicolor*	Soares et al., (2005)
28.	o-coumaric	*Pleurotus involutus*	Zeng et al., (2006)
29.	o-hydroxybiphenyl	*Allescheriella* sp. *Phlebia* sp. *Stachybotrys* sp.	D'Annibale et al., (2006b)
30.	o-hydroxyphenylacetic acids	*Pleurotus involutus*	Zeng et al., (2006)
31.	Olive-mill Dry Residue (DOR)	*Fusarium* sp.	Sampedro et al., (2007)
32.	p-cresol	*Gliomastix indicus*	Singh et al., (2008)

TABLE 10.1   (Continued)

S. No.	Type of Hydrocarbon	Fungi	References
33.	**Pentachlorophenol**	*Anthracophyllum discolour Bjerkandera adusta Phanerochaete chrysosporium Trametes* sp. *Trametes versicolor*	Ford et al., (2007) Rubilar et al., (2007) Yemendzhiev et al., (2008) Zeng et al., (2008)
34.	**Phenanthrene**	*Aspergillus niger Cunninghamella elegans Naematoloma frowardii Phanerochaete chrysosporium Phanerochaete laevis Pleurotus ostreatus Syncephalastrum racemosum Trametes versicolor*	Bumpus, (1989) Hammel et al., (1992) Bezalel et al., (1996a, b, c) Bogan and Lamar, (1996)
		Cunninghamella elegans	Romero et al., (1998)
		Fusarium solani Fusarium oxysporum	Li et al., (2005b) Chen et al., (2007, 2008)
		Pleurotus ostreatus	Pozdnyakova et al., (2006)
		Thrichoderma sp.	Hadibarata et al., (2007a)
35.	**Phenolic Components**	*Trametes versicolor*	Prabu et al., (2005) Udayasoorian et al., (2005) Yemendzhiev et al., (2008)
		Panus tigrinus	D'Annibale et al., (2006a)
36.	**Pyrene**	*Agrocybe aegerita Aspergillus niger Candida parapsilopsis Crinipellis maxima Crinipellis perniciosa Crinipellis stipitaria Crinipellis zonata Cunninghamella elegans Fusarium oxysporum Kuehneromyces mutablis Marasmiellus ramealis Marasmius rotul Mucor* sp. *Naematoloma frowardii Penicillium janczewskii Penicillium janthinellum Phanerochaete chrysosporium Pleurotus ostreatus Syncephalastrum racemosum Trichoderma harzianum*	Hammel et al., (1986) Bezalel et al., (1996a, b, c)
		Cunninghamella elegans	Cerniglia and yang, (1984)
		Pleurotus ostreatus	Pozdnyakova et al., (2006)

TABLE 10.1   (Continued)

S. No.	Type of Hydrocarbon	Fungi	References
		Aspergillus niger	Arun et al., (2008)
		Coriolus versicolor	Wang et al., (2008b)
		Fusarium sp.	
		Trichoderma sp.	
37.	Perylene	*Pleurotus ostreatus*	Pozdnyakova et al., (2006)
38.	Polychlorophenols (PCBs)	*Phlebia brevispora*	Kamei et al., (2006)
39.	Styrene	*Daldinia concentrica*	Lee et al., (2006)
		Phanerochaete chrysosporium	
		Trametes versicolor	

Aliphatic Hydrocarbons

S. No.	Type of Hydrocarbon	Fungi	References
40.	Digitoxin	*Absidia coerulea*	Wang et al., (2008a)
		Curvularia lunata	
41.	Hexachlorocyclohexane (HCH)	*Bjerkandera adusta*	Quintero et al., (2007)
42.	Imidazolium Compounds (ICs)	*Gliocladium roseum*	Zabielska-Matejuk and
		Penicillium brevicompactum	Czaczyk, (2006)
		Penicillium funiculosum	
		Phialophora fastigiata	
		Verticillium lecanii	
43.	n-alkanes	*Aspergillus niger*	Elshafie et al., (2007)
		Aspergillus ochraceus	
		Aspergillus sp.	
		Penicillium sp.	
		Trichoderma asperellum	
44.	n-eicosane	*Trichoderma* sp.	Hadibarata et al., (2007b)
45.	Perchloroethylene (PCE)	*Trametes versicolor*	Marco-Urrea et al., (2006)
		Ganoderma lucidum	Marco-Urrea et al., (2007)
		Irpex lacteus	
46.	Quaternary Ammonium Compounds (QACs)	*Gliocladium roseum*	Zabielska-Matejuk and
		Penicillium brevicompactum	Czaczyk, (2006)
		Penicillium funiculosum	
		Phialophora fastigiata	
		Verticillium lecanii	
47.	Sesquiterpene botrydienediol	*Botrytis cinerea*	Daoubi et al., (2006)
48.	Trichloroethylene (TCE)	*Ganoderma lucidum*	Marco-Urrea et al., (2007)
		Irpex lacteus	

(II) Heavy metals

S. No.	Type of Heavy Metal	Fungi	References
1.	Cadmium	*Pleurotus ostreatus*	Favero et al., (1990a, b)
			Kocaoba and Arisoy, (2011)
			Tay et al., (2011)
			Frutos, (2016)
		Rhizopus oligosporus	Aloysius et al., (1999)
		Aspergillus niger	Kapoor et al., (1999)

TABLE 10.1   (Continued)

S. No.	Type of Heavy Metal	Fungi	References
		Mucor rouxii	Yan and Viraraghavan, (2000)
		Glicocladium roseum *Talaromyces helices*	Massaccesi et al., (2002)
		Pleurotus sapidus	Yalçinkaya et al., (2002)
		Aspergillus terreus	Massaccesi et al., (2002) Joshi et al., (2011)
		Phomopsis sp.	Saiano et al., (2005)
		Aspergillus sp.	Srivastava and Thakur, (2006)
		Saprolegnia delica *Trichoderma viride*	Ali and Hashem, (2007)
		Pleurotus florida	Das et al., (2007)
		Saccharomyces cerevisiae	Chen and Wang, (2007) Yu et al., (2007) Damodaran et al., (2011)
		Trichoderma viridae	Joshi et al., (2011)
		Pleurotus eryngii	Joo et al., (2011) Ozdemir et al., (2012)
		Pleurotus platypus	Vimala and Das, (2011b) Lamrood and Ralegankar, (2013)
		Rhizophagus irregularis *Funneliformis mosseae*	Hassan et al., (2013)
		Agaricus bisporus *Calocybe indica* *Pleurotus floridianus* *Pleurotus sajor-caju*	Lamrood and Ralegankar, (2013)
		Agaricus bisporus *Lactarius piperatus*	Nagy et al., (2013)
2.	**Chromium**	*Aspergillus terreus*	Dias et al., (2002)
		Penicillium chrysogenum	Tan et al., (2003)
		Mucor hiemalis	Tewari et al., (2005)
		Rhizopus sp.	Ahmed et al., (2005)
		Saccharomyces cerevisiae	Chen and Wang, (2007)
		Pleurotus ostreatus	Javaid and Bajwa, (2007) Javaid and Bajwa, (2008) Javaid et al., (2011) Kocaoba and Arisoy, (2011) Arbanah et al., (2012) Carol et al., (2012) Puentes-Cárdenas et al., (2012) Arbanah et al., (2013)
		Agaricus bitorquis *Ganoderma lucidum* *Pleutrotus sajor-caju*	Hanif et al., (2011)

TABLE 10.1 (Continued)

S. No.	Type of Heavy Metal	Fungi	References
		Trichoderma longibrachiatum	Joshi et al., (2011)
		Aspergillus flavus Aspergillus niger	Dwivedi et al., (2012)
		Carum copticum	Kashefi et al., (2012)
		Pleurotus eous	Suseem and Mary Saral, (2014)
		Agaricus bitorquis	Hanif et al., (2015)
		Pleurotus cornucopiae	Xu et al., (2016)
3.	Cobalt	*Rhizopus delemar*	Tsekova and Petrov, (2002)
		Pleurotus ostreatus	Xiangliang et al., (2009)
		Pleurotus eryngii	Özdemir et al., (2012)
4.	Copper	*Ganoderma lucidum*	Muraleedharan and Venkobachar, (1990)
		Aspergillus niger	Kapoor et al., (1999) Mukhopadhyay et al., (2007) Iskandar et al., (2011) Iram et al., (2015)
		Rhizopus delemar	Tsekova and Petrov, (2002)
		Pleurotus ostreatus	Javaid and Bajwa, (2008) Tay et al., (2010) Huo et al., (2011) Javaid et al., (2011) Kocaoba and Arisoy, (2011) Arbanah et al., (2012) Oyetayo et al., (2012) Tay et al., (2012, 2016) Lamrood and Ralegankar, (2013) de Almeida and Burgess, (2013) Frutos, (2016)
		Pleurotus cornucopiae	Danış, (2010)
		Fusarium oxisporum	Simonescu et al., (2012)
		Rhizopus oryzae	Fu et al., (2012)
		Agaricus bisporus Calocybe indica Pleurotus platypus	Lamrood and Ralegankar, (2013)
		Flammulina velutipes	Luo et al., (2013)
		Fomes fasciatus	Sutherland and Venkobachar, (2013)
		Funneliformis mosseae Rhizophagus irregularis	Hassan et al., (2013)
		Pleurotus floridianus Pleurotus sajor-caju	Lamrood and Ralegankar, (2013)

TABLE 10.1   (Continued)

S. No.	Type of Heavy Metal	Fungi	References
		Aspergillus brasiliensis *Penicillium cirtinum*	Pereira et al., (2014)
		Agaricus bitorquis	Hanif et al., (2015)
		Aspergillus flavus	Iram et al., (2015)
5.	Gold	*Aspergillus niger* *Mucor rouxii* *Rhizopus arrhizus*	Kapoor and Viraraghavan, (1997)
6.	Iron	*Aspergillus terreus*	Dias et al., (2002)
		Rhizopus delemar	Tsekova and Petrov, (2002)
		Pleurotus ostreatus	Arbanah et al., (2012)
		Agaricus bisporus *Calocybe indica* *Jelly* sp. *Pleurotus platypus*	Lamrood and Ralegankar, (2013)
		Pleurotus floridianus *Pleurotus sajor-caju*	Lamrood and Ralegankar, (2013)
7.	Lead	*Aspergillus niger*	Kapoor et al., (1999) Iskandar et al., (2011) Dwivedi et al., (2012) Iram et al., (2015)
		Pleurotus ostreatus	Osman and Bandyopadhyay, (1999) Xiangliang et al., (2005) Tay et al., (2009) Liew et al., (2010) Oyetayo et al., (2012) de Almeida and Burgess, (2013) Dulay et al., (2015) Frutos, (2016) Jiang et al., (2017)
		Mucor rouxii	Yan and Viraraghavan, (2000)
		Aspergillus terreus	Massaccesi et al., (2002) Joshi et al., (2011)
		Penicillium chrysogenum	Deng et al., (2005)
		Penicillium oxalicum	Svecova et al., (2006)
		Saccharomyces cerevisiae	Chen and Wang, (2007) Damodaran et al., (2011)
		Saprolegnia delica *Trichoderma viride*	Ali and Hashem, (2007)
		Pleurotus eryngii	Joo et al., (2011)
		Aspergillus flavus	Dwivedi et al., (2012)
		Punica granatum	Alam et al., (2012)
		Pleurotus sajor-caju	Majeed et al., (2012, 2014)

TABLE 10.1 (Continued)

S. No.	Type of Heavy Metal	Fungi	References
		Agaricus bisporus *Calocybe indica* *Jelly* sp. *Pleurotus platypus*	Lamrood and Ralegankar, (2013)
		Pleurotus ferulae	Adebayo, (2013)
		Pleurotus florida	Prasad et al., (2013) Dulay et al., (2015)
		Pleurotus eous	Suseem and Mary Saral, (2014)
		Agaricus bitorquis	Hanif et al., (2015)
		Aspergillus flavus	Iram et al., (2015)
		Pleurotus cystidiosus *Pleurotus djamour* *Pleurotus salmoneo-stramineus*	Dulay et al., (2015)
8.	Manganese	*Pleurotus ostreatus*	Oyetayo et al., (2012)
		Aspergillus brasiliensis *Penicillium cirtinum*	Pereira et al., (2014)
		Pleurotus eryngii	Wu et al., (2016)
9.	Mercury	*Pleurotus ostreatus*	Mandal et al., (1998) Bressa et al., (1988) Nnorom et al., (2012)
		Pleurotus sapidus	Yalçinkaya et al., (2002)
		Pleurotus sajor-caju *Trametes versicolor*	Arica et al., (2003)
		Tolypocladium sp.	Svecova et al., (2006)
		Pleurotus tuber-regium	Nnorom et al., (2012)
10.	Nickle	*Aspergillus niger*	Kapoor et al., (1999) Dwivedi et al., (2012)
		Mucor rouxii	Yan and Viraraghavan, (2000)
		Aspergillus terreus	Dias et al., (2002)
		Penicillium chrysogenum	Tan et al., (2003) Deng et al., (2005)
		Saccharomyces cerevisiae	Chen and Wang, (2007)
		Pleurotus ostreatus	Javaid and Bajwa, (2008) Javaid et al., (2011) Tay et al., (2012, 2016)
		Aspergillus flavus	Dwivedi et al., (2012)
		Agaricus bisporus *Calocybe indica* *Jelly* sp. *Pleurotus floridianus* *Pleurotus platypus* *Pleurotus sajor-caju*	Lamrood and Ralegankar, (2013)
		Pleurotus eous	Suseem and Mary Saral, (2014)

TABLE 10.1 (Continued)

S. No.	Type of Heavy Metal	Fungi	References
11.	Zinc	*Rhizopus arrhizus*	Zhou, (1999)
		Mucor rouxii	Yan and Viraraghavan, (2000)
		Penicillium chrysogenum	Tan et al., (2003)
		Phomopsis sp.	Saiano et al., (2005)
		Saprolegnia delica *Trichoderma viride*	Ali and Hashem, (2007)
		Pleurotus ostreatus	Javaid and Bajwa, (2008) Javaid et al., (2011) Arbanah et al., (2012) Oyetayo et al., (2012)
		Pleurotus sajor-caju	Jibran et al.,(2011)
		Agaricus bisporus *Calocybe indica* *Jelly* sp. *Pleurotus floridianus* *Pleurotus platypus* *Pleurotus sajor-caju*	Lamrood and Ralegankar, (2013)
		Funneliformis mosseae *Rhizophagus irregualris*	Hassan et al., (2013)
		Aspergillus brasiliensis *Penicillium cirtinum*	Pereira et al., (2014)

(III) Dyes

S. No.	Type of Dye	Fungi	References
1.	Acid Black 52	*Trametes trogii*	Park et al., (2007)
2.	Acid Blue 129	*Trametes trogii*	Zeng et al., (2011)
3.	Acid Blue 40	*Cerrena unicolor*	Michniewicz et al., (2008)
4.	Acid Blue 62	*Cerrena unicolor*	Michniewicz et al., (2008)
5.	Acid Blue 80	*Lentinus polychrous*	Ratanapongleka and Phetsom, (2014)
6.	Acid Orange 7	*Trametes versicolor*	Lin et al., (2003)
7.	Acid Red 1	*Trametes trogii*	Zeng et al., (2011)
8.	Acid Violet 43	*Trametes trogii*	Park et al., (2007)
9.	Acid Yellow	*Pestalotiopsis* sp.	Wikee et al., (2019)
10.	Amido Black	*Peniophora* sp.	Shankar and Nill, (2015)
11.	Amido Black 10B	*Trametes* sp.	Yang et al., (2009)
12.	Aniline Blue	*Trametes* sp.	Li et al., (2014)
13.	Anthraquinone Blue	*Trametes trogii* *Thapsia villosa* *Trametes versicolor*	Levin et al., (2010)

TABLE 10.1 (Continued)

S. No.	Type of Dye	Fungi	References
14.	**Astrazon Blue**	*Funalia trogii Phanerochaete chrysosporium Pleurotus florida Pleurotus ostreatus Pleurotus sajor-caju Trametes versicolor*	Yesilada et al., (2003)
15.	**Astrazon Red FBL**	*Funalia trogii Phanerochaete chrysosporium Pleurotus florida Pleurotus ostreatus Pleurotus sajor-caju Trametes versicolor*	Yesilada et al., (2003)
16.	**Azo Dyes**	*Funalia trogii*	Apohan and Ozfer, (2005)
17.	**Basic Fuchsin**	*Aspergillus niger Phanerochaete chrysosporium*	Rani et al., (2014)
18.	**Black-B**	*Fusarium* sp.	Muthezhilan et al., (2014)
19.	**Blue M2R**	*Fusarium* sp.	Muthezhilan et al., (2014)
20.	**Brilliant Green**	*Peniophora* sp.	Shankar and Nill, (2015)
		Lenzites elegans	Pandey et al., (2018)
21.	**Bromo Cresol Purple**	*Pestalotiopsis* sp.	Wikee et al., (2019)
22.	**Bromophenol Blue**	*Funalia trogii Pleurotus sajor-caju Trametes versicolor*	Yesilada and Ozcan, (1998)
		Trametes sp.	Yang et al., (2009)
		Trametes trogii	Grassi et al., (2011)
		Pleurotus ostreatus	Zhuo et al., (2019)
23.	**Brown GR**	*Fusarium* sp.	Muthezhilan et al., (2014)
24.	**Cibacron Blue C-R**	*Bjerkandera adusta*	Robinson and Nigam, (2008)
25.	**Cibacron Blue P-3RGR**	*Coriolus versicolor Phanerochaete chrysosporium Pleurotus ostreatus*	Asgher et al., (2005)
26.	**Cibacron Red C-2G**	*Bjerkandera adusta*	Robinson and Nigam, (2008)
27.	**Cibacron Yellow C-2R**	*Bjerkandera adusta*	Robinson and Nigam, (2008)
28.	**Congo Red**	*Trametes trogii Trametes versicolor*	Birhanli and Yesilada, (2006)
		Trametes sp.	Yang et al., (2009)
		Thapsia villosa Trametes trogii Trametes versicolor	Levin et al., (2010)
		Marasmius cladophyllus	Ngieng et al., (2013)

TABLE 10.1   (Continued)

S. No.	Type of Dye	Fungi	References
		Trametes sp.	Li et al., (2014)
		Curvularia sp.	Senthilkumar et al., (2015)
		Myrothecium verrucaria	Sun et al., (2017)
		Lenzites elegans	Pandey et al., (2018)
29.	**Coomassie Brilliant Blue G250**	*Trametes* sp.	Yang et al., (2009)
30.	**Coralene Dark Red**	*Pleurotus florida* *Pleurotus ostreatus* *Pleurotus sapidus*	Kunjadia et al., (2016)
31.	**Coralene Golden Yellow**	*Pleurotus florida* *Pleurotus ostreatus* *Pleurotus sapidus*	Kunjadia et al., (2016)
32.	**Coralene Navy Blue**	*Pleurotus florida* *Pleurotus ostreatus* *Pleurotus sapidus*	Kunjadia et al., (2016)
33.	**Cotton Blue**	*Coriolopsis* sp.	Chen and Ting, (2015)
		Penicillium simplicissimum	Chen et al., (2019)
34.	**Cresol Red**	*Trametes* sp.	Yang et al., (2009)
35.	**Crystal Violet**	*Funalia trogii* *Phanerochaete chrysosporium Trametes versicolor*	Yesilada et al., (1995)
		Trametes sp.	Yang et al., (2009)
		Pycnoporus sanguineus	Sulaiman et al., (2013)
		Coriolopsis sp.	Chen and Ting, (2015)
		Peniophora sp.	Shankar and Nill, (2015)
		Pleurotus ostreatus	Kunjadia et al., (2016)
		Myrothecium verrucaria	Sun et al., (2017)
		Penicillium simplicissimum	Chen et al., (2019)
36.	**Direct Black 22**	*Cerrena unicolor*	Michniewicz et al., (2008)
37.	**Drimarene Blue CL-BR**	*Pleurotus ostreatus* *Trametes trogii* *Trametes versicolor*	Erkurt et al., (2007)
38.	**Drimarene Blue X3LR**	*Funalia trogii*	Unyayar et al., (2005)
39.	**Drimarene Orange K-GL**	*Phanerochaete chrysosporium* *Pleurotus ostreatus*	Asgher et al., (2005)
40.	**Dylon Navy 17**	*Trametes versicolor*	Sasmaz et al., (2011)
41.	**Fast Blue RR Salt**	*Trametes* sp.	Yang et al., (2009)
42.	**Gentian Violet**	*Thapsia villosa* *Trametes trogii* *Trametes versicolor*	Levin et al., (2010)
		Trametes trogii	Grassi et al., (2011)

TABLE 10.1 (Continued)

S. No.	Type of Dye	Fungi	References
43.	**Indigo Carmine**	*Trametes trogii*	Grassi et al., (2011)
		Funalia trogii	Birhanli et al., (2013)
		Trametes sp.	Li et al., (2014)
		Ganoderma sp.	Lu et al., (2016)
44.	**Indigo Dye**	*Ganoderma weberianum*	Tian et al., (2013)
45.	**Indigo Jean**	*Ganoderma* sp.	Ma et al., (2014)
46.	**Malachite Green**	*Trametes* sp.	Yang et al., (2009)
		Thapsia villosa *Trametes trogii* *Trametes versicolor*	Levin et al., (2010)
		Trametes trogii	Grassi et al., (2011)
		Aspergillus niger *Phanerochaete chrysosporium*	Rani et al., (2014)
		Coriolopsis sp.	Chen and Ting, (2015)
		Lenzites elegans	Pandey et al., (2018)
		Pleurotus ostreatus	Zhuo et al., (2019)
		Lasiodiplodia sp.	Arunprasath et al., (2019)
		Penicillium simplicissimum	Chen et al., (2019)
47.	**Methyl Green**	*Ganoderma* sp.	Lu et al., (2016)
48.	**Methyl Orange**	*Lentinus polychrous*	Ratanapongleka and Phetsom, (2014)
		Peniophora sp.	Shankar and Nill, (2015)
		Myrothecium verrucaria	Sun et al., (2017)
		Pleurotus ostreatus	Zhuo et al., (2019)
49.	**Methyl Red**	*Marasmius cladophyllus*	Ngieng et al., (2013)
		Myrothecium verrucaria	Sun et al., (2017)
		Trametes versicolor	Dayi et al., (2019)
50.	**Methyl Violet**	*Coriolopsis* sp.	Chen and Ting, (2015)
		Penicillium simplicissimum	Chen et al., (2019)
51.	**Methylene Blue**	*Irpex lacteus*	Malachova et al., (2013)
		Trametes sp.	Li et al., (2014)
		Peniophora sp.	Shankar and Nill, (2015)
52.	**Nigrosin**	*Aspergillus niger* *Phanerochaete chrysosporium*	Rani et al., (2014)
53.	**Nitrosulfonazo III**	*Pestalotiopsis* sp.	Wikee et al., (2019)
54.	**Orange 3R**	*Fusarium* sp.	Muthezhilan et al., (2014)
55.	**Orange G**	*Oudemansiella mucida*	Eichlerova et al., (2005)
		Dichomitus squalens	Eichlerová et al., (2007)
		Trametes sp.	Yang et al., (2009)
		Marasmius cladophyllus	Ngieng et al., (2013)

TABLE 10.1 (Continued)

S. No.	Type of Dye	Fungi	References
56.	Orange II	*Funalia trogii* *Pleurotus sajor-caju* *Trametes versicolor*	Yesilada and Ozcan, (1998)
57.	Orange M2R	*Fusarium* sp.	Muthezhilan et al., (2014)
58.	Procion BluePX-5R	*Coriolus versicolor* *Phanerochaete chrysosporium* *Pleurotus ostreatus*	Asgher et al., (2005)
59.	Reactive Black	*Lentinus polychrous*	Ratanapongleka and Phetsom, (2014)
60.	Reactive Black 5	*Ganoderma lucidum*	Murugesan et al., (2007)
		Trametes trogii	Park et al., (2007)
		Trametes trogii	Zeng et al., (2011)
		Trametes versicolor	Asgher et al., (2017)
		Phlebia sp. *Pope formosus*	Bulla et al., (2017)
		Pestalotiopsis sp.	Wikee et al., (2019)
61.	Reactive Blue 171	*Funalia trogii*	Birhanli et al., (2013)
62.	Reactive Blue 19	*Trametes trogii*	Park et al., (2007)
		Phlebia sp. *Pope formosus*	Bulla et al., (2017)
63.	Reactive Blue 21	*Trametes versicolor*	Asgher et al., (2017)
64.	Reactive Blue 220	*Trametes versicolor*	Dayi et al., (2019)
65.	Reactive Blue 4	*Trametes versicolor*	Yemendzhiev et al., (2009)
		Trametes trogii	Zeng et al., (2011)
66.	Reactive Blue 49	*Trametes trogii*	Park et al., (2007)
67.	Reactive Blue 81	*Cerrena unicolor*	Michniewicz et al., (2008)
68.	Reactive Green 19	*Lentinus polychrous*	Ratanapongleka and Phetsom, (2014)
69.	Reactive Orange 16	*Trametes trogii*	Park et al., (2007)
		Ganoderma sp.	Ma et al., (2014)
		Lentinus polychrous	Ratanapongleka and Phetsom, (2014)
70.	Reactive Red 195A	*Trametes versicolor*	Asgher et al., (2017)
71.	Reactive Red 198	*Trametes versicolor*	Sasmaz et al., (2011)
72.	Reactive Violet 1	*Ganoderma cupreum*	Gahlout et al., (2013)
		Trametes versicolor	Asgher et al., (2017)
73.	Reactive Yellow 145A	*Trametes versicolor*	Asgher et al., (2017)
74.	Red BSID	*Fusarium* sp.	Muthezhilan et al., (2014)
75.	Rem Blue RR	*Trametes versicolor*	Sasmaz et al., (2011)
76.	Rem Red RR	*Trametes versicolor*	Sasmaz et al., (2011)
77.	Rem Yellow RR	*Trametes versicolor*	Sasmaz et al., (2011)
78.	Remazol Black B	*Bjerkandera adusta*	Robinson and Nigam, (2008)

TABLE 10.1 (Continued)

S. No.	Type of Dye	Fungi	References
79.	**Remazol Brilliant Blue R (RBBR)**	*Fomes fomentarius*	Eichlerova et al., (2005)
		Oudemansiella mucida	Eichlerova et al., (2005)
		Dichomitus squalens	Eichlerová et al., (2007)
		Ganoderma lucidum	Murugesan et al., (2007)
		Trametes trogii	Zeng et al., (2011)
		Marasmius cladophyllus	Ngieng et al., (2013)
		Pseudolagaro basidiumacaciicola	Adak et al., (2016)
		Ganoderma sp.	Lu et al., (2016)
		Pleurotus ostreatus	Zhuo et al., (2019)
80.	**Remazol Brilliant Yellow 3GL**	*Coriolus versicolor Ganoderma lucidum Phanerochaete chrysosporium*	Asgher et al., (2005)
81.	**Remazol Red RB**	*Bjerkandera adusta*	Robinson and Nigam, (2008)
82.	**Simulated Effluent (containing Acid Red 27 or Basic Green 4 or Acid Violet 17)**	*Calosoma bulleri*	Chhabra et al., (2015)
83.	**Textile Dye Decolourization**	*Aspergillus foetidus Aspergillus niger Aspergillus sojae Geotrichum candidium Penicillium* sp. *Pycnoporus cinnabarinus Trichoderma* sp. *Trichoderma viride*	Jebapriya and Gnanadoss, (2013)
		Bjerkandera adusta Ceriporia metamorphosa Ganoderma sp.	Ma et al., (2014)
84.	**Turquoise Blue HFG**	*Cestraeus plicatilis*	Akdogan and Topuz, (2014)
85.	**Water Blue**	*Lentinus polychrous*	Ratanapongleka and Phetsom, (2014)
86.	**Xylidine**	*Thapsia villosa Trametes trogii Trametes versicolor*	Levin et al., (2010)
		Trametes trogii	Grassi et al., (2011)
87.	**Yellow MR**	*Fusarium* sp.	Muthezhilan et al., (2014)

(IV) Pesticides

S. No.	Type of Pesticide	Fungi	References
1.	**Alachlor**	*Phanerochaete chrysosporium*	McFarland et al., (1996)
2.	**Aldrin**	*Aspergillus niger*	Korte and Porter, (1970)
		Trichoderma viridae	Patil et al., (1970)
3.	**Anthracene**	*Acremonium* sp.	Ma et al., (2014)

TABLE 10.1 (Continued)

S. No.	Type of Pesticide	Fungi	References
4.	**Atrazine**	*Cerrena maxima* *Coriolopsis fulvocinerea* *Coriolus hirsutus*	Koroleva et al., (2001)
		Phanerochaete chrysosporium	Reddy and Mathew, (2001)
		Agrocybe semiorbicularis *Auricularia auricola* *Coriolus versicolor* *Dichotomitus squalens* *Flammulina velupites* *Hypholoma fasciculare* *Phanerochaete velutina* *Pleurotus ostreatus* *Stereum hirsutum*	Bending et al., (2002)
		Pleurotus ostreatus	Pereira et al., (2013)
5.	**Bensulfuron-methyl**	*Penicillium pinophilum*	Peng et al., (2012)
6.	**Benzo[k]fluoranthene**	*Fusarium flocciferum* *Pleurotus ostreatus* *Trametes versicolor* *Trichoderma* sp.	Baldrian et al., (2008)
7.	**Benzo[a]anthracene**	*Fusarium flocciferum* *Pleurotus ostreatus* *Trametes versicolor* *Trichoderma* sp.	Baldrian et al., (2008)
		Pleurotus ostreatus	Bhattacharya et al., (2014)
8.	**Benzo[a]pyrene**	*Fusarium flocciferum* *Pleurotus ostreatus* *Trametes versicolor* *Trichoderma* sp.	Baldrian et al., (2008)
9.	**Benzo[ghi]perylene**	*Fusarium flocciferum* *Pleurotus ostreatus* *Trametes versicolor* *Trichoderma* sp.	Baldrian et al., (2008)
10.	**Carbamate**	*Sphingomona syanoikuyae*	Cases et al., (2005)
11.	**Carbofuran**	*Aspergillus niger*	Hultberg and Bodin, (2018)
12.	**Carfentrazone-ethyl**	*Aspergillus niger*	Hultberg and Bodin, (2018)
13.	**Chlorpyrifos**	*Cladosporium cladosporioides* *Ganoderma* sp.	Chen et al., (2015)
		Streptomyces sp.	Fuentes et al., (2013)
		Aspergillus terreus	Silambarasan and Abraham, (2013)
		Aspergillus flavus	Kurniati et al., (2014)
		Pleurotus coccineus *Pleurotus gigantea* *Trametes versicolor*	Gouma et al., (2019)
14.	**Chlortoluron**	*Mortierella* sp.	Badawi et al., (2009)

TABLE 10.1 (Continued)

S. No.	Type of Pesticide	Fungi	References
15.	**Chrysene**	*Fusarium flocciferum* *Pleurotus ostreatus* *Trametes versicolor* *Trichoderma* sp.	Baldrian et al., (2008)
16.	**Citalopram**	*Bjerkandera adusta* *Bjerkandera* sp. *Phanerochaete chrysosporium*	Rodarte-Morales et al., (2011)
17.	**Cyanazine**	*Aspergillus niger*	Hultberg and Bodin, (2018)
18.	**Cypermethrin**	*Pseudomonas auroginosa*	Bhosle and Nasreen, (2013)
19.	**DDD (Dichlorodiphenyldichloroethane)**	*Trichoderma* sp.	Ortega et al., (2011)
20.	**Dibenzo[a,h]anthracene**	*Fusarium flocciferum* *Pleurotus ostreatus* *Trametes versicolor* *Trichoderma* sp.	Baldrian et al., (2008)
21.	**Dichloro Diphenyl Trichloroethane (DDT)**	*Trichoderma viride*	Matsumura and Boush, (1968) Patil et al., (1970)
		Phanerochaete chrysosporium	Reddy and Mathew, (2001)
		Boletus edulis *Gymnopilus viscidus* *Laccaria bicolor* *Laccaria scaburm*	Huang et al., (2007)
		Pleurotus ostreatus	Purnomo et al., (2010b)
22.	**2,4,Dichlorophenoxy acetic acid**	*Aspergillus niger*	Faulkner and Woodcock, (1964)
23.	**Dieldrin and endrin**	*Phlebia brevispora*	Kamei et al., (2010)
		Mucor racemosus	Kataoka et al., (2010)
		Pachycephala aurea *Phlebia acanthocystis* *Phlebia brevispora*	Xiao et al., (2011a)
		Mucor sp. *Trichoderma* sp.	Velázquez-Fernández et al., (2012)
		Cordyceps brongniartii *Cordyceps militaris*	Xiao and Kondo, (2013)
24.	**Difenoconazole**	*Aspergillus niger*	Hultberg and Bodin, (2018)
		Fusarium oxysporum *Lecanicillium saksenae* *Lentinula edodes* *Penicillium brevicompactum*	Shi et al., (2012)
25.	**Diuron**	*Phanerochaete chrysosporium*	Fratila-Apachitei et al., (1999)
		Beauveria bassiana *Cunninghamella elegans* *Mortirella isabellina*	Tixier et al., (2000)

TABLE 10.1 (Continued)

S. No.	Type of Pesticide	Fungi	References
		Agrocybe semiorbicularis *Auricularia auricola* *Coriolus versicolor* *Dichotomitus squalens* *Flammulina velupites* *Hypholoma fasciculare* *Phanerochaete velutina* *Pleurotus ostreatus* *Stereum hirsutum*	Bending et al., (2002)
		Aspergillus niger *Beauveria bassina* *Cunninghamella elegans* *Mortierella isabellina*	Tixier et al., (2002)
		Mortierella sp.	Badawi et al., (2009)
		Coriolus versicolor *Dichotomitus squalens* *Flammulina velupites* *Pleurotus ostreatus* *Stereum hirsutum*	Bending et al., (2010)
		Ganoderma lucidum	Da Silva Coelho et al., (2010)
		Aspergillus niger	Marco-Urrea et al., (2015)
26.	Dromecarb	*Aspergillus versicolor*	Knowles and Benezet, (1981)
27.	Endosulfan	*Aspergillus niger*	Bhalerao and Puranik, (2007)
		Trametes hirsute	Kamaei et al., (2011)
		Aspergillus fumigates *Candida* sp. *Mucor* sp. *Pencillium* sp.	Mohanasrinivasan et al., (2013)
		Pleurotus eryngii *Coprinus comatus*	Wang et al., (2017)
28.	Endosulfansulphage	*Trametes hirsute*	Kamaei et al., (2011)
29.	Fludioxonil	*Aspergillus niger*	Hultberg and Bodin, (2018)
30.	Fluoranthene	*Acremonium* sp.	Ma et al., (2014)
31.	Fluorene	*Acremonium* sp.	Ma et al., (2014)
32.	Fluoxetine	*Bjerkandera adusta* *Bjerkandera* sp. *Phanerochaete chrysosporium*	Rodarte-Morales et al., (2011)
33.	Folpet	*Gongronella* sp. *Rhizopus stolonifer*	Martins et al., (2014)
34.	Glyphosate	*Trametes versicolor*	Pizzul et al., (2009)
35.	HCH (Lindane)	*Phanerochaete chrysosporium*	Reddy and Mathew, (2001)
		Pleurotus ostreatus	Rigas et al., (2005)
		Ganoderma australe	Rigas et al., (2007)
36.	Heptachlor	*Pleyrotus acanthocystis*	Xiao et al., (2011b)
		Aspergillus niger	Bhalerao, (2012)
		Phanerochaete ostreatus	Purnomo et al., (2013)

TABLE 10.1 (Continued)

S. No.	Type of Pesticide	Fungi	References
37.	Isoproturon	*Alternaria* sp. Basidiomycete strain Gr177 *Mortierella* sp. *Mucor* sp. *Phoma eupyrena*	Rønhede et al., (2005)
		Cunninghamella elegans	Hangler et al., (2007)
		Mortierella sp.	Badawi et al., (2009)
38.	Linuron	*Mortierella* sp.	Badawi et al., (2009)
		Pleurotus coccineus *Pleurotus gigantean* *Trametes versicolor*	Gouma et al., (2019)
39.	Malathion	*Fusarium oxysporum*	Peter et al., (2015)
40.	Metalaxyl	*Agrocybe semiorbicularis* *Auricularia auricola* *Coriolus versicolor* *Dichotomitus squalens* *Flammulina velupites* *Hypholoma fasciculare* *Phanerochaete velutina* *Pleurotus ostreatus* *Stereum hirsutum*	Bending et al., (2002)
		Gongronella sp. *Rhizopus stolonifer*	Martins et al., (2014)
41.	Methyl parathion	*Aspergillus sydowii* *Penicillium decaturense*	Alvarenga et al., (2014)
42.	Metribuzin	*Pleurotus coccineus* *Pleurotus gigantean* *Trametes versicolor*	Gouma et al., (2019)
43.	Monocrotophos (MCP)	*Aspergillus flavus* *Fusarium pallidoroseum* *Macrophomina* sp.	Jain et al., (2014)
44.	Naphthalene	*Acremonium* sp.	Ma et al., (2014)
45.	Organopollutants	*Phanerochaete chrysosporium*	Sasek, (2003)
46.	Parathion	*Bjerkandera adusta* *Phanerochaete chrysosporium* *Pleurotus ostreatus*	Jauregui et al., (2003)
47.	Pendimethalin	*Fusarium oxysporum* *Lecanicillium saksenae* *Lentinula edodes* *Penicillium brevicompactum*	Shi et al., (2012)
48.	Pentachlorophenol	*Lentinus edodes*	Pletsch et al., (1999)
		Trametes versicolor	Tuomela et al., (1999)
		Anthrocophyllum discolor	Rubilar et al., (2007)
49.	Phenanthrene	*Acremonium* sp.	Ma et al., (2014)
		Phanerochaetes ordida	Turlo, (2014)
50.	Phenmedipham	*Aspergillus niger*	Hultberg and Bodin, (2018)

TABLE 10.1   (Continued)

S. No.	Type of Pesticide	Fungi	References
51.	**Polychlorinated biphenyls**	*Doratomyces nanus* *Doratomyces purpureofuscus* *Doratomyces verrucisporus* *Myceliophthora thermophila* *Phoma eupyren* *Thermoascus crustaceus*	Mouhamadou et al., (2013)
52.	**Polychlorinated dibenzofurans**	*Phanerochaete chrysosporium*	Wu et al., (2013)
53.	**Pyrene**	*Fusarium flocciferum* *Pleurotus ostreatus* *Trametes versicolor* *Trichoderma* sp.	Baldrian et al., (2008)
		Pseudo trametes gibbosa	Wen et al., (2011)
54.	**Pyrethrin (OPs)**	*Sphingomonas yanoikuyae*	Cases et al., (2005)
55.	**Simazine**	*Phanerochaete chrysosporium*	Mougin et al., (1997)
		Penicillium steckii	Kodama et al., (2001)
		Fusarium verticillioides	Urlacher et al., (2004) Alzahrani, (2009)
56.	**Sulfamethoxazole**	*Bjerkandera adusta* *Bjerkandera* sp. *Phanerochaete chrysosporium*	Rodarte-Morales et al., (2011)
57.	**Terbuthylazine**	*Agrocybe semiorbicularis* *Auricularia auricola* *Coriolus versicolor* *Dichotomitus squalens* *Flammulina velupites* *Hypholoma fasciculare* *Phanerochaete velutina* *Pleurotus ostreatus* *Stereum hirsutum*	Bending et al., (2002)
		Fusarium oxysporum *Lecanicillium saksenae* *Lentinula edodes* *Penicillium brevicompactum*	Shi et al., (2012)
58.	**2,4,5- Trichlorophenol**	*Clitocybe maxima*	Zhou et al., (2015)
59.	**Trinexapac ethyl**	*Aspergillus niger*	Hultberg and Bodin, (2018)
60.	**Vydate**	*Trichoderma harzianum* *Trichoderma viride*	Helal and Abo-El-Seoud, (2015)

(V) Suspended matter pollutants

S. No.	Type of solid waste	Fungi	References
1.	**Benzo alpha pyrene,** **Phenanthrene and Pyrene**	*Phanerochaete* *chrysosporium*	Magan et al., (2010)
2.	**Cellulose and Hemicellulose**	Brown-rot fungi	Hibbett and Donoghue, (2001)
3.	**Fly Ash**	*Aspergillus niger*	Wu and Ting, (2006)

TABLE 10.1 (Continued)

4.	Industrial effluents of unwanted chemicals	*Candida tropicalis* *Candida utilis* *Saccharomyces carlsbergensis* *Saccharomyces cerevisiae*	Erum and Ahmed, (2011)
5.	Lignin	*Phanerochaete chrysosporium*	Aust, (1995)
6.	Lignocellulose	*Agaricus bisporus* *Lentinus oloides*	Durrant et al., (1991)
7.	Organic Pollutants	*Cunninghamella* sp. *Mucor* sp. *Rhizopus* sp.	Verdin et al., (2004) Casieri et al., (2010)
8.	Pentachlorophenol (PCP) and Creosote	*Phanerochaete chrysosporium* *Phanerochaete sordid* *Trametes hirsuta*	Lamar et al., (1994)
9.	Phenolic Azo	*Pyricularia oryzae*	Chivukula and Renganathan, (1995)
10.	Plant Materials	*Clitocybe* sp. *Collybia* sp. *Mycena* sp.	Osono and Takeda, (2002)
11.	Polycyclic Aromatic Hydrocarbons	*Pleurotus pulmonarius*	Lau et al., (2003)
12.	Polyester Polyurethane	*Fusarium solani*	Zafar et al., (2013)
13.	Polyethylene	*Penicillium simplicissimum*	Yamada-Onodera et al., (2001)
14.	Pulp and Paper Mill Wastes	*Corius versicolor* *Trametes versicolor*	Murugesan, (2003)
15.	Rice Husks	*Moniliophthora roreri*	Solís et al., (2016)
16.	Toluene and 4- Nitrotoluene	*Agrocybe aegerita*	Kinne et al., (2010)
17.	Uranium and Thorium	*Rhizopus arrhizus*	Sears et al., (1984)
18.	Willow Sawdust	*Abortiporus biennis*	Alexandropoulou et al., (2017)
19.	Xenobiotics	*Acremonium* sp. *Aspergillus* sp. *Fusarium* sp. *Geomyces* sp. *Microsporum* sp. *Paecilomyces* sp. *Penicillium* sp.	Field et al., (1993) Pinedo-Rivilla et.al., (2009)

(VI) Microbiological and oxygen depleting pollutants

S. No.	Type of Pollutant	Fungi	References
1.	*Escherichia coli*	Garden giant mushrooms	Stamets, (2005)
		Fungal nanoparticles	Gadd, (2016) Dutta and Hyder, (2017)

10.6   MECHANISMS OF MYCOREMEDIATION OF DIFFERENT AQUATIC POLLUTANTS

A lot of work has been carried on mycoremediation potential of fungi across the world. Successful mycoremediation of several groups of aquatic pollutants like chemical pollutants, suspended matter pollutants, microbiological pollutants, nutrient pollutants, groundwater pollutants, and oxygen depleting aquatic pollutants have been carried out till date (Bhattacharya et al., 2011; Singh and Gauba, 2014; Marco-Urrea et al., 2015; Deshmukh et al., 2016; Gadd, 2016; Bhandari, 2017; Jain et al., 2017; Prakash, 2017; Ceci et al., 2018; Hasan and AI-Jawhari, 2018; Kumar et al., 2018; Mohapatra et al., 2018; Pandey et al., 2018; Rodríguez-Rodríguez et al., 2018; Spina et al., 2018; Yakop et al., 2019). The mycoremediation mechanisms of these aquatic pollutants have been reviewed and highlighted in the subsequent section.

10.6.1   MYCOREMEDIATION OF CHEMICAL, NUTRIENT, AND GROUND WATER POLLUTANTS

Various agro-industrial chemical wastes like polyaromatic hydrocarbon, heavy metals, dyes, and pesticides percolate into groundwater. The diffusion of these recalcitrant chemical pollutants into groundwater is a major environmental threat. The persistence of these pollutants has led to global-scale pollution of water bodies and groundwater resources.

In stressed polluted environments, fungi alleviate pollution stress and clean up polluted environs by diverse mycoremediation mechanisms (Varjani and Patel, 2017). In order to acquire information about pollutant degradation pathways, it is essential to experiment with a single fungus on a particular pollutant. However, such experiments cannot predict fungal mycoremediation performance in natural environmental water systems. Compared to aquatic ecosystems, the majority of fungi have been used for bioremediation of terrestrial ecosystems. Nevertheless, the remediation of pollutants in terrestrial ecosystems will prevent pollutant runoff and will consecutively save aquatic environs from pollution.

10.6.1.1   POLYAROMATIC HYDROCARBON (PAH)

Polyaromatic hydrocarbons (PAH) are a class of intricate organic compounds composed of polycondensed highly stable aromatic rings that have accumulated in the environment due to anthropogenic activities such as incomplete burning of carbon material like coal, oil, gas, garbage, and other organic materials. They are potent environmental pollutants having deleterious effects on human and environmental health (Gupte et al., 2016). They are hydrophobic in nature and can accumulate in fatty tissues with ease and vertically spread through the food chain, which further hampers their degradation (Varjani, 2017).

Fungi have been reported to play an imperative role in the bioremediation of PAH. Fungi can easily degrade low-molecular-weight PAH's; however, high-molecular-weight hydrocarbons are recalcitrant and resist both chemicals as well as microbial degradation (Prakash, 2017). Diverse fungi have evolved effective PAH degradation pathways (Pinedo-Rivilla et al., 2009; El-Enshasy et al., 2017; Prakash, 2017). In general, two pathways have been identified. The first one involves the cytochrome 450 system, which initiates PAH degradation by catalyzing the oxidation of PAH's into arene oxides. The second one entails extracellular enzymatic systems (laccase, lignin peroxidase, manganese-dependent peroxidase, and tyrosinase), which carry out nonspecific oxidation reaction leading to the production of hydroxylated aromatic compounds, which are further broken down into carbon dioxide (Muncnerova and Augustin, 1994; Hammel, 1995; Bogan and Lamar, 1996; Rhodes, 2015; Hasan and AI-Jawhari, 2018; Rodríguez-Rodríguez et al., 2018).

10.6.1.2   HEAVY METALS

Heavy metals pollution of aquatic environs is a major environmental threat. Several conventional methods of heavy metal removal like chemical precipitation, coagulation, electrodialysis, ion exchange are still widely used. Nowadays, researchers across the world are looking for new sustainable techniques for mitigation of such persistent environmental contaminants. Among the techniques employed for heavy metal remediation, mycoremediation is the most economical, effective, and sustainable technique (El-Enshasy et al., 2017; Hassan et al., 2017). Fungi have immense potential to remove a wide range of heavy metals.

Different fungi possess a different mechanism for the removal of heavy metals from contaminated sites. Microorganisms reduce metal toxicity through biosorption and bioaccumulation. Fungi sequester heavy metals from aqueous solutions by the process of biosorption (Prakash, 2017). Fungal cell wall adsorbs heavy metals, followed by intracellular accumulation and sequestration. Interaction of fungi and metals leads to the production of metal-binding peptides, which actively transport heavy metals to the intracellular compartments (Zeng et al., 2010). Several mycorrhizal and endophytic fungi equip plants with the ability to grow in metal-polluted sites through their potential to accumulate and immobilize heavy metal (Hassan et al., 2017; Kumar, 2017). Thus fungal biomass acts like a metal sink and alleviates heavy metal stress (Gadd, 2016).

10.6.1.3 DYES

Synthetic dyes that are mainly used in textile, cosmetics, paper, pharmaceutical, and food commerce are recalcitrant in nature. Effluents from these industries contain a range of pollutants like dyes, inorganic salts, etc. Dyes have an intricate chemical structure which confer them resistance against degradation. Recently, fungal mediated decolorization, degradation, and removal of dyes from industrial wastewaters has emerged as an effectual substitute over conventional approaches of dye removal (Bhattacharya et al., 2011; Senthilkumar et al., 2014).

Fungi remove dyes from aquatic environs by three main mechanisms: biosorption, bioaccumulation, and biodegradation (El-Enshasy et al., 2017). Fungal cells adsorb and accumulate dye by the process of biosorption and bioaccumulation. This is followed by biodegradation of dyes by various extracellular and intracellular fungal enzymes like laccases and peroxidases (Diwanian et al., 2010; Sen et al., 2016; El-Enshasy et al., 2017). *Schizophyllum commune* by means of its ligninolytic enzyme system, decolorize, and degrade solar brilliant red 80 dye (Asgher et al., 2013). Type of degrading enzyme involved and removal rate depend on the chemical structure of dye (Abadulla et al., 2000). Several workers explained decolourization mechanism of different industrial dyes by fungal extracellular enzymes in detail (Bhattacharya et al., 2011; Asgher et al., 2013; Imran et al., 2015; El-Enshasy et al., 2017; Kumar et al., 2018).

10.6.1.4 PESTICIDES

Chemical pollutants can traverse protracted distances in surface or groundwater and the pollution can persist for several decades owing to their slow natural degradation (Aravinna et al., 2017). A survey of over 160 water samples collected from 23 European countries, examined for the incidence of pesticides, revealed the presence of eight pesticides of concentration above the European threshold value of 0.1 μg/l and 29% of water samples were labeled as unsafe (Loos et al., 2010). Leaching and prolonged persistence of pesticides has led to global-scale pollution of water bodies and groundwater resources, and perpetuation of good water quality is a big challenge.

Numerous fungi have been used for the biodegradation of hazardous pesticides. Detailed pathways of pesticide mycoremediation have been postulated by many workers (see Bumpus and Aust, 1987; Reddy and Gold, 2000; Bending et al., 2001; Huang et al., 2007; Carvalho et al., 2011; Xiao et al., 2011; Velázquez-Fernández et al., 2012; Spina et al., 2015, 2018; Bhandari, 2017; Hultberg and Bodin, 2018; Pandey et al., 2018; Gouma et al., 2019). After the biosorption and bioaccumulation of pesticides, various intracellular and extracellular fungal enzymes acts and bring about pesticide biodegradation (Spina et al., 2018). Fungi isolated from contaminated areas have been shown to be efficient mycoremediators (Oliveira et al., 2015). For instance, aquatic microbes associated with submerged leaves that are exposed to nicosulfuron are more efficient in its degradation that than microbes from a less contaminated site (Carles et al., 2017).

Insecticide dichlorodiphenyltrichloroethane (DDT) used since the 1940's was banned in 1972, on account of its detrimental effects on living organisms. Several fungi have been reported to degrade this deadly pesticide, with many other persistent pesticides. Fungi like *Daedalea dickinsii, Gloeophyllum trabeum*, and *Phanerochaete chrysosporium* have been reported to convert DDT into DDE [1,1-dichloro-2,2-bis(p-chlorophenyl) ethylene] by dehydrochlorination and then hydrogenate this DDE to DDD [1,1-dichloro-2,2-bis-(4-chlorophenyl) ethane]. In a few cases, this DDD is the end product of DDT degradation or sometimes DDD is further degraded into DBP (4,4 dichlorobenzophenone) or DBH (4,4-dichlorobenzhydrol) or DDA [bis (4-chlorophenyl) acetic acid] or DDM (di-1,1-chlorophenyl-2-chloroethene), by oxidative dechlorination. DBP is further degraded by *Aspergillus niger* into 4-chlorobenzophenone (CBP) and 4-chloromethylbenzophenone and these residual products are less harmful as compared to DDT (Bumpus and Aust, 1987; Purnomo et al., 2008, 2010; Pandey et al., 2018; Spina et al., 2018).

10.6.2 MYCOREMEDIATION OF SUSPENDED MATTER POLLUTANTS

Suspended matter pollution of aquatic environs cause many environmental and public health problems (Habib et al., 2013; Jain et al., 2017). Incineration and landfilling are most common methods of solid waste disposal but, these are expensive and also lead to the production of numerous secondary pollutants. In contrast to this, mycoremediation of suspended matter by fungi through their hydrolytic enzymes has proven satisfactory (Deshmukh et al., 2016).

Cellulose is a main component of solid organic pollutants and cellulolytic activity of several fungi like *Aspergillus, Chaetomium, Phanerochaete,*

Pleurotus, *Polyporus*, and *Trichoderma* have been used for their bioremediation through bioconversion composting (Pinedo-Rivilla et al., 2009; Deshmukh et al., 2017; Jain et al., 2017). Several workers reported that *Penicillium frequentans*, *Penicillium simplicissimum*, *Phanerochaete chrysosporium*, and *Talaromyces wortmannii* can degrade polyethylene and these species have prospective polyethylene bioremediation potential (Argolo et al., 2006; Hossain, 2006; Seneviratne et al., 2006; Pinedo-Rivilla et al., 2009; Sowmya et al., 2015). White rot fungi have nylon degrading potential (Friedrich et al., 2007). A group of fungi, belonging to genera *Aspergillus*, *Mucor*, *Paecilomyces*, *Penicillium*, *Rhizopus*, *Scopulariopsis*, and *Trichoderma*, *Verticillium* biodegrade leather (Jain et al., 2017). Another fungus, *Resinicium bicolor* has rubber detoxification and degradation capability. An endophytic fungus *Pestalotiopsis microspora* can degrade plastic (Russell et al., 2011). Various strains of *Aspergillus*, *Penicillium*, *Saccharomyces*, and *Yarrowia* have been found to be involved in bioleaching of different metals like Au, Cu, Cd, Sn, Ni, Al, Zn, Pb, etc., from E-waste (discarded electronic waste) through the production of complexing agents like citric acid, gluconic acid, oxalic acid and tartaric acid (Jain et al., 2017).

10.6.3 *MYCOREMEDIATION OF MICROBIOLOGICAL AND OXYGEN DEPLETING POLLUTANTS*

In aquatic ecosystems, microbiological pollution and oxygen depleting pollution are interrelated and interdependent. Microbiological pollution leads to oxygen depletion, which in turn promotes the growth of harmful anaerobic microbes that cause several dreadful diseases. Fungi that act as mycoremediators of microbiological pollutants in turn help in restoring the aquatic ecosystems from oxygen depleting pollutants.

Fungi have proven effective in mitigating several microbiological pollutants. *Escherichia coli*, a major pathogenic microbiological pollutant, was eradicated from aquatic environs through mycoremediation technique (Dutta and Hyder, 2017). Several fungi have been used for the production of nanoparticles (For example, nanosilver) which are antimicrobial in nature (Gadd, 2016).

Mycofiltration (filtration through fungal mycelia) is a promising eco-technique for the management of high-quality water (Rhodes, 2015). This approach wherein cultivated fungi are added to surface water management practices was invented by Paul Stamets in 1970's. Garden giant mushrooms were employed in mycofiltration to decontaminate fecal coliform-contaminated water (Stamets, 2005). Around aquatic ecosystems, fungal mycelia act as a myco-filter system for harmful compounds and microorganisms (Varjani and Patel, 2017).

10.6.4 MYCOREMEDIATION POTENTIAL OF MYCORRHIZAL FUNGI AND MUSHROOMS

The use of mycorrhizal fungi in remediation is known as mycorrhizoremediation (Kumar, 2017). This technology has been used for recuperation of contaminated ecosystems and several workers have reported that native microbes (including non-symbiotic fungi) act in synergy with the mycorrhizal fungi in degradation of pollutants. Mycorrhizal fungi have role in metal immobilization through their protective metal-binding ability (Gadd, 2016), which protect aquatic environs by locking metals and other perilous pollutants.

The process of mycoremediation by employing arbuscular mycorrhizal fungi (AMF) is an economic and ecofriendly technique. The use of AMF for mycoremediation purposes like sequestration of heavy metals have been reported by several workers (Barea et al., 2003; Gadd, 2016), and the molecular aspects of this mycoremediation have also been studied. AMF bind and accumulate heavy metals within the plant roots and extramatrical fungal mycelium, thus restrict their translocation into aboveground plant parts (Hassan et al., 2017; Kumar, 2017).

The mycoremediation potential of aboveground fruiting bodies (mushrooms) of fungi has also been confirmed. The degradation of several pollutants like biodegradable plastic, dyes, and heavy metals have been carried out by means of mycoremediation potential of several macro-fungi belonging to genus *Agaricus, Boletus, Hygrophorus, Lepiota, Lepis, Pleurotus, Psalliota, Rhizopogon, Russula, Suillus, Trametes* (Luz et al., 2013; Kulshreshtha et al., 2014; Rhodes, 2015; Deshmukh et al., 2016). Biosorbent prepared from aged mycelia and spent mushroom substrate (SMS) (produced after harvesting mushrooms) can be utilized for the remediation of the polluted sites (Kapahi and Sachdeva, 2017) through the production of some extracellular enzymes viz., cellulases, ligninase (lignin peroxidase, manganese dependent peroxidase, and laccase), oxidases, pectinases, peroxidases, and xylanases (Casieri et al., 2010; Kulshreshtha et al., 2014). The majority of these mushrooms have been used for soil systems. However, pure inoculum can be produced by culturing these fungi and can be applied in aquatic systems as well. Only a small fraction of mycorrhizal fungi and mushrooms have so far been studied with reference to their remediation potential, and further studies surely warrant the discovery of many novel proficient mycoremediators.

10.7 FUTURE PERSPECTIVES AND CONCLUSION

In aquatic ecosystems, fungi play an imperative function of remediation of range of hazardous compounds. The prospective role of fungi as most efficient remediators lies in:

1. **Large-Scale Application and Commercialization:** Till date, majority of the studies on mycoremediation has been carried out only on small laboratory-scale. Hence, the mycoremediation potential of fungi is yet to be fully utilized commercially. Therefore, further research is pre-requisite to optimize the protocol with reference to natural variables and their applicability in large-scale mycoremediation of vast contaminated aquatic ecosystems.

2. **Use of Biotechnological Tools:** For a thorough understanding and utilization of mycoremediation potential of fungi fully, genomic level studies of these fungi is obligatory. Whole genome studies through various biotechnological tools can facilitate the exploration of mycoremediation pathways. Particularly the proper identification and characterization of fungal genes involved in mycoremediation will enable researchers to enhance their expression and thus will further abet in mycoremediation studies. For example, more efficient fungal strains with superior performance under adverse conditions may be developed by using technique of genetic engineering. Such biotechnological approaches may further promote the commercialization of mycoremediation.

3. **Application of Omics Approach:** With the advent and advancement of metagenomics, metatranscriptomics, metaproteomics, and metabolomics; new fungal species can be identified, isolated, and exploited for more efficient remediation of aquatic environs. Furthermore, an improved perception of fungal community structure and function through omics approach will formulate mycoremediation as the most preferred technique in the future.

4. **Nanoparticle Formation:** Fungi can be used for the production of nanoparticles that have role in metal remediation and antimicrobial

treatments. In the field of bioremediation, the full potential of these nanoparticles is yet to be realized.

5. **Further Research:** Although lots of work has been done in the field of mycoremediation, still some gray areas like fungal physiology, ecology, pollutant remediation pathways; protracted lag-phase, excessive slush production, level of toxicity of end-product, designing of improved expression systems for bioremediation, etc., necessitate further comprehensive research in this direction.

6. **Direct Use of Fungal Enzymes:** Direct use of cell-free fungal enzymes in pollutant degradation is an efficient prospective substitute as it overcomes several drawbacks allied to the growth of whole-cell fungal cultures. Moreover, discovery of novel efficient fungal enzymes will revolutionize the field of mycoremediation.

7. **Use of Synthetic Microbial Community:** Synthetic microbial community is formed by amalgamation of two or more defined microbial populations in a controlled environment, wherein merged populations can perform complex functions more efficiently. Usage of different fungi in consortium as synthetic microbial communities should be further investigated in order to deal efficiently with the multifaceted aquatic pollution.

In conclusion, fungi, and their enzymes act as persuasive remediation agents of aquatic pollutants. The main advantage of mycoremediation is the reduction of pollutants present in the aquatic environs by an economical and eco-friendly technique. This chapter provides information about role of various fungi, their enzymes, metabolic capabilities, and mechanisms in mycoremediation of range of aquatic pollutants. Mycoremediation is an efficient and environmentally practicable tool against aquatic pollution and offers new perspectives of its use, coupled with advanced biotechnological tools. Advanced studies are required to shed light on the molecular mechanism of fungal metabolic pathways, for further exploration of the genes and enzymes involved, so as to further improve mycoremediation strategies for efficient restoration and protection of degraded aquatic environs. This chapter provides substantial information about mycoremediators, and in the future, it will facilitate researchers, policymakers, and funding agencies in taking up aquatic ecosystem management projects primarily based on mycoremediation.

KEYWORDS

- **aquatic pollutants**
- **biodegradation**
- **fungal enzymes**
- **mycofiltration**
- **mycoremediation**
- **mycorrhizal fungi**

REFERENCES

Abadulla, E.; Tzanov, T.; Costa, S.; Robra, K. H.; Cavaco–Paula, A.; Gubitz, G. M. Decolorization and detoxification of textile dyes with a laccase from *Trametes hirsute*. *Appl. Environ. Microbiol.* **2000**, *66*, 3357–3362.

Adak, A.; Tiwari, R.; Singh, S.; Sharma, S.; Nain, L. Laccase production by a novel white–rot fungus *Pseudolagaro basidiumacaciicola* LA 1 through solid–state fermentation of *Parthenium* biomass and its application in dyes decolorization. *Waste Biomass Valor.* **2016**, *7*, 1427–1435.

Adebayo, A. O. Investigation on *Pleurotus ferulae* potential for the sorption of Pb(II) from aqueous solution. *Bull. Chem. Soc. Ethiop.* **2013**, *27*, 25–34.

Ahmed, I.; Zafar, S.; Ahmad, F. Heavy metal biosorption potential of *Aspergillus* and *Rhizopus* sp. isolated from wastewater treated soil. *J. Appl. Sci. Environ. Manage.* **2005**, *9*, 123–126.

Akcil, A.; Erust, C.; Ozdemiroglu, S.; Fonti, V.; Beolchini, F. A review of approaches and techniques used in aquatic contaminated sediments: metal removal and stabilization by chemical and biotechnological processes. *J. Clean. Prod.* **2015**, *86*, 24–36.

Akdogan, H. A.; Topuz, M. C. Decolorization of turquoise blue HFG by *Coprinus plicatilis* for water bioremediation. *Biorem J.* **2014**, *18*, 287–294.

Alam, m.; Nadeem, R.; Jilani, M. I. Pb(II) removal from wastewater using pomegranate waste biomass. *Int. J. Chem. Biochem. Sci.* **2012**, *1*, 24–29.

Alexandropoulou, M.; Antonopoulou, G.; Fragkou, E.; Ntaikou, I.; Lyberatos, G. Fungal pretreatment of willow sawdust and its combination with alkaline treatment for enhancing biogas production. *J. Environ. Manag.* **2017**, *203,* 704–713.

Al–Hawash, A. B.; Alkooranee, J. T.; Abbood, H. A.; Zhang, J.; Sun, J.; Zhang, X.; Ma, F. Isolation and characterization of two crude oil–degrading fungi strains from Rumaila oil field. *Iraq. Biotechnol. Rep.* **2018a**, *17*, 104–109.

Al–Hawash, A. B.; Zhang, X.; Ma, F. Removal and biodegradation of different petroleum hydrocarbons using the filamentous fungus *Aspergillus* sp. RFC–1. *Microbiology open.* **2018b**, 619.

Ali, E. H.; Hashem, M. Removal efficiency of the heavy metals Zn(II), Pb(II) and Cd(II) by *Saprolegnia delica* and *Trichoderma viride* at different pH values and temperature degrees. *Mycobio.* **2007**, *35,* 135–144.

Aloysius, R.; Karim, M. I. A.; Ariff, A. B. The mechanisms of cadmium removal from aqueous solution by nonmetabolizing free and immobilized live biomass of *Rhizopus oligosporus*. *World J. Microbiol. Biotech.* **1999**, *15*, 571–578.

Alzahrani, A. M. Insects cytochrome P450 enzymes: Evolution, functions and methods of analysis. *Glob. J. Mol. Sci.* **2009**, *4*, 167–179.

Apohan, E.; Ozfer Y. Role of white rot fungus *Funalia trogii* in detoxification of textile dyes. *J. Basic. Microbiol.* **2005**, *45*, 99–105.

Aravinna, P.; Priyantha, N.; Pitawala, A.; Yatigammana, S. K. Use pattern of pesticides and their predicted mobility into shallow groundwater and surface water bodies of paddy lands in Mahaweli river basin in Sri Lanka. *J. Environ. Sci. Health. Part B.* **2017**, *52*, 37–47.

Arbanah, M.; Miradatul Najwa, M. R.; Ku Halim, K. H. Biosorption of Cr(III), Fe(II), Cu(II), Zn(II) ions from liquid laboratory chemical waste by *Pleurotus ostreatus*. *Int. J. Biotechnol. Wellness. Ind.* **2012**, *1*, 152–162.

Arbanah, M.; Miradatul Najwa, M. R.; Ku Halim, K. H. Utilization of *Pleurotus ostreatus* in the removal of Cr(VI) from chemical laboratory waste. *Int. Refreed. J. Eng. Sci.* **2013**, *2*, 29–39.

Argolo, E.; Lins, M. C. M.; Palha, M. F.; Bastos de Almeida, Y. M.; Lima, M. Biodegradation of polymeric films by action of the *Talaromyces wortmanii* and *Phanerochaete chrysosporium* fungi. *Braz. Arch. Biol. Technol.* **2006**, *49*, 11–19.

Arica, M. Y.; Arpa, C.; Kaya, B.; Bektaş, S.; Denizli, A.; Genç, O. Comparative biosorption of mercuric ions from aquatic systems by immobilized live and heat inactivated *Trametes versicolor* and *Pleurotus sajur–caju*. *Bioresour. Technol.* **2003**, *89*, 145–154.

Arun, A.; Raja, P.P.; Arthi, R.; Ananthi, M.; Kumar, K.S.; Eyini, M. Polycyclic aromatic hydrocarbons (PAHs) biodegradation by basidiomycetes fungi, *Pseudomonas* isolate, and their cocultures: comparative *in vivo* and *in silico* approach. *Appl. Biochem. Biotechnol.* **2008**, *151*, 132–142.

Arunprasath, T.; Sudalai, S.; Meenatchi, R.; Jeyavishnu, K.; Arumugam, A. Biodegradation of triphenylmethane dye malachite green by a newly isolated fungus strain. *Biocat. Agri. Biotech.* **2019**, *17*, 672–679.

Asgher, M.; Noreen, S.; Bilal, M. Enhancing catalytic functionality of *Trametes versicolor* IBL–04 laccase by immobilization on chitosan microspheres.*Chem. Eng. Res. Des.* **2017**, *119*, 1–11.

Asgher, M.; Shah, S. A. H.; Ali, M.; Legge, R. L. Decolorization of some reactive textile dyes by white rot fungi isolated in Pakistan. *World J. Microbiol. Biotechnol.* **2005**, *22*, 89–93.

Asgher, M.; Yasmeen, Q.; Iqbal, H. M. N. Enhanced decolorization of Solar brilliant red 80 textile dye by an indigenous white rot fungus *Schizophyllum commune* IBL–06. *Saudi J. Biol. Sci.* **2013**, *20*, 347–352.

Atagana, H. I. Fungal bioremediation of PAHs in the presence of heavy metals in soil. *Proceedings of the International Conference on Remediation of Chlorinated and Recalcitrant Compounds, United States.* **2006**.

Aust, S. D. Mechanisms of degradation by white rot fungi. *Environ. Health Perspect.* **1995**, *103*, 59–61.

Badawi, N.; Ronhede, S.; Olsson, S.; Kragelund, B. B.; Johnsen, A. H.; Jacobsen, O. S. Metabolites of the phenylurea herbicides chlorotoluron, diuron, isoproturon and linuron produced by the soil fungus *Mortierella* sp. *Environ. Pollut.* **2009**, *157*, 2806–2812.

Bai, Y.; Wang, Q.; Liao, K.; Jian, Z.; Zhao, C.; Qu, J. Fungal community as a bioindicator to reflect anthropogenic activities in a river ecosystem. *Front. Microbiol.* **2018**, *9*, 3152.

Baldrian, P. Wood–inhabiting ligninolytic basidiomycetes in soils: Ecology and constraints for applicability in bioremediation. *Fungal. Ecol.* **2008**, *1*, 4–12.

Barea, J. M.; Vivas, A.; Marulanda, A.; Gomez, M.; Azcon, R. Physiological characteristics (SDH and ALP activities) of Arbuscular mycorrhizal colonization as affected by *Bacillus thuringiensis* inoculation under two phosphorus levels. *Soil Biol. Biochem.* **2003**, *35*, 987–999.

Bath, H. K.; Arora, D. S. Remediation of chlorobenzoic acids by some white–rot fungi. *Proceedings of the International Conference on Remediation of Chlorinated and Recalcitrant Compounds, United States.* **2006**.

Bending, G. D.; Friloux, M.; Walker, A. Degradation of contrasting pesticides by white rot fungi and its relationship with ligninolytic potential. *FEMS. Microbiol. Lett.* **2002**, *212*, 9–63.

Bending, M. P.; Anderron, A.; Ander, P.; Stenström, J.; Torstensson, L. Establishment of white–rot fungus *Phanerochaete chrysosporium* on unsterile straw in solid substrate fermentation system intended for degradation of pesticides. *World J. Microbiol. Biotechnol.* **2001**, *17*, 627–633.

Bezalel, L.; Hadar, Y.; Cerniglia, C. E. Enzymatic mechanisms involved in phenanthrene degradation by the white rot fungus *Pleurotus ostreatus*. *Appl. Environ. Microbiol.* **1996c**, *63*, 2495–2501.

Bezalel, L.; Hadar, Y.; Fu, P. P.; Freeman, J. P.; Cerniglia, C. E. Metabolism of phenanthrene by the white rot fungus *Pleurotus ostreatus*. *Appl. Environ. Microbiol.* **1996a**, *62*, 2547–2553.

Bezalel, L.; Hadar, Y.; Fu, P. P.; Freeman, J. P.; Cerniglia, C. E. Initial oxidation products in the metabolism of pyrene, anthracene, fluorene, and dibenzothiophene by the white rot fungus *Pleurotus ostreatus*. *Appl. Environ. Microbiol.* **1996b**, *62*, 2554–2559.

Bhalerao, T. S. Bioremediation of endosulfan–contaminated soil by using bioaugmentation treatment of fungal inoculant *Aspergillus niger*. *Turk. J. Biol.* **2012**, *36*, 561–567.

Bhalerao, T. S.; Puranik, P. R. Biodegradation of organochlorine pesticide, endosulfan, by a fungal soil isolate, *Aspergillus niger*. *Int. Biodeter. Biodegrad.* **2007**, *59*, 315–321.

Bhandari, G. Mycoremediation: An eco–friendly approach for degradation of pesticides. In: R. Prasad (ed.), Mycoremediation and Environmental Sustainability, Fungal Biology, *Springer International Publishing AG*. **2017**, 119–131.

Bhattacharya, S.; Das, A.; Mangai, G.; Vignesh, K.; Sangeetha, J. Mycoremediation of congo red dye by filamentous fungi. *Brazilian J. Microbiol.* **2011**, *42*, 1526–1536.

Bhattacharya, S.; Das, A.; Prashanthi, K.; Palaniswamy, M.; Angayarkanni, J. Mycoremediation of benzo[a]pyrene by *Pleurotus ostreatus* in the presence of heavy metals and mediators. *3 Biotech.* **2014**, *4*, 205–211.

Bhosle, N.P.; Nasreen, S. Remediation of cypermethrin–25 EC by microorganisms. *Eur. J. Exp. Biol.* **2013**, *3*, 144–152.

Birhanli, E.; Erdogan, S.; Yesilada, O.; Onal, Y. Laccase production by newly isolated white rot fungus *Funalia trogii*: effect of immobilization matrix on laccase production. *Biochem. Eng. J.* **2013**, *71*, 134–139.

Birhanli, E.; Yesilada, O. Increased production of laccase by pellets of *Funalia trogii* ATCC 200800 and *Trametes versicolor* ATCC 200801 in repeated–batch mode. *Enzym. Microb. Technol.* **2006**, *39*, 1286–1293.

Bogan, B. W.; Lamar, R. T. One–electron oxidation in the degradation of creosote polycyclic aromatic hydrocarbons by *Phanerochaete chrysosporium*. *Appl. Environ. Microbiol.* **1995**, *61*, 2631–2635.

Bogan, B. W.; Lamar, R. T. Polycyclic aromatic hydrocarbon–degrading capabilities of *Phanerochaete laevis* HHB–1625 and its extracellular ligninolytic enzymes. *Appl. Environ. Microbiol.* **1996**, *62*, 1597–1603.

Bressa, G.; Coma, L.; Costa, P. Bioaccumulation of Hg in the mushroom *Pleurotus ostreatus*. *Ecotoxicol. Environ. Safe.* **1988**, *16*, 85–89.

Bulla, L. M. C.; Polonio, J. C.; Portela–Castro, A. L. B.; Kava, V.; Azevedo, J. L.; Pamphile, J. A. Activity of the endophytic fungi *Phlebia* sp. and *Paecilomycesformosus* in decolourisation and the reduction o reactive dyes' cytotoxicity in fish erythrocytes. *Environ. Monit. Assess.* **2017**, *189*, 1–11.

Bumpus, J. A. Biodegradation of polycyclic aromatic hydrocarbons by *phanerochaete chrysosporium*. *Appl. Environ. Microbiol.* **1989**, *55*, 154–158.

Bumpus, J. A.; Tien, M.; Wright, D.; Aust, S. D. Oxidation of persistent environmental pollutants by a white rot fungus. *Science.* **1985**, *4706*, 1434–1436.

Bumpus, J. A.; Aust, S. D. Biodegradation of DDT (1,1,1–trichloro–2,2–bis–(4–chlorophenyl) ethane) by the white–rot fungus *Phanerochaete chrysosporium*. *Appl. Environ. Microb.* **1987**, *53*, 2001–2008.

Carles, L.; Rossi, F.; Joly, M.; Besse–Hoggan, P.; Batisson, I.; Artigas, J. Biotransformation of herbicides by aquatic microbial communities associated to submerged leaves. *Environ. Sci. Pollut. Res.* **2017**, *24*, 3664–3674.

Carol, D.; Kingsley, S. J.; Vincent, S. Hexavalent chromium removal from aqueous solutions by *Pleurotus ostreatus* spent biomass. *Int. J. Eng. Sci. Technol.* **2012**, *4*, 7–22.

Carvalho, M. B.; Tavares, S.; Medeiros, J.; Núñez, O.; Gallart–Ayala, H.; Leitão, M. C.; Galceran, M. T.; Hursthouse, A.; Pereira, C. S. Degradation pathway of pentachlorophenol by *Mucor plumbeus* involves phase II conjugation and oxidation–reduction reactions. *J. Hazard. Mater.* **2011**, *198*, 133–142.

Cases, I.; de Lorenzo, V. Promoters in the environment: Transcriptional regulation in its natural context. *Nat. Rev. Microbiol.* **2005**, *3*, 105–118.

Casieri, L.; Anastasi, A.; Prigione, V.; Varese, G. C. Survey of ectomycorrhizal, litter–degrading, and wood–degrading basidiomycetes for dye decolorization and ligninolytic enzyme activity. *Anton. Leeuw.* **2010**, *98*, 483–504.

Ceci, A.; Pinzari, F.; Russo, F.; Persiani, A. M.; Gadd, G. M. Roles of saprotrophic fungi in biodegradation or transformation of organic and inorganic pollutants in co–contaminated sites. *Appl. Microbiol. Biotechnol.* **2018**.

Cerniglia, C.; Yang, S. Stereoselective metabolism of anthracene and phenanthrene by the fungus *Cunninghamella elegans*. *Appl. Environ. Microbiol.* **1984**, *47*, 119–124.

Chen, C.; Wang, J. Influence of metal ionic characteristics on their biosorption capacity by *Saccharomyces cerevisiae*. *Appl. Microbiol. Biotechnol.* **2007**, *74*, 911–917.

Chen, F.; Liang, L.; Tang, Y.; Mao, L. Research on anthracene removal from water by immobilized *Fusarium oxysporum*. *Zhongguo Jishui Paishui.* **2007**, *23*, 77–80.

Chen, F.; Tang, Y.; Mao, L.; Liang, L. Research on biodegradation characteristics of phenanthrene by *Fusarium solani* strain. *Jiangsu Keji Daxue Xuebao, Ziran Kexueban.* **2008**, *22*, 72–76.

Chen, M.; Xu, P.; Zeng, G.; Yang, C.; Huang, D.; Zhang, J. Bioremediation of soils contaminated with polycyclic aromatic hydrocarbons, petroleum, pesticides, chlorophenols and heavy metals by composting: applications, microbes and future research needs. *Biotechnol. Adv.* **2015**, *33*, 745–755.

Chen, S. H.; Ting, A. S. Y. Biodecolorization and biodegradation potential of recalcitrant triphenylmethane dyes by *Coriolopsis sp.* isolated from compost. *J Environ Manage.* **2015,** *150,* 274–280.

Chen, S. H.; Cheow, Y. L.; Ng, S. L; Ting, A. S. Y. Biodegradation of triphenylmethane dyes by non–white rot fungus *Penicillium simplicissimum*: Enzymatic and toxicity studies. *Int. J. Environ. Res.* **2019,** 1–10.

Chhabra, M.; Mishra, S.; Sreekrishnan, T. R. Immobilized laccase mediated dye decolorization and transformation pathway of azo dye acid red 27. *J. Environ. Health. Sci.* **2015,** *13,* 38–46.

Chiu, S. W.; Law, S. C.; Ching, M. L.; Cheung, K. W.; Chen, M. J. Themes for mushroom exploitation in the 21st century: Sustainability, waste management, and conservation. *J. Gen. Appl. Microbiol.* **2000,** *46,* 269–282.

Chivukula, M.; Renganathan, V. Phenolic azo dye oxidation by laccase from *Pyricularia oryzae. Appl. Environ. Microbiol.* **1995,** *61,* 4347–4377.

Collins, P. J.; Kotterman, M.; Field, J. A.; Dobson, A. Oxidation of anthracene and benzo [a] pyrene by laccases from *Trametes versicolor. Appl. Environ. Microbiol.* **1996,** *62,* 4563–4567.

Da Silva Coelho, J.; de Oliveira, A. L.; de Souza, C. G. M.; Bracht, A.; Peralta, R. M. Effect of the herbicides bentazon and diuron on the production of ligninolytic enzymes by *Ganoderma lucidum. Int. Biodeter. Biodegrad.* **2010,** *64,* 156–161.

Damisa, D.; Oyegoke, T. S.; Ijah, U. J. J.; Adabara, N. U.; Bala, J. D,; Abdulsalam, R. Biodegradation of petroleum by fungi isolated from unpolluted tropical soil. *Int. J. Appl. Biol. Pharm. Technol.* **2013,** *4,* 136–140.

Damodaran, D.; Suresh, G.; Mohan, R. B. Bioremediation of soil by removing heavy metals using *Saccharomyces cerevisiae.* IACSIT Press, Singapore. **2011**.

Danış, Ü. Biosorption of copper(II) from aqueous solutions by *Pleurotus cornucopiae.* BALWOIS 2010, Ohrid, Republic of Macedonia, 25–29 May **2010**.

D'Annibale, A.; Quaratino, D.; Federici, F.; Fenice, M. Effect of agitation and aeration on the reduction of pollutant load of olive mill wastewater by the white–rot fungus *Panus tigrinus. Biochem. Eng. J.* **2006a,** *29,* 243–249.

D'Annibale, A.; Rosetto, F.; Leonardi, V.; Federici, F.; Petruccioli, M. Role of autochthonous filamentous fungi in bioremediation of a soil historically contaminated with aromatic hydrocarbons. *Appl. Environ. Microbiol.* **2006b,** *72,* 28–36.

Daoubi, M.; Duran–Patron, R.; Hernandez–Galan, R.; Benharref, A.; Hanson–James R.; Collado, I. G. The role of botrydienediol in the biodegradation of the sesquiterpenoid phytotoxinbotrydial by *Botrytis cinerea. Tetrahedron.* **2006,** *62,* 8256–8261.

Das, N.; Charumathi, D.; Vimala, R. Effect of pretreatment on Cd2+ biosorption by mycelia biomass of *Pleurotus florida. Afr. J. Biotechnol.* **2007,** *6,* 2555–2558.

Dayi, B.; Kyzy, A. D.; Akdogan, H. A. Characterization of recuperating talent of white–rot fungi cells to dye–contaminated soil/water. *Chinese. J. Chem. Eng.* **2019,** *27,* 634–638.

de Almeida, L. K.; Burgess, J. E. Biosorption and bioaccumulation of copper and lead by *Phanerochaete* and *Pleurotus ostreatus.* **2013**.

Deng, S.; Ting, Y. P. Characterization of PEI–modified biomass and biosorption of Cu(II), Pb(II) and Ni(II). *Water. Res.* **2005,** *39,* 2167–2177.

Deshmukh, R.; Khardenavis, A. A.; Purohit, H. J. Bioprocess for solid waste management. In: H. J. Purohit et al., (eds.), Optimization and Applicability of Bioprocesses, *Springer Nature Singapore Pte Ltd.* **2017,** 73–99.

Deshmukh, R.; Khardenavis, A. A.; Purohit, H. J. Diverse metabolic capacities of fungi for bioremediation. *Indian J. Microbiol.* **2016**, *56*, 247–264.

Dias, M. A. et al., Removal of heavy metals by an *Aspergillus terreus* strain immobilized in a polyurethane matrix. *Letters Appl. Microbiol.* **2002**, *34*, 46–50.

Diwanian, S.; Kharb, D.; Raghukumar, C.; Kuhad, R. C. Decolorization of synthetic dyes and textile effluents by basidiomycetous fungi. *Water Air Soil Pollut.* **2010**, *210*, 409–419.

Dulay, R. M. R.; De Castro, M. A. E. G.; Coloma, N. B.; Bernardo, A. P.; Cruz, A. G. D.; Tiniola, R. C.; Kalaw, S. P.; Reyes, R. G. Effects and myco–remediation of lead (Pb) in five *Pleurotus* mushrooms. *Int. J. Biol. Pharm. Allied Sci.* **2015**, *4*, 1664–1677.

Durrant, A. J.; Wood, D. A.; Cain, R. B. Lignocellulose biodegradation by *Agaricus bisporus* during solid substrate fermentation. *J. Gen. Microbiol.* **1991**, *137*, 1–755.

Dutta, S. D.; Hyder, M. S. Mycoremediation – a potential tool for sustainable management. *Conference Paper.* **2017**.

Dwivedi, S.; Mishra, A.; Saini, D. Removal of heavy metals in liquid media through fungi isolated from wastewater. *Int. J. Sci. Res.* **2012**, *1*, 181–185.

Eichlerova, I.; Homolka, L.; Lisa, L.; Nerud, F. Orange G and Remazol Brilliant Blue R decolorization by white rot fungi *Dichomitus squalens*, *Ischnoderma resinosum* and *Pleurotus*. *Chemosphere.* **2005**, *60*, 398–404.

Eichlerová, I.; Homolka, L.; Benada, O.; Kofroňova, O.; Hubálek, T.; Nerud, F. Decolorization of Orange G and Remazol Brilliant Blue R by the white rot fungus *Dichomitus squalens*: Toxicological evaluation and morphological study. *Chemosphere.* **2007**, *69*, 795–802.

El–Enshasy, H. A.; Hanapi, S. Z.; Abdelgalil, S. A.; Malek, R. A.; Pareek, A. Mycoremediation: Decolourization potential of fungal ligninolytic enzymes. In: R. Prasad (ed.), Mycoremediation and Environmental Sustainability, Fungal Biology, *Springer International Publishing AG*, **2017**, 69–104.

Elshafie, A.; AlKindi, A. Y.; Al–Busaidi, S.; Bakheit, C.; Albahry, S. N. Biodegradation of crude oil and n–alkanes by fungi isolated from Oman. *Marine. Poll. Bull.* **2007**, *54*, 1692–1696.

Erkurt, E. A.; Unyayar, A.; Kumbur, H. Decolorization of synthetic dyes by white rot fungi, involving laccase enzyme in the process. *Process Biochem.* **2007**, *42*, 1429–1435.

Erum, S.; Ahmed, S. Comparison of dye decolorization efficiencies of indigenous fungal isolates. *Afr. J. Biotechnol.* **2011**, *10*, 3399–3411.

Faulkner, J. K.; Woodcock, D. Metabolism of 2, 4–dichlorophenoxyacetic acid ('2, 4–D') by *Aspergillus niger* van Tiegh. *Nature.* **1964**, *203*, 865.

Favero, N.; Bressa, G.; Costa, P. Response of *Pleurotus ostreatus* to cadmium exposure. *Ecotoxicol. Environ. Safe.* **1990a**, *20*, 1–6.

Favero, N.; Costa, P.; Paolo Rocco, G. Role of copper in cadmium metabolism in the basidiomycetes *Pleurotus ostreatus*. *Comp. Biochem. Physiol. Part. C. Comp. Pharmacol.* **1990b**, *97*, 297–303.

Field, J. A.; Dejong, E.; Feijoocosta, G.; Debont, J. A. M. Screening for ligninolytic fungi applicable to the biodegradation of xenobiotics. *Trends biotechnol.* **1993**, *11*, 44–49.

Ford, C. I.; Walter, M.; Northcott, G. L.; Hong J.; Cameron, K. C.; Trower, T. Fungal inoculum properties: Extracellular enzyme expression and pentachlorophenol removal by New Zealand *Trametes* sp. in contaminated field soils. *J. Environ. Quality.* **2007**, *36*, 1749–1759.

Fratila–Apachitei, L. E.; Hirst, J. A.; Siebel, M. A.; Gijzen, H. J. Diuron degradation by *Phanerochaete chrysosporium* BKM–F–1767 in synthetic and natural media. *Biotechnol Lett.* **1999**, *21*, 147–154.

284 *Freshwater Pollution and Aquatic Ecosystems*

Friedrich, J.; Zalar, P.; Mohorcic, M.; Klun, U.; Krzan, A. Ability of fungi to degrade synthetic polymer nylon–6. *Chemosphere,* **2007,** *67,* 2089–2095.

Frutos, I.; García–Delgado, C.; Gárate, A.; Eymar, E. Biosorption of heavy metals by organic carbon from spent mushroom substrates and their raw materials. *Int. J. Environ. Sci. Technol.* **2016,** *13,* 2713–2720.

Fu, Y. Q.; Li, S.; Zhu, H. Y.; Jiang, R.; Yin, L. F. Biosorption of copper(II) from aqueous solution by mycelial pellets of *Rhizopus oryzae. Afr. J. Biotechnol.* **2012,** *11,* 1403–1411.

Fuentes, M. S.; Briceño, G. E.; Saez, J. M.; Benimeli, C. S.; Diez, M. C.; Amoroso, M. J. Enhanced removal of a pesticides mixture by single cultures and consortia of free and immobilized *Streptomyces* strains. *Bio Med.* **2013.**

Gadd, G. M. Fungi and industrial pollutants. In: Environmental and Microbial Relationships, 3rd Edition, The Mycota IV I. S. Druzhinina and C. P. Kubicek (Eds.), *Springer International Publishing Switzerland.* **2016.**

Gahlout, M.; Gupte, S.; Gupte, A. Optimization of culture condition for enhanced decolorization and degradation of azo dye reactive violet 1 with concomitant production of ligninolytic enzymes by *Ganoderma cupreum* AG–1. *3 Biotech.* **2013,** *3,* 143–152.

Gao, H.; Xie, Y.; Hashim, S.; Khan, A. A.; Wang, X.; Xu, H. Application of microbial technology used in bioremediation of urban polluted river: A case study of Chengnan River, China. *Water.* **2018,** *10,* 643.

Gazem, M. A. H.; Nazareth, S. Sorption of lead and copper from an aqueous phase system by marine–derived *Aspergillus* species. *Ann. Microbiol.* **2013,** *63,* 503–511.

Gesinde, A. F.; Agbo, E. B.; Agho, M. O.; Dike, E. F. C. Bioremediation of some Nigerian and Arabian crude oils by fungal isolates. *Int. Jor. P. App. Scs.* **2008,** *2,* 37–44.

Gouma, S.; Anastasia, A.; Papadaki.; Markakis, G,; Magan, N.; Goumas, D. Studies on pesticides mixture degradation by white rot fungi. *JEE.* **2019,** *20,* 16–26.

Grassi, E.; Scodeller, P.; Filiel, N.; Carballo, R.; Levin, L. Potential of *Trametes trogii* culture fluids and its purified laccase for the decolorization of different types of recalcitrant dyes without the addition of redox mediators. *Int. Biodeterior. Biodegrad.* **2011,** *65,* 635–643.

Guiraud, P.; Bonnet, J.L.; Boumendjel, A.; Kadri–Dakir, M.; Dusser, M.; Bohatier, J.; Steiman, R. Involvement of *Tetrahymena pyriformis* and selected fungi in the elimination of anthracene, and toxicity assessment of the biotransformation products. *Ecotox. Environ. Safe.* **2008,** *69,* 296–305.

Gupte, A.; Tripathi, A.; Rudakiya, D.; Patel, H.; Gupte, S. Bioremediation of polycyclic aromatic hydrocarbon (PAHs): A perspective. *The Open Biotech. J.* **2016,** *10,* 363–378.

Habib, K.; Schmidt, J. H.; Christensen, P. A historical perspective of global warming potential from municipal solid waste management. *Waste Manag.* **2013,** *33,* 1926–1933.

Hadibarata, T.; Tachibana, S.; Itoh, K. Biodegradation of neicosane by fungi screened from nature. *Pak. J. Biol. Sci.* **2007b,** *10,* 1804–1810.

Hadibarata, T.; Tachibana, S.; Itoh, K. Biodegradation of phenanthrene by fungi screened from nature. *Pak. J. Biol. Sci.* **2007a,** *10,* 2535–2543.

Hadibarata, T.; Zubir, M. M.; Rubiyabto, T. Z.; Chuang, T. Z.; Yusoff, A. R.; Fulazzaky, M. A.; Seng, B.; Nugroho, A. E. Degradation and transformation of anthracene by white–rot fungus *Armillaria* sp. F022. *Folia Microbiol.* **2013,** *58,* 385–391.

Haemmerli, S. D.; Leisola, M. S. A.; Sanglard, D.; Fiechter, A. Oxidation of benzo[a]pyrene by extracellular ligninases of *Phanerochaete chrysosporium*: Veratryl alcohol and stability of ligninase. *J. Biol. Chem.* **1986,** *261,* 6900–6903.

Hammel, K. E.; Gai, W. Z.; Green, B.; Moen, M. A. Oxidative degradation of phenanthrene by the ligninolytic fungus *Phanerochaete chrysosporium*. *Appl. Environ. Microbiol.* **1992,** *58,* 1832–1838.

Hammel, K. E.; Kalyanaraman, B.; Kirk, T. K. Oxidation of polycyclic aromatic biological hydrocarbons and dibenzo[p]dioxins by *Phanerochaete chrysosporium* ligninase. *J. Chem.* **1986,** *261,* 16948–16952.

Hammel, K. E. Mechanisms for polycyclic aromatic hydrocarbon degradation by lignolytic fungi. *Environ. Heal. Pers.* **1995,** *103,* 41–43.

Hangler, M.; Jensen, B.; Rønhede, S. R. Inducible hydroxylation and demethylation of the herbicide isoproturon by *Cunninghamella elegans*. *FEMS Microbiol. Letters.* **2007,** *268,* 254–260.

Hanif, M. A.; Bhatti, H. N. Remediation of heavy metals using easily cultivable, fast growing, and highly accumulating white rot fungi from hazardous aqueous streams. *Desalination and Water Treatment.* **2015,** *53,* 238–48.

Hanif, M. A.; Bhatti, H. N.; Bhatti, I. A.; Asghar, M. Biosorption of Cr(III) and Cr(VI) by newly isolated white rot fungi: Batch and column studies. *Asian J. Chem.* **2011,** *23,* 3375–3383.

Hasan, I. F.; Al–Jawhari. Role of filamentous fungi to remove petroleum hydrocarbons from the environment. In: V. Kumar et al., (eds.), Microbial Action on Hydrocarbons. *Springer Nature Singapore.* **2018,** 567–580.

Hassan, Z.; Ali, S.; Rizwan, M.; Ibrahim, M.; Nafees, M.; Waseem, M. Role of bioremediation agents (bacteria, fungi, and algae) in alleviating heavy metal toxicity. In: V. Kumar et al., (eds.), Probiotics in Agroecosystem, *Springer Nature Singapore Pte Ltd.* **2017,** 517–537.

Helal, I. M.; Abo–El–Seoud, M. A. Fungal biodegradation of pesticide vydate in soil and aquatic system. In: 4th international conference on radiation sciences and applications. **2015,** 13–17.

Hibbett, D. S.; Donoghue, M. J. Analysis of character correlations among wood decay mechanisms, mating systems, and substrate ranges in homobasidiomycetes. *Syst. Biol.* **2001,** *50,* 215–242.

Hossain, S. K. M. Aerobic biodepolymerization studies of polyethylene using *Phanerochaete chrysosporium*. *Indian J. Environ. Protection,* **2006,** *26,* 1006–1011.

Huang, Y.; Zhao, X.; Luan, S. Uptake and biodegradation of DDT by 4 ectomycorrhizal fungi. *Sci. Total. Environ.* **2007,** *385,* 235–241.

Hultberg, M.; Bodin, H. Effects of fungal–assisted algal harvesting through biopellet formation on pesticides in water. *Biodegrad.* **2018,** *29,* 557–565.

Huo, C. L.; Shang, Y. Y.; Zheng, J. J.; He, R. X.; He, X. S. The adsorption effect of three mushroom powder on Cu2+ of low concentration. In: International symposium on water resource and environmental protection, 20–22 May **2011.**

Ikehata, K. Use of fungal laccases and peroxidases for enzymatic treatment of wastewater containing synthetic dyes. In: Sanjay K. Sharma (ed.) Green Chemistry for Dyes Removal from Wastewater. *Res. Trends Appl.* **2015,** 203–260.

Imran, M.; Crowley, D. E.; Khalid, A.; Hussain, S.; Mumtaz, M. W.; Arshad, M. Microbial biotechnology for decolorization of textile wastewaters. *Rev. Environ. Sci. Biotechnol.* **2015,** *14,* 73–92.

Iram. S.; Shabbir, R.; Zafar, H.; Javaid, M. Biosorption and bioaccumulation of copper and lead by heavy metal–resistant fungal isolates. *Arab J. Sci. Eng.* **2015,** *40,* 1867–1873.

Iskandar, N. L.; Zainudin, N. A. I. M.; Tan, S. G. Tolerance and biosorption of copper (Cu) and lead (Pb) by filamentous fungi isolated from a freshwater ecosystem. *J. Environ. Sci.* **2011,** *23,* 824–830.

Isola, D.; Selbmann, L.; de Hoog, G. S.; Fenice, M.; Onofri , S.; Prenafeta–Boldu F. X.; Zucconi, L. Isolation and screening of black fungi as degraders of volatile aromatic hydrocarbons. *Mycopathologia.* **2013,** *175,* 369–379.

Jain, A.; Yadav, S.; Nigam, V. K.; Sharma, S. R. Fungal–mediated solid waste management: A review. In: R. Prasad (ed.), Mycoremediation and Environmental Sustainability, Fungal Biology, *Springer International Publishing AG,* **2017,** 153–170.

Jain, R.; Garg, V.; Yadav, D. *In vitro* comparative analysis of monocrotophos degrading potential of *Aspergillus flavus, Fusarium pallidoroseum* and *Macrophomina* sp. *Biodegrad.* **2014,** *25,* 437–446.

Janicki, T.; Długoński, J.; Krupiński, M. Detoxification and simultaneous removal of phenolic xenobiotics and heavy metals with endocrine–disrupting activity by the non–ligninolytic fungus *Umbelopsis isabellina. J. Hazard. Mater.* **2018.**

Jauregui, J.; Valderrama, B.; Albores, A.; Vazquez–Duhalt, R. Microsomal transformation of organophosphorus pesticides by white rot fungi. *Biodegrad.* **2003,** *14,* 397–406.

Javaid, A.; Bajwa, R. Biosorption of Cr(III) ions from tannery wastewater by *Pleurotus ostreatus. Mycopathologia.* **2007,** *5,* 71–79.

Javaid, A.; Bajwa, R. Biosorption of electroplating heavy metals by some basiodiomycetes. *Mycopathologia.* **2008,** *6,* 1–6.

Javaid, A.; Bajwa, R.; Shafique, U.; Anwar. J. Removal of heavy metals by adsorption on *Pleurotus ostreatus. Biomass. Bioenergy.* **2011,** *35,* 1675–1682.

Jebapriya, G. R.; Gnanadoss, J. J. Bioremediation of textile dye using white–rot fungi: A review. *Int. J. Curr. Res. Rev.* **2013,** *5,* 1–13.

Jiang, Y.; Hao, R.; Yang, S. Equilibrium and kinetic studies on biosorption of Pb(II) by common edible macrofungi: a comparative study. *Can. J. Microbiol.* **2016,** *62,* 329–337.

Jiang, Y.; Hao, R.; Yang, S. Natural bioaccumulation of heavy metals onto common edible macrofungi and equilibrium and kinetic studies on biosorption of Pb(II) to them. *Acta. Nat. Univ. Pekin.* **2017,** *53,* 125–134.

Jibran, A. K.; Milsee.; Mol, J. P. *Pleurotus sajor–caju* Protein: A potential biosorptive agent. *Adv. Bio. Tech.* **2011,** *4,* 25–27.

Johannes, C.; Majcherczyk, A. Natural mediators in the oxidation of polycyclic aromatic hydrocarbons by laccase mediator systems. *Appl. Environ. Microbiol.* **2000,** *66,* 524–528.

Joo, J. H.; Hussein, K. A.; Hassan, S. H. A. Biosorptive capacity of Cd(II) and Pb(II) by lyophilized cells of *Pleurotus eryngii. Korean J. Soil. Sci. Fert.* **2011,** *44,* 615–624.

Joshi, P. K. et al., Bioremediation of heavy metals in liquid media through fungi isolated from contaminated sources. *Indian J. Microbiol.* **2011,** *51,* 482–487.

Junghanns, C.; Moeder, M.; Krauss, G.; Martin, C.; Schlosser, D. Degradation of the xenoestrogen nonylphenol by aquatic fungi and their laccases. *Microbiol.* **2005,** *151,* 45–57.

Kabenge, I.; Katimbo, A.; Kiggundu, N.; Banadda, N. Bioremediation technology potential for management of soil and water pollution from anticipated rapid industrialization and planned oil and gas sector in Uganda: A review. *J. Environ. Protection.* **2017,** *8,* 1393–1423.

Kamaei, I.; Takagi, K.; Kondo, R. Degradation of endosulfan and endosulfansulphate by white–rot fungus *Trametes hirsuta. J. Wood Sci.* **2011,** *57,* 317.

Kamei, I.; Kondo, R. Biotransformation of dichloro–, trichloro–, and tetrachlorodibenzo–p–dioxin by the white–rot fungus *Phlebialindtneri. Appl. Microbiol.Biotechnol.* **2005,** *68,* 560–566.

Kamei, I.; Sonoki, S.; Haraguchi, K.; Kondo, R. Fungal bioconversion of toxic polychlorinated biphenyls by white–rot fungus *Phlebia brevispora*. *Appl. Microbiol. Biotechnol.* **2006**, *73*, 932–940.

Kamei, I.; Takagi, K.; Kondo, R. Bioconversion of dieldrin by wood–rotting fungi and metabolite detection. *Pest. Manag. Sci.* **2010**, *66*, 888–891.

Kapahi, M.; Sachdeva, S. Mycoremediation potential of *Pleurotus* species for heavy metals: a review. *Bioresour. Bioprocess.* **2017**, *4*, 32.

Kapoor, A.; Viraraghavan, T. Fungi as biosorbents. In: Biosorbents for metal ions. Wase, D. A. J.; Forster, C. F. (Ed) Taylor and Francis, London. **1997**, 67–86.

Kapoor, A.; Viraraghavan, T.; Cullimore, D. R. Removal of heavy metals using fungus *Aspergillus niger*. *Bioresour. Tech.* **1999**, *70*, 95–104.

Kashefi, M.; Salaryan, P.; Hazeri, N.; Valizadeh, J.; Abdi, A.; Mohammadnia, M. S. Biosorption of Cr (VI) from aqueous solutions using *Carum copticum* stem. *Int. J. Chem. Biochem. Sci.* **2012**, *1*, 48–53.

Kasumova, S.Y. Culturing micromycetes on the medium with naphtalan petroleum. *Khabarlar–Azarbaycan Milli Elmlar Akademiyasi, Biologiya Elmlari.* **2006**, *3*, 154–160.

Kasumova, S.Y.; Babaeva, I. Several physiological and biochemical properties of naphthalan petroleum fungi–destructors. *Khabarlar–Azarbaycan Milli Elmlar Akademiyasi, Biologiya Elmlari,* **2007**, *5*, 123–128.

Kataoka, R.; Takagi, K.; Kamei, I.; Kiyota, H.; Sato, Y. Bio–degradation of dieldrin by a soil fungus isolated from a soil with annual endosulfan applications. *Environ. Sci. Techno.* **2010**, *44*, 6343–6349.

Katapodis, P.; Moukouli, M.; Christakopoulos, P. Biodegradation of indole at high concentration by persolvent fermentation with the thermophilic fungus *Sporotrichum thermophile*. *Int. Biodeterior. Biodegrad.* **2007**, *60*, 267–272.

Kinne, M.; Zeisig, C.; Ullrich, R.; Kayser, G.; Hammel, K. E.; Hofrichter, M. Stepwise oxygenations of toluene and 4–nitrotoluene by a fungal peroxygenase. *Biochem. Biophys. Res. Commun.* **2010**, *397*, 18–21.

Knowles, C. O.; Benezet, H. J. Microbial degradation of the carbamate pesticides desmedipham, phenmedipham, promecarb, and propamocarb. *Bull. Environ. Contam. Toxicol.* **1981**, *27*, 529–533.

Kocaoba, S.; Arısoy, M. The use of a white rot fungi (*Pleurotus ostreatus*) immobilized on Amberlite XAD–4 as a new biosorbent in trace metal determination. *Bioresour. Technol.* **2011**, *102*, 8035–8039.

Kodama, T.; Ding, L.; Yoshida, M.; Yajima, M. Biodegradation of anstriazine herbicide, simazine. *J. Mol. Catal. B Enzym.* **2001**, *11*, 1073–1078.

Kohlmeier, S.; Smits, T. H. M.; Ford, R. M.; Keel, C.; Harms, H.; Wick, L. Y. Taking the fungal highway: Mobilization of pollutant–degrading bacteria by fungi. *Environ. Sci. Technol.* **2005**, *39*, 4640–4646.

Koroleva, O. V.; Stepanova, E. V.; Landesman, E. O.; Vasilchenko, L. G.; Khromonygina, V. V.; Zherdev, A. V.; Rabinovich, M. L. *In vitro* degradation of the herbicide atrazine by soil and wood decay fungi controlled through ELISA technique. *Toxicol. Environ. Chem.* **2001**, *80*, 175–188.

Korte, F.; Porter, P. E. Minutes of the fifth meeting of the IUPAC terminal pesticide residues. Erbach, West Germany, **1970**.

Kulshreshtha, S.; Mathur, N.; Bhatnagar, P. Mushroom as a product and their role in mycoremediation. In: AMB Express, *Springer.* **2014**, *4*, 1–7.

Kumar, R.; Dhiman, N.; Negi, S.; Prasher, I. B.; Prakash, C. Role of fungi in dye removal. In: V. Kumar et al., (eds.), Phytobiont and Ecosystem Restitution, *Springer Nature Singapore Pte Ltd.* **2018,** 403–421.

Kumar, V. V. Mycoremediation: A step toward cleaner environment. In: R. Prasad (ed.), Mycoremediation and Environmental Sustainability, Fungal Biology, *Springer International Publishing AG.* **2017,** 171–187.

Kunjadia, P. D.; Sanghvi, G. V.; Kunjadia, A.P.; Mukhopadhyay, P. N.; Dave, G. S. Role of ligninolytic enzymes of white rot fungi (*Pleurotus* spp.) grown with azo dyes. *Springerplus.* **2016,** *5,* 1487–1495.

Kurniati, E.; Arfarita, N.; Imai, T.; Higuchi, T.; Kanno, A.; Yamamoto, K.; Sekine, M. Potential bioremediation of mercury–contaminated substrate using filamentous fungi isolated from forest soil. *J. Environ. Sci.* **2014,** *26,* 1223–1231.

Lamar, R. T.; Davis, M. W.; Dietrich, D. M.; Glaser, J. A. Treatment of a pentachlorophenol– and creosote–contaminated soil using the lignin–degrading fungus *Phanerochaete sordida*: a field demonstration. *Soil Bioil. Biochem.* **1994,** *26,* 1603–1611.

Lamar, R. T.; White, R. B. Mycoremediation: commercial status and recent developments. In: V. S. Magar et al., (eds), *Proc. Sixth Int. Symp. on In Situ and On–Site Bioremediation, San Diego, CA.* **2001,** *6,* 263–278.

Lamrood, P. Y.; Ralegankar, S. D. Biosorption of Cu, Zn, Fe, Cd, Pb and Ni by non treated biomass of some edible mushrooms. *Asian. J. Exp. Biol.* **2013,** *4,*190–195.

Lau, K. L.; Tsang, Y. Y.; Chiu, S. W. Use of spent mushroom compost to bioremediate PAH– contaminated samples. *Chemosphere.* **2003,** *52,* 1539–1546.

Lee, J. W.; Lee, S. M.; Hong, E. J.; Jeung, E. B.; Kang, H. Y.; Kim, M. K.; Choi, I. G. Estrogenic reduction of styrene monomer degraded by *Phanerochaete chrysosporium* KFRI 20742. *J. Microbiol.* **2006,** *44,* 177–184.

Levin, L.; Melignani, E.; Ramos, A. M. Effect of nitrogen sources and vitamins on ligninolytic enzyme production by some white–rot fungi: Dye decolorization by selected culture filtrates. *Bioresour. Technol.* **2010,** *101,* 4554–4563.

Li, H. X.; Zhang, R. J.; Tang, L.; Zhang, J. H.; Mao, Z. G. *In vivo* and *in vitro* decolorization of synthetic dyes by laccase from solid state fermentation with *Trametes* sp SYBC–L4. *Bioprocess Biosyst. Eng.* **2014,** *37,* 2597–2605.

Li, H.; Lan, W.; Lin, Y. Biotransformation of 1,2,3,4–tetrahydronaphthalene by marine fungus *Hypoxylono ceanicum. Fenxi. Ceshi. Xuebao.* **2005a,** *24,* 45–47.

Li, P.; Li, H.; Stagnitti, F.; Wang, X.; Zhang, H.; Gong, Z.; Liu, W.; Xiong, X.; Li, L.; Austin, C.; Barry, D. A. Biodegradation of pyrene and phenanthrene in soil using immobilized fungi *Fusarium* sp. *Bull. Environ. Contam.Tox.* **2005b,** *75,* 443–450.

Li, Y. Q.; Liu, H. F.; Tian, Z. L.; Zhu, L. H.; Wu, Y. H.; Tang, H. Q. Diesel pollution biodegradation: synergetic effect of *Mycobacterium* and filamentous fungi. *Biomed. Environ. Sci.* **2008,** *21,* 181–187.

Liew, H. H.; Tay, C. C.; Yong, S. K.; Surif, S.; Abdul Talib, S. Biosorption characteristics of lead [Pb(II)] by *Pleurotus ostreatus* biomass. In: Abstracts of the proceedings of international conference on science and social research (CSSR), Kuala Lumpur, **2010.**

Lin, J. P.; Lian, W.; Xia, L. M.; Cen, P. L. Production of laccase by *Coriolus versicolor* and its application in decolorization of dyestuffs: (II) decolorization of dyes by laccase containing fermentation broth with or without self–immobilized mycelia. *J. Environ. Sci.* **2003,** *15,* 5–8.

Lisowska, K.; Bizukojc, M.; Dlugonski, J. An unstructured model for studies on phenanthrene bioconversion by filamentous fungus *Cunninghamella elegans. Enzyme. Microb. Tech.* **2006,** *39,* 1464–1470.

Loos, R.; Locoro, G.; Comero, S.; Contini, S.; Schwesig, D.; Werres, F.; Balsaa, P.; Gans, O.; Weiss, S.; Blaha, L. Pan–European survey on the occurrence of selected polar organic persistent pollutants in ground water. *Water Res.* **2010**, *44*, 4115–4126.

Lu, R.; Ma, L.; He, F.; Yu, D.; Fan, R.; Zhang, Y.; Long, Z.; Zhang, X.; Yang, Y. White–rot fungus *Ganoderma* sp. En3 had a strong ability to decolorize and tolerate the anthraquinone, indigo and triphenylmethane dye with high concentrations. *Bioprocess. Biosyst.* Eng. **2016**, *39*, 381–390.

Luo, D.; Yf, X.; Tan, Z. L.; Li, X. D. Removal of Cu2+ ions from aqueous solution by the abandoned mushroom compost of *Flammulina velutipes*. *J. Environ. Biol.* **2013**, *4*, 359–365.

Luz, J. M. R.; Paes, S. A.; Nunes, M. D.; da Silva, M. C. S.; Kasuya, M. C. M. Degradation of oxo–biodegradable plastic by *Pleurotus ostreatus*. *PLoS One.* **2013**, *8*, 69386.

Ma, L.; Zhuo, R.; Liu, H.; Yu, D.; Jiang, M.; Zhang, X.; Yang, Y. Efficient decolorization and detoxification of the sulfonatedazo dye Reactive Orange 16 and simulated textile wastewater containing Reactive Orange 16 by the white–rot fungus *Ganoderma* sp. En3 isolated from the forest of Tzu–chin Mountain in China. *Biochem. Eng. J.* **2014**, *82*, 1–9.

Ma, X.; Ling Wu, L.; Fam, H. Heavy metal ions affecting the removal of polycyclic aromatic hydrocarbons by fungi with heavy–metal resistance. *Appl. Microbiol. Biotechnol.* **2014**, *98*, 9817–9827.

Magan, N.; Fragoeiro, S.; Bastos, C. Environmental factors and bioremediation of xenobiotics using white rot fungi. *Mycobiology.* **2010**, *38*, 238–248.

Majeed, A.; Jilani, M. I.; Nadeem, R.; Hanif, M. A.; Ansari, T. M. Novel studies for the development of hybrid biosorbent. *Int. J. Chem. Biochem. Sci.* **2012**, *2*, 78–82.

Majeed, A.; Jilani, M. I.; Nadeem, R.; Hanif, M. A.; Ansari, T. M. Adsorption of Pb(II) using novel *Pleurotus sajor–caju* and sunflower hybrid biosorbent. *Environ. Prot. Eng.* **2014**, *40*, 5–15.

Malachova, K.; Rybkova, Z.; Sezimova, H.; Cerven, J.; Novotny, C. Biodegradation and detoxification potential of rotating biological contactor (RBC) with *Irpex lacteus* for remediation of dye–containing wastewater. *Water res.* **2013**, *47*, 7143–7148.

Mandal, T. K.; Baldrian, P.; Gabriel, J.; Nerud, F.; Zadrazil F. Effect of mercury on the growth of wood–rotting basidiomycetes *Pleurotus ostreatus*, *Pycnoporus cinnabarinus* and Serpula *lacrymans*. *Chemosphere.* **1998**, *36*, 435–440.

Marco–Urrea, E.; Caminal, G.; Gabarrell, X.; Vicent, T.; Reddy, C. A. Aerobic degradation/ mineralization of trichloroethylene and perchloroethylene by white–rot fungi. *Proceedings of the International In Situ and On–Site Bioremediation Symposium, United States.* **2007**.

Marco–Urrea, E.; Gabarrell, X.; Sarra, M.; Caminal, G.; Vicent, T.; Reddy, C. A. Novel aerobic perchloroethylene degradation by the white–rot fungus *Trametes versicolor.* *Environ. Sci. Technol.* **2006**, *40*, 7796–7802.

Marco–Urrea, E.; Garcia–Romera, I.; Aranda, E. Potential of non–ligninolytic fungi in bioremediation of chlorinated and polycyclic aromatic hydrocarbons. *New biotechnol.* **2015**, *32*, 620–628.

Martins, T. M.; Núñez, O.; Gallart–Ayala, H.; Leitão, M. C.; Galceran, M. T.; Pereira, C. S. New branches in the degradation pathway of monochlorocatechols by *Aspergillus nidulans*: a metabolomics analysis. *J. Hazard.Mater.* **2014**, *268*, 264–272.

Massaccesi, G. et al., Cadmium removal capacities of filamentous soil fungi isolated from industrially polluted sediments, La Plata, Argentina. *World J. Microbio. Biotech.* **2002**, *18*, 817–820.

Matsumura, F.; Boush, G. M.; Tai, A. Breakdown of dieldrin in the soil by a microorganism. *Nature.* **1968**, *219*, 965–967.

McFarland, M.; Salladay, D.; Ash, D.; Baiden, E. Composting treatment of alachlor impacted soil amended with the white rot fungus *Phanerochaete chrysosporium*. *Hazard. Waste. Hazard. Mater.* **1996,** *13,* 363–373.

Michniewicz, A.; Ledakowicz, S.; Ullrich, R.; Hofrichter, M. Kinetics of the enzymatic decolorization of textile dyes by laccase from *Cerrena unicolor. Dyes Pigm.* **2008,** *77,* 295–302.

Mishra, A.; Malik, A. Novel fungal consortium for bioremediation of metals and dyes from mixed waste stream. *Bioresour. Technol.* **2014,** *171,* 217–226.

Mishra, R.; Sarma, V. V. Mycoremediation of heavy metal and hydrocarbon pollutants by endophytic fungi. In: R. Prasad (ed.), Mycoremediation and Environmental Sustainability, Fungal Biology, *Springer International Publishing AG.* **2017,** 133–151.

Mohanasrinivasan, V.; Suganthi, V.; Selvarajan, E.; Subathra–Devi C.; Ajith, E.; Muhammed, F. N. P.; Sreeram, G. Bioremediation of endosulfan contaminated soil. *Res. J. Chem. Environ.* **2013,** *17,* 93–101.

Mohapatra, D.; Rath, S. K.; Mohapatra, P. K. Bioremediation of insecticides by white–rot fungi and its environmental relevance. In: R. Prasad (ed.), Mycoremediation and Environmental Sustainability, Fungal Biology, *Springer International Publishing AG.* **2018,** 181–212.

Mollea, C.; Bosco, F.; Ruggeri, B. Fungal biodegradation of naphthalene: Microcosms studies. *Chemosphere.* **2005,** *60,* 636–643.

Mougin, C.; Laugero, C.; Asther, M.; Chaplain, V. Biotransformation of triazine herbicides and related degradation products in liquid culture by the white rot fungus *Phanerochaete chrysosporium. Pest. Sci.* **1997,** *49,* 169–177.

Mouhamadou, B.; Faure, M.; Sage, L.; Marcais, J.; Souard, F.; Geremia, R. A. Potential of autochthonous fungal strains isolated from contaminated soils for degradation of polychlorinated biphenyls. *Fungal Biol.* **2013,** *117,* 268–274.

Mukhopadhyay, M.; Noronha, S. B.; Suraishkumar, G. K. Kinetics modeling for the biosorption of copper by pretreated *Aspergillus niger biomass. Bioresour. Technol.* **2007,** *98,* 1781–1787.

Muncnerova, D.; Augustin, J. Fungal metabolism and detoxification of polycyclic aromatic hydrocarbons: a review. *Bioresour. Technol.* **1994,** *48,* 97–106.

Muraleedharan, T. R.; Venkobachar, C. Mechanism of biosorption of Cu2+ by *Ganoderma lucidum. Biotechnol. Bioeng.* **1990,** *35,* 320–325.

Murugesan, K. Bioremediation of paper and pulp mill effluents. *Indian J. Exp. Biol.* **2003,** *41,* 1239–1248.

Murugesan, K.; Nam, I. H.; Kim, Y. M.; Chang, Y. S. Decolorization of reactive dyes by a thermo stable laccase produced by *Ganoderma lucidum* in solid state culture. *Enzym. Microb. Technol.* **2007,** *40,* 1662–1667.

Muthezhilan, R.; Vinoth, S.; Gopi, K.; Jaffar, H. A. A dye degrading potential of immobilized laccase from endophytic fungi of coastal sand dune plants. *Int. J. Chem. Tech. Res.* **2014,** *6,* 4154–4160.

Nagy, B.; Măicăneanu, A.; Indolean, C.; Mânzatu, C.; ilaghi–Dumitrescu, M. C. Comparative study of Cd(II) biosorption on cultivated *Agaricus bisporus* and wild *Lactarius piperatus* based biocomposites. Linear and nonlinear equilibrium modelling and kinetics. *J. Taiwan. Inst. Chem. E.* **2013.**

Nam, I.; Kim, Y.; Murugesan, K.; Jeon, J.; Chang, Y.; Chang, Y. Bioremediation of PCDD/Fs–contaminated municipal solid waste incinerator fly ash by a potent microbial biocatalyst. *J. Hazard. Mater.* **2008,** *157,* 114–121.

Ngieng, N. S.; Zulkharnain, A.; Roslan, H. A.; Husaini, A. Decolorization of synthetic dyes by endophytic fungal flora isolated from Senduduk Plant (*Melastoma malabathricum*). *Biotech.* **2013**, 260–730.

Nnorom, I. C.; Jarzyńska, G.; Falandysz, J.; Drewnowska, M.; Okoye, I.; Oji–Nnorom, C. G. Occurrence and accumulation of mercury in two species of wild grown *Pleurotus* mushrooms from southeastern Nigeria. *Ecotoxicol. Environ. Safe.* **2012**, *84*, 78–83.

Ogbo, E. M.; Okhuoya, J. A. Biodegradation of aliphatic, aromatic, resinic and asphaltic fractions of crude oil contaminated soils by *Pleurotus tuber–regium* Fr. Singer – A white rot fungus. *Afr. J. Biotechnol.* **2008**, *7*, 4291–4297.

Oliveira, B. R.; Penetra, A.; Cardoso, V. V.; Benoliel, M. J.; Barreto Crespo, M. T.; Samson, R. A.; Pereira, V. J. Biodegradation of pesticides using fungi species found in the aquatic environment. *Environ. Sci. Pollut. Res. Int.* **2015**, *22*, 11781–11791.

Ortega, N. O.; Nitschke, M.; Mouad, A. M.; Landgraf, M. D.; Rezende, M. O. O.; Seleghim, M. H. R.; Sette, L. D.; Porto, A. L. M. Isolation of Brazilian marine fungi capable of growing on DDD Pesticide. *Biodegrad.* **2011**, *22*, 43–50.

Osman, M. S.; Bandyopadhyay, M. Bioseparation of lead ions from wastewater by using a fungus *P. ostreatus*. *J. Civil. Eng.* **1999**, *27*, 183–196.

Osono, T.; Takeda, H. Comparison of litter decomposing ability among diverse fungi in a cool temperate deciduous forest in Japan. *Mycologia.* **2002**, *94*, 421–427.

Owa, F. D. Water pollution: Sources, effects, control and management. *Mediterranean Journal of Social Sciences. Rome: MCSER Publishing.* **2014**.

Oyetayo, V. O.; Adebayo, A.O.; Ibileye, A. Assessment of the biosorption potential of heavy metals by *Pleurotus tuber–regium*. *Int. J. Adv. Biol. Res.* **2012**, *2*, 293–297.

Özdemir, S.; Okumuşa, V.; Kılınçb, E.; Bilgetekinc, H.; Dündara, A.; Ziyadanog'ullarıb, B. *Pleurotus eryngii* immobilized Amberlite XAD–16 as a solid–phase biosorbent for preconcentrations of Cd2+and Co2+ and their determination by ICP–OES. *Talanta.* **2012**, *99*, 502–506.

Pandey, C.; Prabha, D.; Negi, Y. K. Mycoremediation of common agricultural pesticides. In: Prasad, R. (ed.), Mycoremediation and Environmental Sustainability, Fungal Biology, *Springer International Publishing AG.* **2018**, 155–179.

Pandey, R. K.; Tewari, S.; Tewari, L. Lignolytic mushroom *Lenzites elegans* WDP2: Laccase production, characterization, and bioremediation of synthetic dyes. *Ecotox. Environ. safe.* **2018**, *158*, 50–58.

Park, C.; Lee, M.; Lee, B.; Kim, S. W.; Chase, H. A.; Lee, J.; Kim, S. Biodegradation and biosorption for decolorization of synthetic dyes by *Funalia trogii*. *Biochem. Eng. J.* **2007**, *36*, 59–65.

Passarini, M. R.; Rodrigues, M. V.; da Silva, M.; Sette L. D. Marine–derived filamentous fungi and their potential application for polycyclic aromatic hydrocarbon bioremediation. *Mar. Pollut. Bull.* **2011**, *62*, 364–370.

Patil, K. C.; Matsumura, F.; Boush, G. M. Degradation of endrin, aldrin, and DDT by soil microorganisms. *J. App. Microbiol.* **1970**, *19*, 879–881.

Peng, X.; Huang, J.; Liu, C.; Xiang, Z.; Zhou, J.; Zhong, G. Biodegradation of bensulphuron–methyl by a novel *Penicillium pinophilum* strain BP–H–02. *J. Hazard. Mater.* **2012**, *213*, 216–222.

Pereira, A. R. B. et al., Removal of trace element by isolates of *Aspergillus brasiliensis* EPAMIG 0084 and *Penicillium cirtinum* EPAMIG 0086 in biofilters. *African J. Biotech.* **2014**, *13*, 3759–3773.

Pereira, P. M.; Sobral Teixeira, R. S.; de Oliveira, M. A. L.; da Silva, M.; Ferreira, V. S. Optimized atrazine degradation by *Pleurotus ostreatus* INCQS 40310: an alternative for impact reduction of herbicides used in sugarcane crops. *J. Microb. Biochem. Technol. S.* **2013,** *12*, 006.

Peter, L.; Gajendiran, A.; Mani, D.; Nagaraj, S.; Abraham, J. Mineralization of malathion by *Fusarium oxysporum* strain JASA1 isolated from sugarcane fields. *Environ. Prog. Sustainable. Energy.* **2015,** *34*, 112–116.

Pinedo–Rivilla, C.; Aleu, J.; Collado, I. G. Pollutants biodegradation by fungi. *Curr. Org. Chem.* **2009,** *13*, 1194–1214.

Pizzul, L.; Castillo, M. D. P.; Stenström, J. Degradation of glyphosate and other pesticides by ligninolytic enzymes. *Biodegrad.* **2009,** *20*, 751–759.

Pletsch, M.; de Araujo, B.; Charlwood, B. Novel biotechnological approaches in environmental remediation research. *Biotechnol. Advances.* **1999,** *17*, 679–687.

Pointing, S. B. Feasibility of bioremediation by white–rot fungi. *Appl. Microbiol. Biotechnol.* **2001,** *57*, 20–33.

Pozdnyakova, N. N.; Rodakiewicz–Nowak, J.; Turkovskaya, O. V.; Haber, J. Oxidative degradation of polyaromatic hydrocarbons catalyzed by blue laccase from *Pleurotus ostreatus* D1 in the presence of synthetic mediators. *Enzyme microbial technol.* **2006,** *39*, 1242–1249.

Prabu, P. C.; Udayasoorian, C. Biodecolorization of phenolic paper mill effluent by ligninolytic fungus *Trametes versicolor. J. Biol. Sci.* **2005,** *5*, 558–561.

Prakash, V. Mycoremediation of environmental pollutants. *Int. J. Chem. Tech. Res.* **2017,** *10*, 149–155.

Prasad, A. S. A.; Varatharaju, G.; Anushri, C.; Dhivyasree, S. Biosorption of lead by *Pleurotus florida* and *Trichoderma viride. Br. Biotechnol. J.* **2013,** *3*, 66–78.

Puentes–Cárdenas, I. J.; Pedroza–Rodríguez, A. M.; Navarrete–López, M.; Villegas–Garrido, T. L.; Cristiani–Urbina, E. Biosorption of trivalent chromium from aqueous solutions by *Pleurotus ostreatus* biomass.

Purnomo, A. S.; Kamei, I.; Londo, R. Degradation of 1,1,1–trchloro–2,2–bis(4–chlorophenyl) ethane (DDT) by brown–rot fungi. *J. Biosci. Bioeng.* **2008,** *105*, 614–621.

Purnomo, A. S.; Mori, T.; Kondo, R. Involvement of Fenton reaction in DDT degradation by brown rot fungi. *Int. Biodeterior. Biodegrad.* **2010a,** *64*, 560–565.

Purnomo, A. S.; Mori, T.; Putra, S. R.; Kondo, R. Biotransformation of heptachlor and heptachlor epoxide by white–rot fungus *Pleurotus ostreatus. Int. Biodeter. Biodegrad.* **2013,** *82*, 40–44.

Purnomo, A.S.; Mori, T.; Kameil, Nishii, T.; Kondo, R. Application of mushroom waste medium from *Pleurotus ostreatus* for bioremediation of DDT–contaminated soil. *Int. Biodeterior. Biodegrad.* **2010b,** *64*, 397–402.

Quintero, J. C.; Lu–Chau, T. A.; Moreira, M. T.; Feijoo, G.; Lema, J. M. Bioremediation of HCH present in soil by the white–rot fungus *Bjerkandera adusta* in a slurry batch bioreactor. *Int. Biodeterior. Biodegrad.* **2007,** *60*, 319–326.

Rani, B.; Kumar, V.; Singh, J.; Bisht, S.; Teotia, P.; Sharma, S.; Kela, R. Bioremediation of dyes by fungi isolated from contaminated dye effluent sites for bio–usability. *Braz. J. Microbiol.* **2014,** *45*, 1055–1063.

Ratanapongleka, K.; Phetsom, J. Decolorization of synthetic dyes by crude laccase from *Lentinus polychrous* Lev. *Int. J. Chem. Eng. Appl.* **2014,** *5*, 26–30.

Reddy, C.; Mathew, Z. Bioremediation potential of white rot fungi. In: Gadd G. (Eds.) Fungi in bioremediation. Cambridge University Press. Cambridge, U.K. **2001**.

Reddy, G. V. B.; Gold, M. H. Degradation of pentachlorophenol by *Phanerochaete chrysosporium:* intermediates and reactions involved. *Microbiol.* **2000**, *146*, 405–413.

Ren, D.; Zhang, X.; Yan, K.; Yuan, S.; Lu, X. Studies on the degradation of indole using white rot fungus. *Fresen. Environ. Bull.* **2006**, *15*, 1238–1243.

Rhodes, C. J. Mycoremediation (bioremediation with fungi) – growing mushrooms to clean the earth. *Chem. Spec. Bioavailab.* **2015**, *26*, 196–198.

Rigas, F.; Dritsa, V.; Marchant, R.; Papadopoulou, K.; Avramides, E. J.; Hatzianestis, I. Biodegradation of Lindane by *Pelourotus ostreatus* via Central Composite Design. *Environ. Int.* **2005**, *31*, 191–196.

Rigas, F.; Papadopoulou, K.; Dritsa, V.; Doulia, D. Bioremediation of a soil contaminated by lindane utilizing the fungus *Ganoderma australe* via response surface methodology. *J. Hazard. Mater.* **2007**, *140*, 325–332.

Robinson, T.; Nigam, P. S. Remediation of textile dye waste water using a white–rot fungus *Bjerkandera adusta* through solid–state fermentation (SSF). *Appl. Biochem. Biotechnol.* **2008**, *151*, 618.

Rodarte–Morales, A. I.; Feijoo, G.; Moreira, M. T., Lema, J. M. Degradation of selected pharmaceutical and personal care products (PPCPs) by white–rot fungi. *World. J. Microbiol. Biotechnol.* **2011**, *27*, 1839–1846.

Rodríguez–Rodríguez, C. E.; Castro–Gutiérrez, V.; Chin–Pampillo, J. S.; Ruiz–Hidalgo, K. On–farm biopurification systems: role of white rot fungi in depuration of pesticide–containing wastewaters. *FEMS Microbiol. Lett.* **2013**, *345*, 1–12.

Rodríguez–Rodríguez, C. E.; Castro–Gutiérrez, V.; Tortella, G. Mycoremediation fungal mediated processes for the elimination of organic pollutants. In: Fuentes et al., Strategies for Bioremediation of Organic and Inorganic Pollutants, *CRC Press, Taylor and Francis Group, New York.* **2018**, 130–150.

Romero, M.; Cazau, M.; Giorgieri, S.; Arambarri, A. Phenanthrene degradation by microorganisms isolated from a contaminated stream. *Environ. Pollut.* **1998**, *101*, 355–359.

Rønhede, S.; Jensen, B.; Rosendahl, S.; Kragelund, B. B.; Juhler, R. K.; Aamand, J. Hydroxylation of the herbicide isoproturon by fungi isolated from agricultural soil. *Appl. Environ. Microbiol.* **2005**, *71*, 7927–7932.

Rosales, E.; Pazos, M. A.; Sanroma, M. Feasibility of solid–state fermentation using spent fungi–substrate in the biodegradation of PAHs. *Clean Soil Air Water.* **2013**, *41*, 610–615.

Rubilar, O.; Feijoo, G.; Diez, C.; Lu–Chau, T. A.; Moreira, M. T.; Lema, J. M. Biodegradation of pentachlorophenol in soil slurry cultures by *Bjerkandera adusta* and *Anthracophyllum discolor*. *Ind. Eng. Chem. Res.* **2007**, *46*, 6744–6751.

Russell, J. R.; Huang, J.; Anand, P.; Kucera, K.; Sandoval, A. G.; Dantzler, K. W.; Hickman, D.; Jee, J.; Kimovec, F. M.; Koppstein, D.; Marks, D. H.; Mittermiller, P. A.; Núñez, S. J.; Santiago, M.; Townes, M. A.; Vishnevetsky, M.; Williams, N. E.; Vargas, M. P. N.; Boulanger, L, A.; Slack, C. B.; Strobel, S. A. Biodegradation of polyester polyurethane by endophytic fungi. *Appl. Environ. Microbiol.* **2011**, *77*, 6076–6084.

Saiano, F.; Ciofalo, M.; Cacciola, S. O.; Ramirez, S. Metal ion adsorption by *Phomopsis* sp. biomaterial in laboratory experiments and real wastewater treatment. *Water. Res.* **2005**, *39*, 2273–2280.

Sampedro, I.; D'Annibale, A.; Ocampo, J. A.; Stazi, S. R.; GarciaRomera, I. Solid–state cultures of *Fusarium oxysporum* transform aromatic components of olive–mill dry residue and reduce its phytotoxicity. *Bioresour. Technol.* **2007**, *98*, 3547–3554.

Sasek, V. Why mycoremediations have not yet come to practice. In: Sasek V. et al., (Eds.) The utilization of bioremediation to reduce soil contamination: Problems and solutions. *Kluwer Academis Publishers.* **2003**, 247–276.

Sasmaz, S.; Gedikli, S.; Aytar, P.; Gungormedi, G.; Cabuk, A.; Hur, E.; Unal, A.; Kolankaya, N. Decolorization potential of some reactive dyes with crude laccase and laccase–mediated system. *Appl. Biochem. Biotechnol.* **2011**, *163*, 346–361.

Schwarzenbach, R. P.; Egli, T.; Hofstetter, T. B.; Von Gunten, U.; Wehrli, B. Global water pollution and human health. *Annu. Rev. Environ. Resour.* **2010**, *35*, 109–136.

Sears, M. E.; Volesky, B.; Neufeld, R. J. Ion exchange/complexation of uranyl ion by *Rhizopus* biosorbent. *Biotechnol. Bioeng.* **1984**, *26*, 1323–1329.

Sen, S. K.; Raut, S.; Bandyopadhyay, P.; Raut, S. Fungal decolouration and degradation of azo dyes: A review. *Fungal Biol. Rev.* **2016**, *30*, 112–133.

Seneviratne, G.; Tennakoon, N. S.; Weerasekara, M. L. M. A. W.; Nandasena, K. A. Polyethylene biodegradation by a developed *Penicillium–Bacillus* biofilm. *Curr. Sci.* **2006**, *90*, 20–21.

Senthilkumar, S.; Perumalsamy, M.; Prabhu, H. J. Decolourization potential of white–rot fungus *Phanerochaete chrysosporium* on synthetic dye bath effluent containing Amido black 10B. *J. Saudi Chem. Soci.* **2014**, *18*, 845–853.

Senthilkumar, S.; Perumalsamy, M.; Prabhu, H. J.; Thajudin, N. Optimization of process parameters for efficient decolorization of Congo red by novel fungus *Curvularia* sp. using BoxBehnken design. *Desalin Water Treat.* **2015**, *54*, 1708–1716.

Shankar, S.; Nill, S. Effect of metal ions and redox mediators on decolorization of synthetic dyes by crude laccase from a novel white rot fungus *Peniophora* sp. (NFCCI–2131). *Appl. Biochem. Biotechnol.* **2015**, *175*, 635–647.

Shi, H.; Pei, L.; Gu, S.; Zhu, S.; Wang, Y.; Zhang, Y.; Li, B. Glutathione S–transferase (GST) genes in the red flour beetle, *Tribolium castaneum,* and comparative analysis with five additional insects. *Genomics.* **2012**, *100*, 327–335.

Shin, E. H.; Choi, H. T.; Song, H. G. Biodegradation of endocrine disrupting bisphenol A by white rot fungus *Irpex lacteus. J. Microbiol. Biotechnol.* **2007**, *17*, 1147–1151.

Silambarasan, S.; Abraham, J. Ecofriendly method for bioremediation of chlorpyrifos from agricultural soil by novel fungus *Aspergillus terreus* JAS1. *Water Air Soil Pollut.* **2013**, *224*, 1369.

Simonescu C. M.; Ferdes, M. Fungal biomass for Cu(II) uptake from aqueous systems. *Pol. J. Environ. Stud.* **2012**, *21*, 1831–1839.

Singh, A.; Gauba, P. Mycoremediation: A treatment for heavy metal pollution of soil. *J. Civil Eng. Environ. Technol.* **2014**, *1*, 59–61.

Singh, R. K.; Kumar, S.; Kumar, S.; Kumar, A. Biodegradation kinetic studies for the removal of p–cresol from wastewater using *Gliomastix indicus* MTCC 3869. *Biochem. Eng. J.* **2008**, *40*, 293–303.

Soares, A.; Jonasson, K.; Terrazas, E.; Guieysse, B.; Mattiasson, B. The ability of white–rot fungi to degrade the endocrine–disrupting compound nonylphenol. *Appl. Microbiol. Biotechnol.* **2005**, *66*, 719–725.

Solís, N.; Chiriboga, C.; Ávalos, R.; Rueda, D.; Albán, C.; Manjunatha, B.; Rueda, B.; Selvanayagam, M. Use of Cellulase enzyme obtained from Monilia (*Moniliophthora roreri*) for treatment of solid waste of cob, rice husks and cocoa shell. *J. Appl. Pharm. Sci.* **2016,** *6,* 66–70.

Sowmya, H. V.; Ramalingappa, B.; Nayanashree, G.; Thippeswamy, B.; Krishnappa, M. Polyethylene degradation by fungal consortium. *Int. J. Environ. Res.* **2015,** *9,* 823–830.

Spina, F.; Cecchi, G.; Landinez–Torres, A.; Pecoraro, L.; Russo, F.; Wu, B.; Cai, L.; Liu, X. Z.; Tosi, S.; Varese, G. C.; Zotti, M.; Persiani, A. M. Fungi as a toolbox for sustainable bioremediation of pesticides in soil and water. *Plant Biosyst.* **2018,** *152,* 474–488.

Spina, F.; Cordero, C.; Schiliro, T.; Sgorbini, B.; Pignata, C.; Gilli, G. Removal of micropollutants by fungal laccases in model solution and municipal wastewater: evaluation of estrogenic activity and ecotoxicity. *J. Cleaner Prod.* **2015,** *100,* 185–194.

Srivastava, S.; Thakur, I. S. Isolation and process parameter optimization of *Aspergillus* sp. for removal of chromium from tannery effluent. *Bioresour. Technol.* **2006,** *97,* 1167–1173.

Stamets, P. Mycelium running: how mushrooms can help save the world. *Ten Speed Press, Berkley.* **2005.**

Steffen K. T.; Schubert S.; Tuomela M.; Hatakka A.; Hofrichter M. Enhancement of bioconversion of high–molecular mass polycyclic aromatic hydrocarbons in contaminated non–sterile soil by litter–decomposing fungi. *Biodegradation.* **2007,** *18,* 359–369.

Sulaiman, S. S.; Annuar, M. S. M.; Razak, N. N. A.; Ibrahim, S.; Bakar, B. Triarylmethane dye decolorization by pellets of *Pycnoporus sanguineus*: Statistical optimization and effects of novel impeller geometry. *Biorem. J.* **2013,** *17,* 305–315.

Sun, J.; Guo, N.; Niu, L. L.; Wang, Q. F.; Zang, Y. P.; Zu, Y. G. Production of laccase by a new *Myrothecium verrucaria* MD–R–16 isolated from Pigeon Pea [*Cajanus cajan* (L.) Millsp.] and its application on dye decolorization. *Molecules.* **2017,** *22,* 673.

Suseem, S. R.; Mary Saral, A. Biosorption of heavy metals using *Pleurotus eous*. *J. Chem. Pharm. Res.* **2014,** *6,* 2163–2168.

Sutherland, C.; Venkobachar, C. Equilibrium modeling of Cu (II) biosorption onto untreated and treated forest macro–fungus *Fomes fasciatus*. *Int. J. Plant Animal Environ Sci.* **2013,** *4,*193–203.

Svecova, L.; Spanelova, M.; Kubal, M.; Guibal, E. Cadmium, lead and mercury biosorption on waste fungal biomass issued from fermentation industry. I. Equilibrium studies. *Sep. Purif. Technol.* **2006,** *52,* 142–153.

Tan, T. W.; Cheng, P. Biosorption of metal ions with *Penicillium chrysogenum*. *Appl. Biochem. Biotechnol.* **2003,** *104,* 119–128.

Tay, C. C.; Liew, H. H.; Abdul–Talib, S.; Redzwan, G. Bi–metal biosorption using *Pleurotus ostreatus* spent mushroom substrate (PSMS) as a biosorbent: isotherm, kinetic, thermodynamic studies and mechanism. Desalination Water Treat. **2016,** *57.*

Tay, C. C.; Liew, H. H.; Yin, C. Y.; Abdul–Talib, S.; Surif, S.; Abdullah, A.; Yong, S. K. Biosorption of cadmium ions using *Pleurotus ostreatus*: growth kinetics, isotherm study and biosorption mechanism. *Kor. J. Chem. Eng.* **2011,** *28,* 825–830.

Tay, C. C.; Liew, H. H.; Yong, S. K.; Surif, S.; Abdul–Talib, S. Biosorption of lead(II) from aqueous solutions by *Pleurotus* as a toxicity biosorbent. In: Environmental science and technology conference (ESTEC2009), Kuala Terengganu Malaysia, Dec 7–8, **2009.**

Tay, C. C.; Redzwan, G.; Liew, H. H.; Yong, S. K.; Surif, S.; Abdul–Talib, S. Fundamental behavior for biosorption of divalence cations by *Pleurotus* mushroom spent–substrate. *Malays. J. Sci.* **2012,** *31,* 40–44.

Tay, C. C; Redzwan, G.; Liew, H. H.; Yong, S. K.; Surif, S.; Abdul–Talib, S. Copper (II) Biosorption characteristic of *Pleurotus* spent mushroom compost. In: International conference on science and social research (CSSR 2010), Kuala Lumpur, Malaysia, Dec 5–7, **2010**.

Tegli, S.; Cerboneschi, M.; Corsi, M.; Bonnanni, M.; Bianchini, R. Water recycle as a must: decolorization of textile wastewaters by plant–associated fungi. *J. Basic Microbiol.* **2014**, *54*, 120–132.

Tewari, N.; Vasudevan, P.; Guha, B. K. Study on biosorption of Cr(VI) by *Mucor hiemalis*. *Biochem. Eng. J.* **2005**, *23*, 185–192.

Thatai, S.; Verma, R.; Khurana, P.; Goel, P.; Kumar, D. Water quality standards, its pollution and treatment methods. In: A New Generation Material Graphene: Applications in Water Technology, *Springer, Cham.* **2019**, 21–42.

Tian, C. E.; Tian, R.; Zhou, Y.; Chen, Q.; Cheng, H. Decolorization of indigo dye and indigo dyecontaining textile effluent by *Ganoderma weberianum*. *Afr. J. Microbiol Res.* **2013**, *7*, 941–947.

Tixier, C.; Bogaerts, P.; Sancelme, M.; Bonnemoy, F.; Twagilimana, L.; Cuer, A. Fungal biodegradation of a phenylurea herbicide, diuron: structure and toxicity of metabolites. *Pest Manag. Sci.* **2000**, *56*, 455–462.

Tixier, C.; Sancelme, M.; Aït–Aïssa, S.; Widehem, P.; Bonnemoy, F.; Cuer, A. Biotransformation of phenylurea herbicides by a soil bacterial strain, *Arthrobacter* sp. N2: structure, ecotoxicity and fate of diuron metabolite with soil fungi. *Chemosphere.* **2002**, *46*, 519–526.

Tsekova, K.; Petrov, G. Removal of heavy metals from aqueous solution using *Rhizopus delemar* mycelia in free and polyurethane bound form. Zeitschrift fuer Naturforschung. C: *J. Biosci.* **2002**, *57*, 629–633.

Tuomela, M.; Lyytikainen, M.; Oivanen, P.; Hatakka, A. Mineralization and conversion of pentachlorophenol (PCP) in soil inoculated with the white rot fungus *Trametes versicolor*. *Soil Biol. Biochem.* **1999**, *31*, 65–74.

Turlo, J. The biotechnology of higher fungi–current state and perspectives. *Folia. Biol. Oecol.* **2014**, *10*, 49–65.

Udayasoorian, C.; Prabu, P. C. Biodegradation of phenols by ligninolytic fungus *Trametes versicolor. J. Biol. Sci.* **2005**, *5*, 824827.

Ulčnik, A.; Cigić, I. K.; Pohleven, F. Degradation of lindane and endosulfan by fungi, fungal and bacterial laccases. *World J. Microbiol. Biotechnol.* **2013**, *29*, 2239–2247.

Unyayar, A.; Mazmanci, M. A.; Erkurt, E. A.; Atacag, H.; Gizir, A. M. Decolorization kinetics of the azodye drimaren blue X3LR by laccase. *React. Kinet. Catal. L.* **2005**, *86*, 99–107.

Urlacher, V. B.; Lutz–Wahl, S.; Schmid, R. D. Microbial P450 enzymes in biotechnology. *Appl. Microbiol. Biotechnol.* **2004**, *64*, 317–325.

Varjani, S. J. Microbial degradation of petroleum hydrocarbons. *Bioresour. Technol.* **2017**, *223*, 277–286.

Varjani, S.; Patel, R. K. Fungi: A remedy to eliminate environmental pollutants. In: R. Prasad (ed.), Mycoremediation and Environmental Sustainability, Fungal Biology, *Springer International Publishing AG.* **2017**, 53–67.

Velázquez–Fernández, J. B.; Martínez–Rizo, A. B.; Ramírez–Sandoval, M.; Domínguez–Ojeda, D. Biodegradation and bioremediation of organic pesticides. In: Pesticides–recent trends in pesticide residue assay. *IntechOpen.* **2012**.

Verdin, A.; Sahraoui, A. L. H.; Durand, R. Degradation of benzo[a]pyrene by mitosporic fungi and extracellular oxidative enzymes. *Int. Biodeter. Biodegr.* **2004**, *53*, 65–70.

Vimala, R.; Charumathi, D.; Nilanjana Das. Packed bed column studies on Cd(II) removal from industrial wastewater by macrofungi *Pleurotus platypus*. *Desalination.* **2011b**, *275,* 291–296.

Volke–Sepúlveda, T. L.; Gutiérrez–Rojas, M.; Favela–Torres, E. Biodegradation of hexadecane in liquid and solid–state fermentations by *Aspergillus niger*. *Bioresour. Technol.* **2003**, *87*, 81–86.

Wang, L.; Wang, M.; Wen, Z.; Liu, Z. Studies on biotransformation identity of digitoxin by three fungi. *Tianjin Keji Daxue Xuebao.* **2008a**, *23*, 8–12.

Wang, X.; Gong, Z.; Li, P.; Zhang, L.; Hu, X. Degradation of pyrene and benzo–pyrene in contaminated soil by immobilized fungi. *Environ. Eng. Sci.* **2008b**, *25*, 677–684.

Wang, Y.; Zhang, B.; Chen, N.; Wang, C.; Feng, S.; Xu, H. Combined bioremediation of soil co–contaminated with cadmium and endosulfan by *Pleurotus eryngii* and *Coprinus comatus. J. Soils. Sediments.* **2017**, *18*, 2136–2147.

Wen, J.; Gao, D.; Zhang, B.; Liang, H. Co–metabolic degradation of pyrene by indigenous white–rot fungus *Pseudotrametes gibbosa* from the northeast China. *Int. Biodeter. Biodegr.* **2011**, *65*, 600–604.

Wikee, S.; Hatton, J.; Turbé–Doan, A.; Mathieu, Y.; Daou, M.; Lomascolo, A.; Kumar, A.; Lumyong, S.; Sciara, G.; Faulds, C.B.; Record, E. Characterization and dye decolorization potential of two laccases from the marine–derived fungus *Pestalotiopsis* sp. *Int. J. Mol. Sci.* **2019**, *20*, 1864.

Wu, H. Y.; Ting, Y. P. Metal extraction from municipal solid waste (MSW) incinerator fly ash–chemical leaching and fungal bioleaching. *Enzym. Microb. Technol.* **2006**, *38*, 839–847.

Wu, J.; Zhao, Y.; Liu, L.; Fan, B.; Li, M. Remediation of soil contaminated with decarbrominated diphenyl ether using white rot fungi. *J. Environ. Eng. Landsc. Manag.* **2013**, *21*, 171–179.

Wu, M.; Xu, Y.; Ding, W.; Li. Y.; Xu, H. Mycoremediation of manganese and phenanthrene by *Pleurotus eryngii* mycelium enhanced by tween 80 and saponin. *Appl. Microbiol. Biotechno.* **2016**, *100*, 7249–7261.

Xiangliang, P.; Jianlong, W.; Daoyong, Z. Biosorption of Co(II) by immobilised *Pleurotus ostreatus. Int. J. Environ. Pollut.* **2009**, *37*, 289–298.

Xiangliang, P.; Jianlong, W.; Daoyong, Z. Biosorption of Pb(II) by *Pleurotus ostreatus* immobilized in calcium alginate gel. *Process Bio. Chem.* **2005**, *40*, 2799–2803.

Xiao, P. F.; Kondo, R. Biodegradation of dieldrin by *Cordyceps* fungi and detection of metabolites. *Appl. Mecha. Mater.* **2013**, *295*, 30–34.

Xiao, P.; Mori, T.; Kamei, I.; Kondo, R. A novel metabolic pathway for biodegradation of DDT by the white rot fungi *Phlebia lindtneri* and *Phlebia brevispora. Biodegrad.* **2011a**, *22*, 859–867.

Xiao, P.; Mori, T.; Kamei, I.; Kondo, R. Metabolism of organochlorine pesticide heptachlor and its metabolite heptachlor epoxide by white rot fungi belonging to genus *Phlebia. FEMS Microbiol. Lett.* **2011**, *314*, 140–146.

Xiao, P.; Mori, T.; KameiI, I.; Kondo, R. Metabolism of organochlorine pesticide heptachlor and its metabolite heptachlor epoxide by white rot fungi, belonging to genus *Phlebia. FEMS. Microbiol. Lett.* **2011b**, *314*, 140–146.

Xu, F.; Liu, X.; Chen, Y.; Zhang, K.; Xu, H. Self–assembly modified–mushroom nano composite for rapid removal of hexavalent chromium from aqueous solution with bubbling fluidized bed. Sci. Rep. **2016**, *6*, 26201.

Xu, X.; Lin, L.; He, B.; Jiang, L.; Ye, J.; Origele, A.; Sun, R. Degradation of three kinds of dioxins by white rot fungi. *Yingyong Yu Huanjing Shengwu Xuebao.* **2006**, *12*, 701–705.

Yakop, F.; Taha, H.; Shivanand, P. Isolation of fungi from various habitats and their possible bioremediation. *Curr. Sci.* **2019,** *116,* 733–740.

Yalçinkaya, Y.; Arica, M. Y.; Soysal, L.; Bektaş, S. Cadmium and mercury uptake by immobilized *Pleurotus sapidus. Turk. J. Chem.* **2002,** *26,* 441–452.

Yamada–Onodera, K.; Mukumoto, H.; Katsuyaya, Y.; Saiganji, A.; Tani, Y. Degradation of poly– ethylene by a fungus, *Penicillium simplicissimum* YK. *Polym. Degrad. Stab.* **2001,** *72,* 323–327.

Yamaguchi, M.; Kamei, I.; Nakamura, M.; Takano, M.; Sekiya, A. Selection of *Pleurotus pulmonarius* from domestic basidiomycetous fungi for biodegradation of chlorinated dioxins as environmentally persistent organopollutants. *Shinrin Sogo Kenkyusho Kenkyu Hokoku.* **2007,** *6,* 231–237.

Yan, G.; Viraraghavan, T. Effect of pretreatment on the bioadsorbents of Hg on *Mucor rouscii. Water Res.* **2000,** *26,* 119–123.

Yang, X. Q.; Zhao, X. X.; Liu, C. Y.; Zheng, Y.; Qian, S. J. Decolorization of azo, triphenylmethane and anthraquinone dyes by a newly isolated *Trametes* sp SQ01 and its laccase. *Process. Biochem.* **2009,** *44,* 1185–1189.

Ye, J. S.; Yin, H.; Qiang, J.; Peng, H.; Qin, H. M.; Zhang, N.; He, B. Y. Biodegradation of anthracene by *Aspergillus fumigatus. J. Hazard. Mater.* **2011,** *185,*174–181.

Yemendzhiev, H.; Gerginova, M.; Krastanov, A.; Stoilova, I.; Alexieva, Z. Growth of *Trametes versicolor* on phenol. *J. Ind. Microbiol. Biotechnol.* **2008,** *35,* 1309–1312.

Yemendzhiev, H.; Alexieva, Z.; Krastanov, A. Decolorization of synthetic dye reactive blue 4 by mycelial culture of white–rot fungi *Trametes versicolor* 1. *Biotechnol. Biotec. Eq.* **2009,** *23,* 1337–1339.

Yesilada, O.; Fiskin, K.; Yesilada, E. The use of white–rot fungus *Funalia trogii* (Malatya) for the decolorization and phenol removal from olive mill wastewater. *Environ. Technol.* **1995,** *16,* 95–100.

Yesilada, O.; Asma, D.; Cing, S. Decolorization of textile dyes by fungal pellets. *Process Biochem.* **2003,** *38,* 933–938.

Yesilada, O.; Ozcan, B. Decolorization of orange II dye with the crude culture filtrate of white rot fungus, *Coriolus versicolor. Turk. J. Biol.* **1998,** *22,* 463–476.

Yoshioka, T.; Komuro, M. The biodegradation of the oil using fungi isolated from polluted oils. *Kaijo Hoan Daigakko Kenkyu Hokoku, Rikogaku–kei.***2006,** *50,* 1–9.

Yu, J.; Tong, M.; Sun, X.; Li, B. Cystine–modified biomass for Cd(II) and Pb(II) biosorption. *J. Hazard Mater.* **2007,** *143,* 277–284.

Zabielska–Matejuk, J.; Czaczyk, K. Biodegradation of new quaternary ammonium compounds in treated wood by mould fungi.*Wood. Sci. Technol.* **2006,** *40,* 461–475.

Zafar, U.; Houlden, A.; Robson, G. D. Fungal communities associated with the biodegradation of polyester polyurethane buried under compost at different temperature. *Appl. Environ. Microbiol.* **2013,** *79,* 7313–7324.

Zeng, B.; Ning, D. L.; Wang, H. Preliminary study on biodegradation of pentachlorophenol by white–rot fungus. *Huanjing Huaxue.* **2008,** *27,* 181–185.

Zeng, R.; Mallik, A. U. Selected ectomycorrhizal fungi of Black Spruce (*Picea mariana*) can detoxify phenolic compounds of *Kalmia angustifolia. J. Chem. Ecol.* **2006,** *32,* 1473–1489.

Zeng, X.; Su, S.; Jiang, X.; Li, L.; Bai, L.; Zhang, Y. Capability of pentavalent arsenic bioaccumulation and biovolatilization of three fungal strains under laboratory conditions. *Clean: Soil, Air, Water.* **2010,** *38,* 238–241.

Zeng, X.; Cai, Y.; Liao, X.; Zeng, X.; Li, W.; Zhang, D. Decolorization of synthetic dyes by crude laccase from a newly isolated *Trametes trogii* strain cultivated on solid agro–industrial residue. *J. Hazard. Mater.* **2011,** *187,* 517–525.

Zhang, J. H.; Xue, Q. H, Gao, H.; Ma, X.; Wang. P. Degradation of crude oil by fungal enzyme preparations from *Aspergillus* spp. for potential use in enhanced oil recovery. *J. Chem. Technol. Biotechnol.* **2016,** *91,* 865–875.

Zhou, J. L. Zn biosorption by *Rhizopus arrhizus* and other fungi. *Appl. Microbiol. Biotech.* **1999,** *51,* 686–693.

Zhou, Z.; Chen, Y.; Liu, X.; Zhang, K.; Xu, H. Interaction of copper and 2,4,5–trichlorophenol on bioremediation potential and biochemical properties in co–contaminated soil incubated with *Clitocybe maxima. RSC Adv.* **2015,** *5,* 42768–42776.

Zhuo, R.; Zhang, J.; Yu, H.; Ma, F.; Zhang, X. The roles of *Pleurotus ostreatus* HAUCC 162 laccase isoenzymes in decolorization of synthetic dyes and the transformation pathways. *Chemosphere.* **2019,** *234,* 733–745.

WEBLINKS

Water Pollution Guide. Retrieved from https://www.water-pollution.org.uk/types/ (accessed on June 1, 2019) **2019.**

CHAPTER 11

Genetically Modified Organisms as Tools for Water Bioremediation

FERNANDA MARIA POLICARPO TONELLI,[1]
MOLINE SEVERINO LEMOS,[1] and
FLÁVIA CRISTINA POLICARPO TONELLI[2]

[1]*Department of Cell Biology, Institute of Biological Sciences,
Federal University of Minas Gerais, Belo Horizonte, Brazil,
E-mail: tonellibioquimica@gmail.com (F. M. P. Tonelli)*

[2]*Department of Biochemistry, Federal University of São João del Rei,
Divinópolis, Brazil*

ABSTRACT

Environmental pollution poses a serious risk to human health and the ecosystem as a whole. Water (surface water and groundwater) can be contaminated directly by pollutants or receive them from the soil through leaching. Human intervention in the environment without concern for sustainable development aggravates the problem as residues from industrial activities, the use of chemical agents in agricultural crops, dumps, mining among other activities that produce toxic waste, are relevant sources of pollutants, mainly pesticides and heavy metals. To deal with this problem and promote the remediation of contaminated waters, different methodologies can be applied. This chapter will focus on the use of recombinant DNA technology to genetically modify organisms to make them suitable to perform bioremediation of contaminated waters.

11.1 INTRODUCTION

Anthropological action on the environment, if not carried out sustainably, can be very harmful. Pollution resulting from environmental exploitation without environmental liability poses severe threats to the entire ecosystem. Human economical activities as mine exploitation, pesticides' use to improve agricultural production, industrial production of fuels, fabrics, medicines, organic solvents, and paints may produce toxic wastes. The environmental contaminants can be organic or inorganic (Jafari et al., 2013; Barrios-Estrada et al., 2018; Bilal et al., 2019a; Mendes et al., 2019; Pesantes et al., 2019; Pu et al., 2019; Rosculete et al., 2019; Vázquez-Luna and Cuevas-Díaz, 2019; Zhang et al., 2019).

From the inorganic pollutants, it is possible to highlight heavy metals, a serious environmental problem once they are persistent on the contaminated environment being able to bioaccumulate and cause severe health damages (as carcinogenic agents) or even death (Alkorta et al., 2004; Karthik et al., 2017; Jacob et al., 2018; Ali et al., 2019). Commonly these metals are released into watercourses in mining regions, and when in contact with living organisms, it causes oxidative stress and ends up inducing cell death (Jaishankar et al., 2014; Jacob et al., 2018).

Among organic pollutants, it is also important to find persistent substances, like aromatic molecules, that are hard to degrade and need human intervention as remediation strategies to be removed from the environment (Parrilli et al., 2010). Organic pollutants can also present high toxicity to live forms threatening the environmental sustainability (Bilial et al., 2019b). Pesticides, for example, used with the intention to avoid damages from crop pests, can contaminate the environment threatening life (Craig, 2018; Li, 2018; Shabbir, 2018). Dyes applied in industries such as textile, plastic, and cosmetic ones can end up being discharged in the wastewater polluting the environment (Bilal et al., 2017; Chatha et al., 2017). The exploitation of offshore oil and activities related to marine transportation contribute to oil and petroleum hydrocarbons pollution of seawater (Carpenter, 2018). Explosives and their metabolites can also pollute the environment in a persistent way, especially in regions of mining activities (Dhanarani et al., 2016). The synthetic chemicals polychlorinated biphenyls (PCBs) are used to offer insulation and heat resistance to products and present serious risks to the ecology and human health being carcinogenic pollutants that can easily accumulate in the environment (Aldhafiri et al., 2018).

It is necessary to developed strategies capable of extracting these pollutants from the contaminated areas, and the genetic manipulation is an elegant solution. When it comes specifically to water, the focus of this book, it is a worldwide concern to provide the population with water of good quality (Albering et al., 2016).

Methods considered traditional (e.g., incineration, electrochemical treatment, reverse osmosis) in water remediation present some disadvantages (e.g., elevated costs to be performed on a large scale, generation of toxic waste, requiring harmful reagents (Dasgupta et al., 2015; Bilal et al., 2018)) that the use of biological agents can help to surpass.

Bioremediation contemplates the use of biological agents, mainly microorganisms, i.e., yeast, fungi, or bacteria to clean up contaminated soil and water (Strong and Burgess, 2008). Among the bioremediation strategies is the use of genetically modified organisms (GMOs), reprogramed to perform efficient remediation of different environments, including superficial and groundwater (Zhao et al., 2017; Liu et al., 2019).

11.2 GENETICALLY MODIFIED PLANTS (GMPS) FOR PHYTOREMEDIATION

Some plant species, like the ones from the Brassicaceae family, can naturally hyperaccumulate some pollutants (in this case, metals). They can be used, without any kind of genetic manipulation, to perform bioremediation: accumulating and transporting well pollutants from root to shoot. However, there is an important disadvantage in this kind of practice: very time-consuming bioremediation because these species can present slow growth in environments containing a high level of contaminants (Jafari et al., 2013).

Genetic manipulation is an appropriate solution to this problem. Transgenesis made it possible to implant desirable DNA sequences into an organism's genome to induce the manifestation of some characteristics not natural from the species or to change undesirable ones. For example, genes from hyperaccumulators (that can survive in a toxic environment)–like selenocysteine methyltransferase from *Astragalus bisulcatus*–can be inserted into a non-hyperaccumulator (that grows fast but cannot survive in the toxic environment) – as *Brassica juncea* – to make the latter survive into a toxic environment and growing fast, being very useful for phytoremediation (Ali et al., 2013) – in this example selenium bioremediation (LeDuc et al., 2004).

Actually, plants present a relevant advantage over microorganisms when it comes to bioremediation of heavy metals. Microorganisms generally do not remove these pollutants from a contaminated environment; they only convert metals into a less toxic form. Plants can, in fact, extract these pollutants from polluted sites (Garbisu et al., 2002; Ojuederie and Babalola, 2017).

The pesticide Atrazine is commonly used as an herbicide to protect crop yields from weed. Transgenic *Medicago sativa* and *Nicotiana tabacum* can be genetically reprogramed to efficiently metabolize this substance by inserting bacterial (*Pseudomonas* sp) atrazine chlorohydrolase gene (*atzA* gene) into the plants genome. The modified plants could capture and transport atrazine and converted the pollutant to hydroxyatrazine proving to be an important tool on bioremediation of soil and water environments contaminated with this chemical (Wang et al., 2005).

A large number of scientists that work on bioremediation field using transgenic plants focus on soil phytoremediation (Connolly et al., 2002; Inui and Ohkawa, 2005; Karavangeli et al., 2005; Martinez et al., 2006; Gasic and Korban, 2007; Korenkov et al., 2007; Gandia-Herrero et al., 2008; Guo et al., 2008; Hsieh et al., 2009; Bhuiyan et al., 2011; Rodríguez-Llorente et al., 2012; Shim et al., 2013; Merlot et al., 2014; Liu et al., 2015; Lu et al., 2015; Das et al., 2016). However, in some of the research papers, the scientists perform experiments on liquid media, allowing the reader to comprehend also the transgenic's behavior dealing with pollutants in an aquatic environment. Normally, these assays are performed to allow the researchers to analyze plants' metabolism of pollutants apart from the metabolism of soil microorganisms.

Transgenic plants to deal with toxic nitro-substituted compounds pollutants were developed, for example, inserting *Enterobacter cloacae* nitroreductase *NfsI* gene into tobacco plants. The genetically modified plants (GMPs) were germinated and grown on liquid media, then transferred to water to be exposed to the pollutant in different concentrations, and finally their extracts were prepared to be analyzed by HPLC. GMPs presented enhanced tolerance, transformation of 2,4,6-trinitrotoluene (TNT) into 4-hydroxylamino-2,6-dinitrotoluene and consequently enhanced capacity to conjugate this metabolite to promote detoxification (Hannink et al., 2007). *NfsA* gene from *Escherichia coli* allowed the conversion of *Arabidopsis thaliana* into a phytoremediator species. Seeds were germinated and grown, and then transferred to liquid media containing TNT where they were let do develop for 7 days in which the pollutant was quantified daily through HPLC. The GMPs differently from the wild type plants could grow almost normally in the presence of the contaminant. The transgenic presented the ability to efficiently uptake TNT and degraded it into 4-amino-2,6-dinitrotoluene (Kurumata et al., 2005).

The genes *XplA, XplB* from *Rhodococcus rhodochorus* were used to engineer *Arabidopsis thaliana* to perform 1,3,5-trinitro-1,3,5-triazine remediation. The GMPs were able to remove saturating levels of the pollutant not only from soil leachate, but also from liquid culture faster than non-modified plants (Jackson et al., 2007).

Nitroglycerin or glycerol trinitrate (GTN), other contaminant that comes from explosives use, could be degraded by the *Nicotiana tabacum* engineered to contain *ONR* gene (pentaerythritol tetranitrate reductase) from *Enterobacter cloaceae*. The transgenic plants were exposed to GTN in sterile water that was daily monitored for the presence of the pollutant and its denitration products popularly known as GDN and GMN. Besides being able to germinate and grow in the presence of the pollutant at concentrations that inhibited germination and growth of wild-type organisms, the GMPs could perform complete denitration of GTN faster than non-transgenic ones (French et al., 1999).

CYP2E1 gene (codifying the cytochrome P450 2E1 enzyme) from *Homo sapiens* was used to generate transgenic *Arabidopsis thaliana* for phytoremediation. As groundwater in the United States is commonly contaminated with chlorinated solvents, scientists developed GMP to deal with trichloroethylene (TCE) pollution. Compared to wild-type plants, transgenic ones enhanced up to 640-fold the capacity to oxidize TCE and also could uptake and debrominate ethylene dibromide (other relevant halogenated hydrocarbon water pollutant) more efficiently (Doty et al., 2000).

CYP1A1, CYP2B6, and *CYP2C19* genes from *Homo sapiens* were used to reprogram *Oryza sativa* to perform phytoremediation of the herbicides atrazine and metolachlor. Transgenic plants in hydroponic culture, differently from control plants, could reduce the amounts of the pesticides atrazine, metolachlor, and simazine efficiently removing these pollutants from water during the incubation period (Kawahigashi et al., 2006).

Nicotiana tabacum was engineered with *MT2* (a cysteine-rich metal-binding protein) gene from *Sedum alfredii* (Cd/Zn co-hyperaccumulator) to present Cu increased tolerance and accumulation in both shoots and roots. The assays to determine the heavy metals concentration were performed after hydroponic culture using ICP-MS (Zhang et al., 2014).

Arabidopsis thaliana could be modified to constitutively overexpress the zinc transporter ZNT1 from the metal hyperaccumulator from Brassicaceae family: *Noccaea cearulescens*. After hydroponic culture and analysis, it was observed that GMPs presented enhanced and accumulation of Zn and Cd when exposed to excessive supply differently from the non-modified plants (Lin et al., 2016).

Nicotiana tabacum modified to overexpress the metallothionein 1 (MT1) of *Elsholtzia haichowensis* (Cu-accumulator) could accumulate more Cu in roots than the wild-type plant. Transgenic organism also presented reduced level of lipid peroxidation and production of hydrogen peroxide once the MT could bind to Cu^{2+} avoiding its harmful activity to induce the production of reactive oxygen species (ROS) (Xia et al., 2012).

γ-Glutamylcysteine synthetase (codified by *gshI* gene) from *Escherichia coli* overexpression in *Brassica juncea* led to plants that could, in hydroponic culture, grow better in Cd presence (enhanced tolerance) and accumulate more of this heavy metal in shoots than wild-type organisms (Zhu et al., 1999a). The *gshII* gene when overexpressed in *Brassica juncea* also conducted to improved Cd tolerance and shoot accumulation, been suitable for this heavy metal phytoremediation (Zhu et al., 1999b).

To generate a transgenic plant, there are different methodologies (e.g., the use of viral vectors (Zaidi and Mansoor, 2017; Keshavareddy et al., 2018)). However, the most commonly used methodology involves the use of bacteria from *Agrobacterium* gender. *Agrobacterium rhizogenes* can be the organism chosen, but *Agrobacterium tumefaciens* (a soil bacterium that naturally infects dicots) is the most common choice. By inserting its DNA (containing the transgenes of interest replacing virulence genes) inside host's plant DNA it is possible to generate GMPs (Cunningham et al., 2018) (Figure 11.1).

11.3 GENETICALLY MODIFIED BACTERIA (GMB) FOR BIOREMEDIATION

Microorganisms such as bacteria have a great potential for decontamination of aquatic environments as they can degrade, accumulate or convert in less toxic forms various pollutants (as aromatic compounds containing or not nitrogen and chlorine, biphenyls, heavy metals-like uranium (U), nickel (Ni), mercurium (Hg), chromium (Cr) and cadmium (Cd)- and hydrocarbons). They have been isolated for the purpose of exploiting their metabolic potential for remediating contaminated sites. For example, *Geobacter metallireducens* and *Shewanella oneidensis* are Fe^{3+}-reducing bacteria and present dissimilatory U^{4+} reduction, conserving energy for anaerobic growth (Lovley et al., 1991) and being suitable for uranium remediation from drainage waters in mining operations and from contaminated groundwater. *Pseudomonas fluorescens 4F39* can accumulate nickel ions. It was possible, for example, to immobilize this organism in beads of agar (biobeads) to increase the binding efficiency and removal capacity (Lopez et al., 2002). *Pseudomonas putida* can deal with Cr^{6+} pollution expressing a chromate reductase, the ChrR; *Escherichia coli* can do the same by expressing an oxygen-insensitive nitroreductase: NfsA (Ackerley et al., 2004). *Methylococcus capsulatus*, methane-oxidizing bacteria, can also reduce Cr^{6+} bioremediating this heavy metal pollution over a wide range of concentrations (Hasin et al., 2010). *Pseudomonas aeruginosa* strain KUCd1 could accumulate high levels of Cd, removing the metal from industrial wastewater (Sinha and Mukherjee, 2009). Not only metals but also other pollutants like pesticides can be naturally degraded by bacteria. *Pseudomonas* sp. strain ADP can metabolize, using as a nitrogen source, atrazine, and even more efficiently the cyanuric acid (major intermediate of the pesticide degradation) (Neumann et al., 2004). *Sphingobium wenxiniae* strain JZ-1 possess a gene cluster (*pbaA1A2B*) which can catalyze 3-phenoxybenzoate degradation (Cheng et al., 2015).

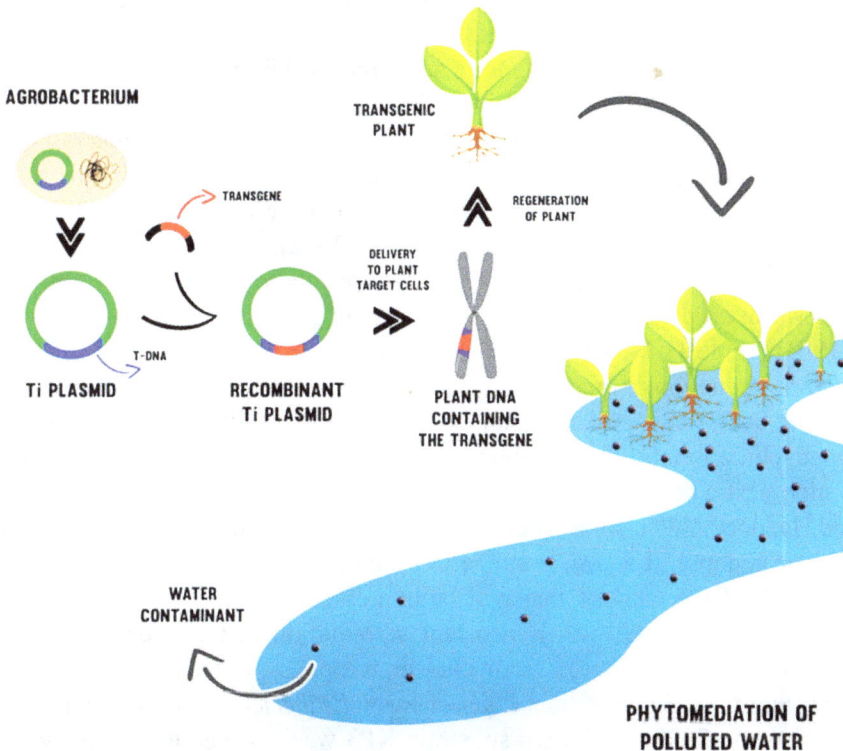

FIGURE 11.1 Use of *Agrobacterium* bacterium to generate GMP to perform bioremediation. The Ti-plasmid from *Agrobacterium* can, in a site known as T-DNA, receive the transgene to be delivered to the host plant genome. This transgene can, for example, codify for enzymes or other proteins involved in remediation of water's processes. The transgene of interest is integrated to the host's DNA generating the GMP suitable for phytoremediation.

However, the development and advances in molecular biology field made possible genome editing and generation of optimized DNA sequences to engineer organisms. *Escherichia coli*, for example, could be modified with a sequence of manipulated DNA to efficiently deal with a wide range of PCBs. This DNA sequence was constructed based on the biphenyl dioxygenases sequence of *Pseudomonas pseudoalcaligenes* KF707 and *Pseudomonas cepacia* LB400 that could metabolize different kinds of PCBs. A chimeric bphA1 gene was developed exchanging fragments between KF707 bphA1 and LB400 bphA1 to codify a biphenyl dioxygenase with an expanded range of substrates (Kimura et al., 1997).

It is feasible to insert more copies of a gene that already belongs to the species to improve some desirable characteristic. For example, *arsR* that codifies for the metalloregulatory protein ArsR could be overexpressed in *Escherichia coli* to enhance arsenic's bioaccumulation, making the manipulated organism suitable for arsenic removal from contaminated water sources (Kostal et al., 2004). Using pGEMT vector, the number of copies of *merB* gene (that codifies for an organomercurial lyase) could be increased in *E. coli* resulting in bacteria capable of tolerating 5 more times organic form of mercury than wild type organisms, being an interesting option for polluted water bodies bioremediation (Murtaza et al., 2002).

It is possible to optimize bacterium conferring to them specific abilities to perform efficient bioremediation generating genetically modified bacteria (GMB) (Table 11.1). It is possible, for example, to program theses GMBs to perform reduction or present high capacity to accumulate some toxic metals, and to deal with contamination from various chemical substances like polycyclic aromatic hydrocarbons from incomplete combustion of organic matter, pesticides, explosives, solvents, etc., (Samanta et al., 2002; Parales and Haddock, 2004; Jafari et al., 2013) by inserting genes from other species in their genome. It is also possible to optimize DNA sequences to improve the work of the enzyme that will be expressed in the transgenic bacteria. *Deinococcus radiodurans* is resistant to radiation and was engineered to perform Hg^{2+} remediation from nuclear weapons production wastes. The gene *merA* (a Hg^{2+} resistance gene) from *Escherichia coli* strain BL308 (Brim et al., 2000). *Ralstonia* sp. strain KP3 was exposed to *Pseudomonas putida* BN210 containing a *clc*-element (self-transferable element (van der Meer et al., 2001)) coding enzyme to promote 3-chlorobenzoate (CBA) degradation. After being in contact in a biofilm reactor to remediate wastewater-containing CBA, KP3 acquired the ability to degrade this chemical and surpassed the capacity of other bacteria present in the reactor (Springael et al., 2002). Fenpropathrin, a pyrethroid ester insecticide, could be metabolized by *Sphingobium* sp. strain JZ-2 due to the enzyme pyrethroid hydrolase; its gene was cloned and inserted into *Sphingobium* sp. BA3. Strain JZ-2 could degrade 60% of the contaminant and the modified BA3 strain, 95% being suitable for pyrethroid insecticides contaminated samples remediation (Duan et al., 2011). *Escherichia coli* could acquire the ability to

TABLE 11.1 Examples of GMB Developed to Possess the Ability to Perform Bioremediation

Source of Transgene	Transgene	Genetically Modified Bacteria (GMB)	Desired Characteristic Presented by the GMB	References
Pyrus calleryana	*PcPCS1*	*Escherichia coli*	Enhanced accumulation and tolerance to Cd, Cu, and Hg	Li et al. (2015)
Yeast and mammalian	CUP1 and HMT-1A metallothioneins genes	*Escherichia coli*	Improved ability to bind Cd^{2+}	Sousa et al. (1998)
Sphingobium sp. strain JZ-2	*pytH*	*Sphingobium* sp. BA3	Efficient metabolism of Fenpropathrin (a pyrethroid ester insecticide)	Duan et al. (2011).
Escherichia coli	*merA*	*Deinococcus radiodurans*	Hg^{2+} increased accumulation	Brim et al. (2000)
Gordonia sp.	*alkB*	*Streptomyces coelicolor* M145	Capacity to degrade n-hexadecane	Gallo et al. (2012)
Artrobacter globiformis KZT1	*fcbA*	*Escherichia coli*	Capacity to degrade chlorobenzoic acid	Tsoi et al. (1991)
Pseudomonas putida strain BNF1	*xylE*	*Acinetobacter* sp. BS3	Capacity to degrade aromatic hydrocarbon compounds	Xie et al. (2014)
Enterococcus sp. L2	*azoA*	*Escherichia coli* DH5α and *Pseudomonas fluorescens* PfO-1	Decolorization of recalcitrant azo dyes	Rathod et al. (2017)
Rhodobacter sphaeroides AS1.1737	Azoreductase gene	*Escherichia coli* JM109	Enhanced C.I. Direct Blue 71 (DB 71) decolorization	Jin et al. (2009)
Plesiomonas sp. strain M6	*mpd*	*Escherichia coli*	Metabolism of methyl parathion to *p*-nitrophenol	Zhongli et al. (2001)
Rhodococcus sp. Strain T1	*feh*	*Escherichia coli*	Metabolism of pesticide fenoxaprop-ethyl into fenoxaprop acid	Hou et al. (2011)

TABLE 11.1 *(Continued)*

Source of Transgene	Transgene	Genetically Modified Bacteria (GMB)	Desired Characteristic Presented by the GMB	References
Pseudomonas pseudomallei	3.0-kb DNA fragment containing DNA sequence codifying enzyme that metabolize the phosphonate herbicide	*Escherichia coli*	Metabolism of Glyphosate	Peñaloza-Vazquez et al. (1995)
Ochrobactrum sp	*tpd*	*Pseudomonas putida* KT2440	Metabolism of the organophosphorous pesticide Triazophos	Gu et al. (2006)
Pseudomonas sp. LB400 and *Pseudomonas alcaligenes* KF707	*bphA1*	*Escherichia coli*	Degradation of polychlorinated biphenyls (PCBs)	Kimura et al. (1997)
Pseudomonas putida BN210	clc element encoding enzyme to promote degradation of 3-chlorobenzoate	*Ralstonia* sp. strain KP3	Degradation of 3-chlorobenzoate	Springael et al. (2002)

degrade the pesticide fenoxaprop-ethyl after being genetically engineered to express a fenoxaprop-ethyl hydrolase from *Rhodococcus* sp. strain T1 (Hou et al., 2011). Glyphosate could be degraded by *Escherichia coli* genetically modified with a 3.0-kb DNA fragment from strains of *Pseudomonas pseudomallei* that naturally metabolize this phosphonate herbicide as phosphorus source (Peñaloza-Vazquez et al., 1995). *Pseudomonas putida* KT2440 could metabolize the organophosphorous pesticide Triazophos after being engineered to contain the codifying sequence of the triazophos hydrolase from *Ochrobactrum* sp. (Gu et al., 2006). Methyl parathion could be metabolized to *p*-nitrophenol by *Escherichia coli* after being genetically modified to express an organophosphate hydrolase from *Plesiomonas* sp. strain M6 (Zhongli et al., 2001). *Escherichia coli* JM109 could decolorize the azo dye (C.I. Direct Blue 71) from wastewater after being engineered with the azoreductase gene from *Rhodobacter sphaeroides* AS1.1737 (Jin et al., 2009). *Escherichia coli* DH5α and *Pseudomonas fluorescens* PfO-1 acquired the ability to promote decolorization of recalcitrant azo dyes after having their DNA engineered to contain the *azo*A gene from *Enterococcus* sp. L2, responsible for expressing the enzyme azoreductase. Overexpressing not only the *azoA* gene but also the *fdh* from *Mycobacterium vaccae* N10 (a formate dehydrogenase coding gene) increased even more the efficiency of RV5R decolorization (Rathod et al., 2017). *Escherichia coli* expressing metallothioneins (MT) from yeast (CUP1) and mammalian (HMT-1A) fused to membrane protein LamB could improve in 15 to 20 folds the natural ability of the bacterium to bind Cd^{2+} (Sousa et al., 1998). *E. coli* overexpressing a phytochelatin synthase from *Pyrus calleryan* presented enhanced accumulation and tolerance to the heavy metals Cd, Cu, and Hg (Li et al., 2015). *Streptomyces coelicolor* M145 acquired the capacity to degrade n-hexadecane thanks to the insertion and overexpression of alkB gene (codifying an alkane monooxygenase) from *Gordonia sp.* turning the bacterium into a suitable part of a bioremediation system (Gallo et al., 2012). *E. coli*, after being modified to contain the gene of a 4-chlorobenzoate-4-hydroxylase from *Artrobacter globiformis* KZT1, became able to degrade 4-chlorobenzoic acid (Tsoi et al., 1991). Aromatic hydrocarbon compounds could be efficiently degraded by *Acinetobacter* sp. BS3 (an alkanes degrading strain) after being genetically modified by receiving the xylE gene (that codifies a catechol 2,3-dioxygenase) from *Pseudomonas putida* strain BNF1 (Xie et al., 2014). A dehalogenase from *Arthrobacter* sp. FG1 when expressed in *Pseudomonas putida* PaW340 after transformation resulted in 4-chlorobenzoic acid biodegradation capacity (Massa et al., 2009).

312   *Freshwater Pollution and Aquatic Ecosystems*

11.4 FUNGAL GENETIC MODIFICATIONS FOR BIOREMEDIATION

Fungi and yeast are naturally an interesting option for bioremediation, especially when it comes to heavy metals. There are species that can naturally: accumulate these metals or change their redox state or even precipitate them extra/intracellularly (Ayangbenro and Babalola, 2017). *Aspergillus versicolor*, for example, can efficiently bioaccumulate Cu^{2+} and Ni^{2+} and remove dye form polluted environment. In fact, Cu^{2+} removal rate can be increased if dye is also present (Taştan et al., 2010). The yeast *Pichia guilliermondii* can accumulate and growth well in a medium containing Cr^{3+} and Cr^{6+} (Ksheminska et al., 2003).

However, fungal species can also be genetically modified to be optimized to bioremediate. A synthetic DNA sequence codifying a histidine oligopeptide (hexa-His) was developed and used to genetically manipulate *Saccharomyces cerevisiae*. With this oligopeptide capable of chelating divalent metals the yeast could present increased capacity to absorb Cu^{2+} 8-times better than a wild type organism (Kuroda et al., 2001). The yeast could also present high biosorption of Pb^{2+} and Cd^{2+} accumulating them on cell wall after being genetically modified with the gene of a synthetic phytochelatins known as EC20 (Fontes et al., 2015). The expression of rat CYP1A from the family of cytochrome P450 in *S. cerevisiae* provided the yeast with the ability to degrade dioxins (Sakaki et al., 2002).

11.5 FUTURE PERSPECTIVES

The biotechnology field has in recombinant DNA technology a very important and strategic tool, and when it comes to bioremediation, the generation of GMOs is a useful and promising strategy. The persistent and highly toxic water pollution needs to receive attention once it threatens human health and other living forms on Earth.

It is expected that the studies of bioremediation using these GMOs continue to be performed, involving the proposal of innovative techniques, as the knowledge over the genomes of different species continues to increase. This knowledge can conduct to the generation of interesting methods to reprogram organisms to deal efficiently, even with mixed contamination.

Field trials are also expected to increase in number to enhance the understanding of GMOs interactions on the ecosystem, making it possible to evaluate better the impacts they can cause, for example, accessing the real risk of horizontal gene transfer. Technical and ethical obstacles need to receive attention and be surpassed in a safe way.

Regarding this kind of concern, it is also desirable that ways of improving GMOs bioremediation security are developed as the use of conditional lethality genes.

11.6 CONCLUSION

Pollution is an increasingly global problem involving substances from diverse chemical nature like heavy metals, dyes, pesticides, explosives, solvents, etc. Bioremediation strategies offer advantages over traditional methods such as lower costs, and the generation of GMOs. It made possible the improvement of preexisting desirable characteristics or the insertion of DNA sequences capable of conferring new abilities to the target organisms to bioremediate environments. This approach made it possible to convert different species of bacteria, yeast, and plants into useful organisms to promote the cleaning up of contaminated aquatic environments.

ACKNOWLEDGMENTS

The authors would like to thank Rafael Severino Lemos for freely converting our sketch into better quality, visually appealing image.

KEYWORDS

- **bioremediation**
- **degradation**
- **fungal genetic modifications**
- **genetically modified bacteria**
- **genetically modified plants**
- **phytoremediation**

REFERENCES

Ackerley, D. F., Gonzalez, C. F., Keylan, M., Blake, R., & Matin, A., (2004). Mechanism of chromate reduction by the *Escherichia coli* protein, NfsA, and the role of different chromate reductases in minimizing oxidative stress during chromate reduction. *Environ. Microbiol., 6*, 851–860.

Albering, H. J., Rila, J. P., Moonen, E. J., Hoogewerff, J. A., & Kleinjans, J. C., (2016). Human health risk assessment in relation to environmental pollution of two artificial freshwater lakes in the Netherlands. *J. Res. Med. Sci., 107*, 27–35.

Aldhafiri, S., Mahmoud, H., Al-Sarawi, M., & Ismail, W. A., (2018). Natural attenuation potential of polychlorinated biphenyl-polluted marine sediments. *Pol. J. Microbiol., 67*, 37–48.

Ali, H., Khan, E., & Ilahi, I., (2019). Environmental chemistry and ecotoxicology of hazardous heavy metals: Environmental persistence, toxicity, and bioaccumulation. *J. Chem.*, 1–14.

Ali, H., Khan, E., & Sajad, M. A., (2013). Phytoremediation of heavy metals-concepts and applications. *Chemosphere, 91*, 869–881.

Alkorta, I., Hernandez-Allica, J., Becerril, J. M., Amezaga, I., Albizu, I., & Garbisu, C., (2004). Recent findings on the phytoremediation of soils contaminated with environmentally toxic heavy metals and metalloids such as zinc, cadmium, lead, and arsenic. *Rev. Environ. Sci. Biotechnol., 3*, 71–90.

Ayangbenro, A. S., & Babalola, O. O., (2017). A new strategy for heavy metal polluted environments: A review of microbial bio-sorbents. *Int. J. Environ. Res. Public Health, 14*, 1–7.

Barrios-Estrada, C., Rostro-Alanis, M. D. J., Muñoz-Gutiérrez, B. D., Iqbal, H. M., Kannan, S., & Parra-Saldivar, R., (2018). Emergent contaminants: Endocrine disruptors and their laccase-assisted degradation: A review. *Sci. Total Environ., 612*, 1516–1531.

Bhuiyan, M. S. U., Min, S. R., Jeong, W. J., Sultana, S., Choi, K. S., Lee, Y., & Liu, J. R., (2011). Overexpression of AtATM3 in *Brassica juncea* confers enhanced heavy metal tolerance and accumulation. *Plant Cell Tissue Organ Cult, 107*, 69–77.

Bilal, M., Adeel, M., Rasheed, T., Zhao, Y., & Iqbal, H. M., (2019a). Emerging contaminants of high concern and their enzyme-assisted biodegradation: a review. *Environ. Int., 124*, 336–353.

Bilal, M., Asgher, M., Iqbal, H. M., Hu, H., & Zhang, X., (2017). Bio-based degradation of emerging endocrine-disrupting and dye-based pollutants using cross-linked enzyme aggregates. *Environ. Sci. Pollut. Res., 24*, 7035–7041.

Bilal, M., Rasheed, T., Nabeel, F., Iqbal, H. M., & Zhao, Y., (2019b). Hazardous contaminants in the environment and their laccase-assisted degradation: A review. *J. Environ. Manag., 234*, 253–264.

Bilal, M., Rasheed, T., Sosa-Hernández, J., Raza, A., Nabeel, F., & Iqbal, H., (2018). Biosorption: An interplay between marine algae and potentially toxic elements: A review. *Marine Drugs, 16*, 65–70.

Brim, H., McFarlan, S. C., Fredrickson, J. K., Minton, K. W., Zhai, M., Wackett, L. P., & Daly, M. J., (2000). Engineering *Deinococcus radiodurans* for metal remediation in radioactive mixed waste environments. *Nat. Biotechnol., 18*, 85–90.

Carpenter, A., (2018). Oil pollution in the North Sea: The impact of governance measures on oil pollution over several decades. *Hydrobiologia*, 1–19.

Chatha, S. A. S., Asgher, M., & Iqbal, H. M., (2017). Enzyme-based solutions for textile processing and dye contaminant biodegradation: A review. *Environ. Sci. Pollut. Res., 24*, 14005–14018.

Cheng, M., Chen, K., Guo, S., Huang, X., He, J., Li, S., & Jiang, J., (2015). PbaR, an IclR family transcriptional activator for the regulation of the 3-phenoxybenzoate 1',2'-dioxygenase gene cluster in *Sphingobium wenxiniae* JZ-1T. *Appl. Environ. Microbiol., 81*, 8084–8092.

Connolly, E. L., Fett, J. P., & Guerinot, M. L., (2002). Expression of the IRT1 metal transporter is controlled by metals at the levels of transcript and protein accumulation. *Plant Cell, 14*, 1347–1357.

Craig, K., (2018). A review of the chemistry, pesticide use, and environmental fate of sulfur dioxide, as used in California. *Rev. Environ. Contam. Toxicol., 246*, 33–64.

Cunningham, F. J., Goh, N. S., Demirer, G. S., Matos, J. L., & Landry, M. P., (2018). Nanoparticle-mediated delivery towards advancing plant genetic engineering. *Trends Biotechnol., 36*, 882–897.

Das, N., Bhattacharya, S., & Maiti, M. K., (2016). Enhanced cadmium accumulation and tolerance in transgenic tobacco overexpressing rice metal tolerance protein gene OsMTP1 is promising for phytoremediation. *Plant Physiol. Biochem., 105*, 297–309.

Dasgupta, J., Sikder, J., Chakraborty, S., Curcio, S., & Drioli, E., (2015). Remediation of textile effluents by membrane-based treatment techniques: A state of the art review. *J. Environ. Manag., 147*, 55–72.

Dhanarani, S., Viswanathan, E., Piruthiviraj, P., Arivalagan, P., & Kaliannan, T., (2016). Comparative study on the biosorption of aluminum by free and immobilized cells of *Bacillus safensis* KTSMBNL 26 isolated from explosive contaminated soil. *J. Taiwan Inst. Chem. Eng., 69*, 61–67.

Doty, S. L., Shang, Q. T., Wilson, A. M., Moore, A. L., Newman, L. A., Strand, S. E., & Gordon, M. P., (2000). Enhanced metabolism of halogenated hydrocarbons in transgenic plants contain mammalian P450 2E1. *Proc. Natal. Acad. Sci. USA, 97*, 6287–6291.

Duan, X. Q., Zheng, J. W., Zhang, J., Hang, B. J., He, J., & Li, S. P., (2011). Characteristics of a 3-phenoxybenzoic acid degrading-bacterium and the construction of an engineering bacterium. *Huan Jing Ke Xue, 32*, 240–246.

Fontes, L. C., Mario, H. B., & Benedito, C., (2015). Potential application of modified *Saccharomyces cerevisiae* for removing lead and cadmium. *J. Bioremed. Biodegrad., 6*, 2–11.

French, C. J., Rosser, S. J., Davies, G. J., Nicklin, S., & Bruce, N. C., (1999). Biodegradation of explosives by transgenic plants expressing pentaerythritol tetranitrate reductase. *Nat. Biotechnol., 17*, 491–494.

Gallo, G., Piccolo, L. L., Renzone, G., La, R. R., Scaloni, A., Quatrini, P., & Puglia, A. M., (2012). Differential proteomic analysis of an engineered *Streptomyces coelicolor* strain reveals metabolic pathways supporting growth on n-hexadecane. *Appl. Microbiol. Biotechnol., 94*, 1289–1301.

Gandia-Herrero, F., Lorenz, A., Larson, T., Graham, I. A., Bowles, J., & Rylott, E. L., (2008). Detoxification of the explosive 2,4,6-trinitrotoluene in *Arabidopsis*: Discovery of bi-functional O and C-glucosyltransferases. *Plant J., 56*, 963–974.

Garbisu, C., Hernandez-Allica, J., Barrutia, O., Alkorta, I., & Becerril, J. M., (2002). Phytoremediation: A technology using green plants to remove contaminants from polluted areas. *Rev. Environ. Health, 17*, 173–188.

Gasic, K., & Korban, S. S., (2007). Transgenic Indian mustard (*Brassica juncea*) plants expressing an *Arabidopsis* phytochelatin synthase (AtPCS1) exhibit enhanced As and Cd tolerance. *Plant Mol. Biol., 64*, 361–369.

Gu, L. F., He, J., Huang, X., Jia, K. Z., & Li, S. P., (2006). Construction of a versatile degrading bacteria *Pseudomonas putida* KT2440-DOP and its degrading characteristics. *Wei Sheng Wu Xue Bao, 46*, 763–766.

Guo, J., Dai, X., Xu, W., & Ma, M., (2008). Overexpressing gsh1 and AsPCS1 simultaneously increases the tolerance and accumulation of cadmium and arsenic in *Arabidopsis thaliana*. *Chemosphere, 72*, 1020–1026.

Hannink, N. K., Subramanian, M., Rosser, S. J., Basran, A., Murray, J. A. H., Shanks, J. V., & Bruce, N. C., (2007). Enhanced transformation of TNT by tobacco plants expressing a bacterial nitroreductase. *Int. J. Phytoremediation, 9*, 385–401.

Hasin, A. A., Gurman, S. J., Murphy, L. M., Perry, A., Smith, T. J., & Gardiner, P. E., (2010). Remediation of chromium (VI) by a methane-oxidizing bacterium. *Environ. Sci. Technol.,* *44*, 400–405.

Hou, Y., Tao, J., Shen, W., Liu, J., Li, J., Li, Y., Cao, H., & Cui, Z., (2011). Isolation of the fenoxaprop-ethyl (FE)-degrading bacterium *Rhodococcus* sp. T1, and cloning of FE hydrolase gene *feh. FEMS Microbiol. Lett.,* *323*, 196–203.

Hsieh, J. L., Chen, C. Y., Chiu, M. H., Chein, M. F., Chang, J. S., Endo, G., & Huang, C. C., (2009). Expressing a bacterial mercuric ion binding protein in plant for phytoremediation of heavy metals. *J. Hazard. Mater., 161*, 920–925.

Inui, H., & Ohkawa, H., (2005). Herbicide resistance in transgenic plants with mammalian P450 monooxygenase genes. *Pest Manag. Sci., 61*, 286–291.

Jackson, E. G., Rylott, E. L., Fournier, D., Hawari, J., & Bruce, N. C., (2007). Exploring the biochemical properties and remediation applications of the unusual explosive-degrading P450 system XplA/B. *Proc. Natl. Acad. Sci. USA, 104*, 16822–16827.

Jacob, J. M., Karthik, C., Saratale, R. G., Kumar, S. S., Prabakar, D., Kadirvelu, K., & Pugazhendhi, A., (2018). Biological approaches to tackle heavy metal pollution: A survey of literature. *J. Environ. Manag., 217*, 56–70.

Jafari, M., Danesh, Y. R., Goltapeh, E. M., & Varma, A., (2013). Bioremediation and genetically modified organisms. In: Goltapeh, E., Danesh, Y., & Varma, A. M., (eds.), *Fungi as Bioremediators* (pp. 433–450). Springer, Berlin.

Jaishankar, M., Tseten, T., Anbalagan, N., Mathew, B. B., & Beeregowda, K. N., (2014). Toxicity, mechanism, and health effects of some heavy metals. *Interdiscip. Toxicol., 7*, 60–72.

Jin, R., Yang, H., Zhang, A., Wang, J., & Liu, G., (2009). Bioaugmentation on decolorization of CI Direct Blue 71 by using genetically engineered strain *Escherichia coli* JM109 (pGEX-AZR). *J. Hazard. Mater., 163*, 1123–1128.

Karavangeli, M., Labrou, N. E., Clonis, Y. D., & Tsaftaris, A., (2005). Development of transgenic tobacco plants overexpressing maize glutathione S-transferase I for chloroacetanilide herbicides phytoremediation. *Biomol. Eng., 22*, 121–128.

Karthik, C., Barathi, S., Pugazhendhi, A., Ramkumar, V. S., Thi, N. B. D., & Arulselvi, P. I., (2017). Evaluation of Cr (VI) reduction mechanism and removal by cellulosimicrobium funkei strain AR8, a novel haloalkaliphilic bacterium. *J. Hazard. Mater., 333*, 42–53.

Kawahigashi, H., Hirose, S., Ohkawa, H., & Ohkawa, Y., (2006). Phytoremediation of herbicide atrazine and metolachlor by transgenic rice plants expressing human CYP1A1, CYP2B6, and CYP2C19. *J. Agric. Food Chem., 54*, 2985–2991.

Keshavareddy, G., Kumar, A. R. V., & Ramu, V. S., (2018). Methods of plant transformation: A review. *Int. J. Curr. Microbiol. App. Sci., 7*, 2656–2668.

Kimura, N., Nishi, A., Goto, M., & Furukawa, K., (1997). Functional analyses of a variety of chimeric dioxygenases constructed from two biphenyl dioxygenases that are similar structurally but different functionally. *J. Bacteriol., 179*, 3936–3943.

Korenkov, V., Hirschi, K., Crutchfield, J. D., & Wagner, G. J., (2007). Enhancing tonoplast Cd/H antiport activity increases Cd, Zn, and Mn tolerance, and impacts root/shoot Cd partitioning in *Nicotiana tabacum* L. *Planta, 226*, 1379–1387.

Kostal, J. R. Y., Wu, C. H., & Chen, W., (2004). Enhanced arsenic accumulation in engineered bacterial cells expression ArsR. *Appl. Environ. Microbiol., 70*, 4582–4587.

Ksheminska, H., Jaglarz, A., Fedorovych, D., Babyak, L., Yanovych, D., Kaszycki, P., & Koloczek, H., (2003). Bioremediation of chromium by the yeast *Pichia guilliermondii*: Toxicity and accumulation of Cr (III) and Cr (VI) and the influence of riboflavin on Cr tolerance. *Microbiol. Res., 158*, 59–67.

Kuroda, K., Shibasaki, S., Ueda, M., & Tanaka, A., (2001). Cell surface-engineered yeast displaying a histidine oligopeptide (hexa-His) has enhanced adsorption of and tolerance to heavy metal ions. *Appl. Microbiol. Biotechnol., 57*, 697–701.

Kurumata, M., Takahashi, M., Sakamoto, A., Ramos, J. L., Nepovim, A., Vanek, T., Hirata, T., & Morikawa, H., (2005). Tolerance to, and uptake and degradation of 2,4,6-trinitrotoluene (TNT) are enhanced by the expression of a bacterial nitro reductase gene in *Arabidopsis thaliana. Z Naturforsch C., 60*, 272–278.

LeDuc, D. L., Tarun, A. S., Montes-Bayon, M., Meija, J., Malit, M. F., Wu, C. P., Abdel, S. M., et al., (2004). Overexpression of selenocysteine methyltransferase in *Arabidopsis* and Indian mustard increases selenium tolerance and accumulation. *Plant Physiol., 135*, 377–383.

Li, H., Cong, Y., Lin, J., & Chang, Y., (2015). Enhanced tolerance and accumulation of heavy metal ions by engineered *Escherichia coli* expressing *Pyrus calleryana* phytochelatin synthase. *J. Basic Microbiol., 55*, 398–405.

Li, Z., (2018). Health risk characterization of maximum legal exposures for persistent organic pollutant (POP) pesticides in residential soil: An analysis. *J. Environ. Manag., 205*, 163–173.

Lin, Y. F., Hassan, Z., Talukdar, S., Schat, H., & Aarts, M. G., (2016). Expression of the ZNT1 zinc transporter from the metal hyperaccumulator *Noccaea caerulescens* confers enhanced zinc and cadmium tolerance and accumulation to *Arabidopsis thaliana. Plos One, 11*, e0149750.

Liu, J., Shi, X., Qian, M., Zheng, L., Lian, C., Xi, Y., & Shen, Z., (2015). Copper-induced hydrogen peroxide upregulation of a metallothionein gene, OsMT2c, from *Oryza sativa* L. confers copper tolerance in *Arabidopsis thaliana. J. Hazard. Mater., 294*, 99–108.

Liu, L., Bilal, M., Duan, X., & Iqbal, H. M. N., (2019). Mitigation of environmental pollution by genetically engineered bacteria: Current challenges and future perspectives. *Sci. Total Environ., 667*, 444–454.

Lopez, A., Lazaro, N., Morales, S., & Margues, A. M., (2002). Nickel biosorption by free and immobilized cells of *Pseudomonas fluorescens* 4F39: A comparative study. *Water Air Soil Pollut., 135*, 157–172.

Lovley, D. R., Phillips, E. J. P., Gorby, Y. A., & Landa, E. R., (1991). Microbial reduction of uranium. *Nature, 350*, 413–416.

Lu, Y., Deng, X., Quan, L., Xia, Y., & Shen, Z., (2015). Metallothioneins BcMT1 and BcMT2 from *Brassica campestris* enhance tolerance to cadmium and copper and decrease production of reactive oxygen species in *Arabidopsis thaliana. Plant Soil, 367*, 507–519.

Martinez, M., Bernal, P., Almela, C., Velez, D., Garcia-Agustin, P., & Serrano, R., (2006). An engineered plant that accumulates higher levels of heavy metals than *Thlaspi caerulescens*, with yields of 100 times more biomass in mine soils. *Chemosphere, 64*, 478–485.

Massa, V., Infantino, A., Radice, F., Orlandi, V., Tavecchio, F., Giudici, R., Conti, F., Urbini, G., Di Guardo, A., & Barbieri, P., (2009). Efficiency of natural and engineered bacterial strains in the degradation of 4- chlorobenzoic acid in soil slurry. *Int. Biodeter. Biodegr., 63*, 112–115.

Mendes, K. F., Régo, A. P. J., Takeshita, V., & Tornisielo, V. L., (2019). *Water Resource Pollution by Herbicide Residues*. Intech Open: https://doi.org/10.5772/intechopen.85159.

Merlot, S., Hannibal, L., Martins, S., Martinelli, L., Amir, H., Lebrun, M., & Thomine, S., (2014). The metal transporter PgIREG1 from the hyperaccumulator *Psychotria gabriellae* is a candidate gene for nickel tolerance and accumulation. *J. Exp. Bot., 65*, 1551–1564.

Mishra, S., Sarma, P. M., & Lal, B., (2004). Crude oil degradation efficiency of a recombinant *Acinetobacter baumannii* strain and its survival in crude oil-contaminated soil microcosm. *FEMS Microbiol. Lett., 2*, 323–331.

Murtaza, I., Dutt, A., & Ali, A., (2002). Biomolecular engineering of *Escherichia coli* organomercurial lyase gene and its expression. *Indian J. Biotech.*, *1*, 117–120.

Neumann, G., Teras, R., Monson, L., Kivisaar, M., Schauer, F., & Heipieper, H. J., (2004). Simultaneous degradation of atrazine and phenol by *Pseudomonas* sp. strain ADP: Effects of toxicity and adaptation. *Appl. Environ. Microbiol.*, *70*, 1907–1912.

Ojuederie, O. B., & Babalola, O. O., (2017). Microbial and plant-assisted bioremediation of heavy metal polluted environments: A review. *Int. J. Environ. Res. Public Health*, *14*, 1504–1511.

Parales, R. E., & Haddock, J. D., (2004). Biocatalytic degradation of pollutants. *Curr. Opin. Biotechnol.*, *15*, 374–379.

Parrilli, E., Papa, R., Tutino, M. L., & Sannia, G., (2010). Engineering of a psychrophilic bacterium for the bioremediation of aromatic compounds. *Bioeng. Bugs.*, *1*, 213–216.

Peñaloza-Vazquez, A., Mena, G. L., Herrera-Estrella, L., & Bailey, A. M., (1995). Cloning and sequencing of the genes involved in glyphosate utilization by *Pseudomonas pseudomallei*. *Appl. Environ. Microbiol.*, *61*, 538–543.

Pesantes, A. A., Carpio, E. P., Vitvar, T., López, M. M. M., & Menéndez-Aguado, J. M., (*2019*). A Multi-index analysis approach to heavy metal pollution assessment in river sediments in the ponce enríquez area, Ecuador. *Water, 11, 590–596.*

Pu, Q., Sun, J. Q., Zhang, F. F., Wen, X. Y., Liu, W. H., & Huang, C. M., (2019). Effects of copper mining on heavy metal contamination in a rice agro-system. *Acta Geochim.*, 1–21.

Rathod, J., Dhebar, S., & Archana, G., (2017). Efficient approach to enhance whole cell azo dye decolorization by heterologous overexpression of *Enterococcus* sp. L2 azoreductase (azoA) and *Mycobacterium vaccae* formate dehydrogenase (fdh) in different bacterial systems. *Int. Biodeter. Biodegr.*, *124*, 91–100.

Rodríguez-Llorente, I. D., Lafuente, A., Doukkali, B., Caviedes, M. A., & Pajuelo, E., (2012). Engineering copper hyperaccumulation in plants by expressing a prokaryotic cop C gene. *Environ. Sci. Technol.*, *46*, 12088–12097.

Rosculete, C. A., Bonciu, E., Rosculete, E., & Olaru, L. A., (2019). Determination of the environmental pollution potential of some herbicides by the assessment of cytotoxic and genotoxic effects on *Allium cepa*. *Int. J. Environ. Res. Public Health*, *16*, 75–85.

Sakaki, T., Shinkyo, R., Takita, T., Ohta, M., & Inouye, K., (2002). Biodegradation of polychlorinated dibenzo-p-dioxins by recombinant yeast expressing rat CYP1A subfamily. *Arch Biochem. Biophys.*, *401*, 91–98.

Samanta, S. K., Singh, O. V., & Jain, R. K., (2002). Polycyclic aromatic hydrocarbons: Environment pollution and bioremediation. *Trends Biotechnol.*, *20*, 243–248.

Shabbir, M., Singh, M., Maiti, S., Kumar, S., & Saha, S. K., (2018). Removal enactment of organophosphorus pesticide using bacteria isolated from domestic sewage. *Bioresour. Technol.*, *263*, 280–288.

Shim, D., Kim, S., Choi, Y. I., Song, W. Y., Park, J., Youk, E. S., Jeong, S. C., Martinoia, E., Noh, E. W., & Lee, Y., (2013). Transgenic poplar trees expressing yeast cadmium factor 1 exhibit the characteristics necessary for the phytoremediation of mine tailing soil. *Chemosphere*, *90*, 1478–1486.

Sinha, S., & Mukherjee, S. K., (2009). *Pseudomonas aeruginosa* KUCD1, a possible candidate for cadmium bioremediation. *Braz. J. Microbiol., 40*, 655–662.

Sousa, C., Kotrba, P., Ruml, T., Cebolla, A., & De Lorenzo, V., (1998). Metalloadsorption by *Escherichia coli* cells displaying yeast and mammalian metallothioneins anchored to the outer membrane protein lam B. *J. Bacteriol.*, *180*, 2280–2284.

Springael, D., Peys, K., Ryngaert, A., Van, R. S., Hooyberghs, L., Ravatn, R., Heyndrickx, M., et al., (2002). Community shifts in seeded 3-chlorobenzoate degrading membrane biofilm reactor: Indications for involvement of in situ horizontal transfer of the clc-element from inoculum to contaminant bacteria. *Environ. Microbiol., 4*, 70–80.

Strong, P. J., & Burgess, J. E., (2008). Treatment methods for wine-related ad distillery wastewaters: A review. *Biorem. Jou., 12*, 7087–7095.

Taştan, B. E., Ertuğrul, S., & Dönmez, G., (2010). Effective bio-removal of reactive dye and heavy metals by *Aspergillus versicolor*. *Bioresour. Technol., 101*, 870–876.

Tsoi, T. V., Zaitsev, G. M., Plotnikova, E. G., Kosheleva, I. A., & Boronin, A. M., (1991). Cloning and expression of the Arthrobacter globiformis KZT1 fcbA gene encoding dehalogenase (4-chlorobenzoate-4-hydroxylase) in *Escherichia coli*. *FEMS Microbiol. Lett., 81*, 165–169.

Van, D. M. J., R., Ravatn, R., & Sentchilo, V., (2009). The element of *Pseudomonas* sp. strain B13 and other mobile degradative elements employing phage-like integrases. *Arch Microbiol., 175*, 79–85.

Vázquez-Luna, D., & Cuevas-Díaz, M. C., (2019). Soil contamination and alternatives for sustainable development. In: Vázquez-Luna, D., & Cuevas-Díaz, M. C., (eds.) *Soil Contamination and Alternatives for Sustainable Development*. Intech Open. https://doi.org/10.5772/intechopen.83720.

Wang, L., Samac, D. A., Shapir, N., Wackett, L. P., Vance, C. P., Olszewski, N. E., & Sadowsky, M. J., (2015). Biodegradation of atrazine in transgenic plants expressing a modified bacterial atrazine chlorohydrolase (atzA) gene. *Plant Biotechnol. J., 3*, 475–486.

Xia, Y., Qi, Y., Yuan, Y., Wang, G., Cui, J., Chen, Y., Zhang, H., & Shen, Z., (2012). Overexpression of *Elsholtzia haichowensis* metallothionein 1 (EhMT1) in tobacco plants enhances copper tolerance and accumulation in root cytoplasm and decreases hydrogen peroxide production. *J. Hazard. Mater., 234*, 65–71.

Xie, Y., Yu, F., Wang, Q., Gu, X., & Chen, W., (2018). Cloning of catechol 2, 3-dioxygenase gene and construction of a stable genetically engineered strain for degrading crude oil. *Indian J. Microbiol., 54*, 59–64.

Zaidi, S. S., & Mansoor, S., (2017). Viral vectors for plant genome engineering. *Front Plant Sci., 8*, 539–545.

Zhang, J., Zhang, M., Tian, S., Lu, L., Shohag, M. J. I., & Yang, X., (2014). Metallothionein 2 (SaMT2) from *Sedum alfredii* Hance confers increased Cd tolerance and accumulation in yeast and tobacco. *Plos One, 9*, e102750.

Zhang, Q., Yu, R., Fu, S., Wu, Z., Chen, H. Y. H., & Liu, H., (2019). Spatial heterogeneity of heavy metal contamination in soils and plants in Hefei, China. *Sci. Rep., 1049*, 2045–2051.

Zhao, Q., Yue, S., Bilal, M., Hu, H., Wang, W., & Zhang, X., (2017). Comparative genomic analysis of 26 *Sphingomonas* and *Sphingobium* strains: Dissemination of bioremediation capabilities, biodegradation potential and horizontal gene transfer. *Sci. Total Environ., 609*, 1238–1247.

Zhongli, C., Shunpeng, L., & Guoping, F., (2001). Isolation of methyl parathion-degrading strain M6 and cloning of the methyl parathion hydrolase gene. *Appl. Environ. Microbiol., 67*, 4922–4925.

Zhu, Y., Pilon-Smits, E. A. H., Jouanin, L., & Terry, N., (1999b). Overexpression of glutathione synthetase in *Brassica juncea* enhances cadmium tolerance and accumulation. *Plant Physiol.*, *119*, 73–79.

Zhu, Y., Pilon-Smits, E. A., Tarun, A. S., Weber, S. U., Jouanin, L., & Terry, N., (1999a). Cadmium tolerance and accumulation in Indian mustard is enhanced by overexpressing γ-glutamylcysteine synthetase. *Plant Physiol.*, *121*, 1169–1177.

CHAPTER 12

Bio-Indicator Species and Their Role in Monitoring Water Pollution

MUNIZA MANZOOR,[1] KULSUM AHMAD BHAT,[2] NAZIYA KHURSHID,[3] ALI MOHD. YATOO,[4] ZARKA ZAHEEN,[4] SHAFAT ALI,[5] MD. NIAMAT ALI,[5] INSHA AMIN,[6] MANZOOR UR RAHMAN MIR,[6] SHAHZADA MUDASIR RASHID,[6] and MUNEEB U. REHMAN[6,7]

[1]*Wildlife Laboratory, Department of Zoology/Cytogenetics and Molecular Biology Laboratory, Center of Research for Development, University of Kashmir, Hazratbal, Srinagar–190006, Jammu and Kashmir, India*

[2]*Wildlife Laboratory, Department of Zoology/Phytochemistry Laboratory, Center of Research for Development, University of Kashmir, Hazratbal, Srinagar–190006, Jammu and Kashmir, India*

[3]*Parasitology Laboratory, Department of Zoology/Microbiology Laboratory, Center of Research for Development, University of Kashmir, Hazratbal, Srinagar–190006, Jammu and Kashmir, India*

[4]*Center of Research for Development/Department of Environmental Science, University of Kashmir, Hazratbal, Srinagar–190006, Jammu and Kashmir, India*

[5]*Cytogenetics and Molecular Biology Laboratory, Center of Research for Development, University of Kashmir, Hazratbal, Srinagar–190006, Jammu and Kashmir, India*

[6]*Division of Veterinary Biochemistry, Faculty of Veterinary Science and Animal Husbandry, SKUAST-Kashmir, Alustang, Shuhama, Srinagar–190006, Jammu and Kashmir, India*

[7]*Department of Clinical Pharmacy, College of Pharmacy, King Saud University, Riyadh–11541, Saudi Arabia, E-mail: muneebjh@gmail.com*

ABSTRACT

The menace of pollution has spread globally with the rapid growth of industries and has mostly affected water. Pollutants of different kinds like physical, chemical, radioactive, and infective microscopic biological substances pass through water resources like oceans, lakes, ponds, rivers, and thus degrade the quality and essence of water, which often leads to the detrimental and negative effect on various aquatic species. Animals, plants, planktons, and microbes act as bio-indicators and are used to monitor the well-being of the natural environment. The bio-indicators discover their main utility in evaluating ecological well-being and bio-geographic alterations taking place in the natural ecosystem. Almost all the biological entities reveal the health of its environment like plankton; these respond quickly to slight fluctuations occurring in the natural environment and thus serve as an important biomarker for reflecting the water quality and in also monitoring the water pollution. Plankton reveals the fitness of aquatic flora, and act as primary warning indicators. In the past, various attempts of examining and monitoring techniques came to the forefront that could improve the capability of perceiving more pollutants in less time, and at lesser concentrations. The technique of bio-monitoring or biological monitoring is an important and essential means through which the quality of water can be assessed, in which morphological, biochemical, and physiological alterations occur in indicators and are being related to particular environmental pollution.

12.1 INTRODUCTION

The rapid growth of industries has tremendously increased the production and use of both natural and chemical substances (Markert et al., 2000; Pereira et al., 2009; Ambasht and Ambasht, 2003), which are directly or indirectly drained into the aquatic bodies in adequate amounts enough to cause detrimental effects to aquatic life in them (Tsui and Chu, 2003; Meng et al., 2008). Due to the many potential impacts and unpredicted reactions, there is a necessity for active monitoring mechanisms. Biological monitoring or biomonitoring is an extremely significant and vital tool that evaluates the changes in our surroundings based on natural and biological responses (Oertel and Salánki, 2003). Comparing the traditional methods of chemical investigation of aquatic milieu with bio-monitoring, it unveils the noticeable predominance that makes bio-monitoring an attractive device for exercising incomparable roles in the estimation of environmental pollution, mainly for the pollution in aquatic bodies (Zhou et al., 2008). Thus, many organisms which are assigned as bio-indicators show the distinct advantages for bio-monitoring of pollution in aquatic ecosystem in comparison with the others. All the ecosystems of the world are adversely affected due to human involvement and interference in the name of development and are shrinking the survival of humans through global warming and climate change. Air, water, and soil, the three main constituents of the universe have become polluted beyond allowable limits (Haribhau, 2012). Good quality air and freshwater have now become essential commodities of life. Natural forests are the main source that renew oxygen in the atmosphere and are capable of safeguarding the hostile and destructive effects to a large extent, but they also fall prey to the humans in the name of development and thus adding to the problem of pollution. Vehicular traffic and industries release different toxic gases and effluents like carbon dioxide, sulfur dioxide, nitrous oxide, and chlorine into the environment, all of which pollute the air and water ecosystems (Srivastava and Kumar, 2012). Moreover, the tools and techniques of advanced agricultural practices eyeing on rapid and instant returns make the use of pesticides, weedicides, fertilizers, and synthetic hormones a norm that further ends up by degrading and polluting the environment. Industrialization, urbanization, and modernization and its subsequent demands on land and other amenities are yet additional reason that contributes to the pollution of the global environment. Different toxic pollutants and contaminants have an adverse impact on the living organisms gradually over a period of time and lead to many interactions (Nwajei et al., 2012; Ranjeeta, 2012). Primarily these interactions may occur at the cellular and biochemical level and often lead to biological impairments such as interruption and disorders of locomotory, respiratory, feeding, excretory, reproductive, neural, and circulatory processes in animals and in the case of plants, the pollutants adversely affect the transpiratory, photosynthetic, respiratory, growth, and reproductive processes. Physical and chemical methods are used to assess the quality of water (Espinal-Carreon et al., 2013) yet biological monitoring has been used to complement all the other techniques (Li et al., 2010; Springer, 2010). Bio-monitoring indicates employing living species to assess the ecological conditions (Gerhardt, 1996), and it essentially involves the selection of either one or a group of organisms (as the bio-monitors) and biological responses shown by the indicators is used to recognize and closely follow changes occurring in the environment. Bioindicators can be used to detect and recognize changes in the quality of water or the variations in the river or stream habitat (Reece and Richardson, 2000). Likewise, biological monitoring offers supplementary and genuine evidence regarding the present and the past developments in environmental activities and behavior (Oertel and Salanki, 2003). Organisms of water, such as microscopic diatoms (John, 2003), macro-invertebrates (Flores and Zafaralla, 2012; Ferreira et al., 2011), and fish (Tejeda-Vera et al., 2007; Trujillo Jimenez et al., 2011), are more frequently used as bio-indicators. Polluted water symbolizes a spring of ailments and diseases in living organisms including humans thus making efficient and reliable monitoring of water a necessity (Bae et al., 2012; Hassan et al., 2012). Fish (Mishra et al., 2000), algae (Zhang et al., 2012), living cells (Wolf et al., 2013), Daphnia (Guilhermino et al., 2000) and various other invertebrates (Beyene et al., 2009) are employed as indicator species since they react to the noxious effects of the metals and other pollutants in water bodies.

A bio-indicator is defined as any organism called an indicator species or group of organisms whose purpose, utility, and populace can disclose the qualitative and quantitative properties of the ecosystem, for example, microscopic water crustaceans and copepods occurring in water bodies can be monitored for biochemical and physiological variations that may specify a problem within the ecosystem. Bio-indicators are used to quantify or measure the cumulative properties of various contaminants in the environment and about how long this problem has been present there (Karr, 1981). Thus, a good bio-indicator is sensitive even to a slight occurrence of the contaminant and also provides ample knowledge regarding the concentration and intensity of the exposure. A bio-indicator can also be used to assess the sterility of the water ecosystem through the use of resilient and impervious microorganism strains (Protak Scientific, 2017). Biological monitors can be defined as the overview of extremely resilient microscopic organisms to a given ecosystem before sterilization and the tests are conducted to quantity the efficacy of the sterilization or purification methods. Thus biological indicators use extremely impervious microbes, so that one can be assured that any purification method that renders them indolent will have also got killed more common than feebler pathogens.

Bio-indicators give a clear picture about the total effects of various impurities or contaminants in the environment and for how long a problem has persisted in the water body. Additionally, short term or long-term stress events or conditions can predict impending conditions and changes by recognizing the deviations in the monitoring species (Cairns and Pratt, 1993). Bio-indicator is a name specified to a living species or group of organisms that express or confirms the facts or information based on the ecosystem or a part of it (Wilkomirski, 2013). Hence, the organisms that can be scrutinized without any effort, difficulty, or exertion about the environmental situations of their locale or territory can be regarded as indicator species (Landres et al., 1988; Cairns and Pratt, 1993; Bartell, 2006). Different kinds of bio-indicators, like lichens, microorganisms, animals, plants, etc., yield certain peculiar molecular signs under different ecological conditions (Posudin, 2014). With the help of these bio-indicators, one can forecast the normal condition of a certain area or the degree or range of contamination (Khatri and Tyagi, 2015). All the living species or populations cannot serve as effective bio-indicators. Physical, chemical, and biotic factors such as light, type of substrate, temperature conditions, and degree of competition may vary among habitats, which can affect the function of bio-indicators. As time proceeds, organisms develop different tactics to maximize growth and reproduction within a particular range of environmental conditions (Devlin et al., 2011). Bio-indicators effectively indicate the environmental conditions because of their restrained tolerance to the changeability of ecosystem properties (Akbulut, 2003; Demir et al., 2014).

The bio-monitoring species vary from several important or key indicator species, while both are equally useful in revealing information about their environmental conditions. These bio-monitors clarify and describe the characteristics of different habitats through their population richness and abundance values. The vital or key indicator species are those species that are indispensable to the environment. The loss or disappearance of such species will lead to modifications in the food web as important components would be lost. The utilization of indicators species is not just limited to a lone organism with a restricted biological tolerance. Societies that involve a wide range of biological tolerances can assist as bio-monitors and signify numerous sources of data to evaluate ecosystem conditions in a "biotic index" (Gokce, 2016).

12.2 BIO-INDICATION-MEANING

Bio-indication is often used frequently but rarely in different ways and means (O'Brien et al., 1993). Overall the purpose of bio-indication is purely the application of living organisms as indicators of environmental status. McGeoch (1998) stated that there are three wide and comprehensive categories or types of bio-indicators that can be recognized based on their biodiversity, environmental, and ecological applications. Even though these classes may be considered to some extent as artificial, and clearly overlap and show similarity, yet, they provide a valuable and beneficial framework which consider the diversity and variances in bio-indicator methodologies. Ecological indicators can also be termed as pointers as they point towards certain stresses and aim to measure either quantitatively or semi-quantitatively or specify the importance of an ecological variable by regularly reviewing the reactions of organism shape, community structure and population size. An important illustration of such a study is the use of the protist communities to quantitatively measure the sea-level variability along different gradients of pollution and then apply the same elsewhere. The different classes of bio-indicators also show bioaccumulation. Bioaccumulation is the property to accumulate different chemicals within living systems (usually a pollutant) to indicate environmental stress, such as the use of shellfish to indicate TBT pollution (Morcillo et al., 1999). Bio-indicators are often regarded as the environmental sentinels and are thus deliberately used to indicate the presence or level of a pollutant, like the use of bioluminescent bacteria which is genetically altered to indicate the occurrence of mutagenic pollutants (Podgórska and Węgrzyn, 2006). There is a slight contrast between environmental indicators and ecological indicators, and the latter mainly focuses on the influences of an environmental aspect or cause rather than the level at which they operate. However, the distinction is subtle and somehow, indicators may fulfill both the roles. This is a very significant reason as the conditions for effectiveness is different. Simply for the environmental indicators, the basic criteria for an active indicator are the capability to evaluate the existence of an environmental factor, whereas in the case of ecological indicators, the main objective is to indicate or validate the impacts on a wider range of living organisms or ecological factors. Valentine et al. (2013) illustrate the use of testate amoebae as a potential indicator to monitor peatland renewal and restoration. Testate amoebae may also be used to indicate environmental factors like water table and pH and therefore also prove to be an effective environmental indicator. These factors are of immense importance as the position of this particular testate amoeba at topmost of the microbial food-web means or represents changes in a wider group of organisms. The last category of bio-indicators is biodiversity indicators which aim to focus on the diversity of one set of organisms as a substitute for the diversity of a wider range of other organisms (McGeoch, 1998).

12.3   AQUATIC BIO-MONITORING

The freshwater qualities were traditionally mainly focused on biological, chemical, physical, and microbial dimensions (Metcalfe, 1989; Ouyang,

2005). In the modern era, coming across to the wide range of composite pollutants and their richness without assessment, traditional methods of monitoring cannot redirect the addition of various environmental factors and long-lasting sustainability of aquatic ecosystems (Li et al., 2010). Moreover, uses of other indicators have been widely accepted and are known to enhance, are complementary to traditional physical and chemical monitoring techniques, and can greatly boost the assessment and management of aquatic ecosystems (Zhou et al., 2008). Biological monitoring or bio-monitoring is the use of biotic organisms or their reactions to regulate and closely monitor the variations in the aquatic ecosystem. There are numerous bio-monitoring methods or techniques that are employed in the water environment. The basic and main bio-monitoring methods include the use of indicator species for biotic indices, quick bio-assessment, and analytical models (Reece and Richardson, 1999). Employing of indicator species, indicator organisms, or bio-indicators is the distinguishing technique in water quality valuation largely due to its capacity to signify the overall status or condition of the environment, detects the trends through their sensitivity to a wide array of contaminants (Li et al., 2010; Gerhardt, 2002; Holt and Miller, 2011).

12.4 CHARACTERISTICS OF A BIO-INDICATOR

A good indicator must have the following characteristics: should be easily recognized by non-specialists, should have wide distribution and high abundance, quick enough to reflect environmental variations, short life span, sessile lifestyle and hence symbolic to native pollution. They may live in sediment deposits where contaminants have a tendency to accumulate. They should be sensitive to a wide range of stressors and must gather a high level of toxins without death in their systems. They should be essential components of the ecosystem and appropriate for scientific experiments (easy sampling) (Gerhardt, 2002; Holt and Miller, 2011).

12.5 TYPES OF BIO-INDICATORS

Bio-indicators are currently used and endorsed by countless organizations (the World Conservation Union, International Union for Conservation of Nature) as a novel way to handle bio-monitoring and assess anthropogenic pressures. Figure 12.1 describes the different categories of bio-indicators.

TYPES OF BIOINDICATORS

MICROBIAL INDICATORS

PLANT INDICATORS

ANIMAL INDICATORS

Micro organisms : used in indicating the maritime or physical natural community health.

Bioluminiscent microorganisms are generally used to test water for natural poisons.

Region or nonappearance of certain plant or vegetative life in a natural group can give basic bits of data about the well being of the environment. • Eg: lichens, planktons.

An expand or decline in a creature populace might show harm to biological community brought on by contamination

Zooplanktons like Alona guttata, Moscyclopesedex, Cyclips, Aheyella.

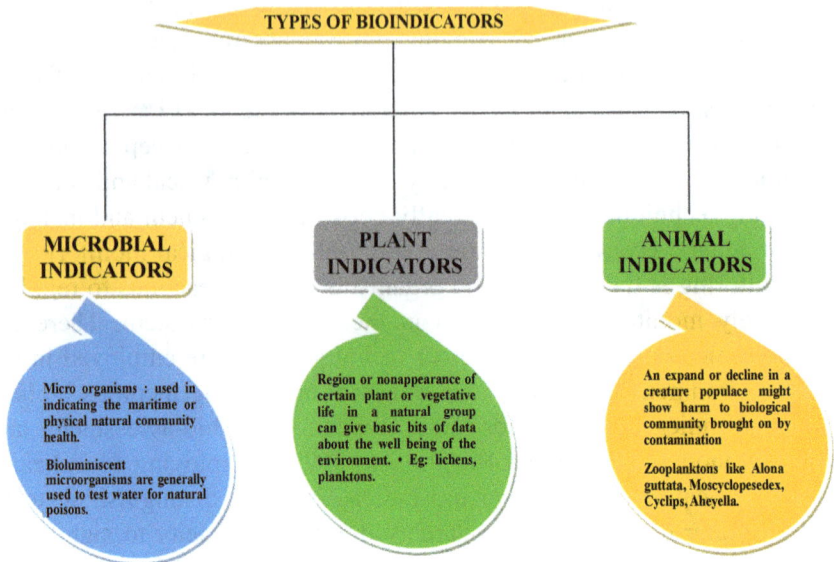

FIGURE 12.1 Different types of biological indicators that help in assessing the health of environment.

12.5.1 MICROBIAL INDICATORS

Different microbial organisms are used as fitness indicators of terrestrial and aquatic ecosystems. Due to their richness and abundance, microorganisms are easy to investigate and easily accessible. Various microorganisms, when exposed to benzene and cadmium pollutants, develop new kinds of molecular proteins known as stress proteins, which can be used as primary threatening signs (Khatri and Tyagi, 2015). These microorganisms are an essential part of marine biomass and contribute for the majority of productivity and nutrient enrichment in oceanic ecosystem. These microbes happen to show very rapid growth, and respond to even very minute concentrations of toxins and other physical, chemical, and biological alterations occurring in different living systems. From a research outlook, they give vital signs of ecological change (Zannatul and Muktadir, 2009; Nkwoji et al., 2010; Hosmani, 2014). Microbial indicators are employed in numerous ways to identify ecological contaminants in water like bioluminescent bacteria. The occurrence of toxic chemicals and pollutants in aquatic ecosystem can be easily gauged either by changes in the digestive system of microbes which is altered by the presence of toxins and other pollutants which may result in variations in the amount of light discharged by the bacteria (Butterworth et al., 2001; Uttah et al., 2008). When comparing it to former existing traditional tests, these tests are rapid to check; however, they can only specify the changes occurring in the organisms due to the presence of these toxins (Malik and Bharti, 2012; Khatri and Tyagi, 2015). One such example is the bacterium *Vogesella indigofera* that responds to heavy metals. The bacterium *V. indigofera* produces blue skin-coloration in the presence of metal contamination, which is a vital indication of external variation that has occurred and is very much conspicuous in nature. The production of pigment can be attributed to the connection between the concentration of chromium and the generation of blue pigmentation by the bacterium. On the other hand, under the surrounding area of hexavalent chromium, the pigment production is blocked, so far the reasons not known (Jain et al., 2010; Aslam et al., 2012; Malik and Bharti, 2012).

12.5.2 PLANT INDICATORS

Plants are regarded as effective and delicate tools for estimation of various environmental stresses. In 21st century, due to rapid industrial development and expansion, the problem of water pollution has intensified (Batiuk et al., 1992; Joanna, 2006). Marine flora provides prized evidences to calculate the status of oceanic ecosystem, as these plants are sedentary and rapidly attain a state of equilibrium with their natural surrounding (Phillips and Rainbow, 1993; Jain et al., 2010). Similarly, the presence or absence of some particular plants or other vegetation types provides sufficient knowledge about the ecological state of a particular place. Lichens that are generally seen growing on the tree trunks and rocks are association of algae and fungi both. Lichens are known to respond to minute ecological changes in terrestrial ecosystem, including changes in the structure of the woods, quality of the air, and climate of the region. Hence these environmental stress can be specified by the vanishing lichens in forests, as triggered by the changes such as increases in the level of sulfur dioxide (SO$_2$), pollutants of sulfur, and nitrogen (Gerhardt, 2002; Holt and Miller, 2010; Khatri and Tyagi, 2015). *Wolffia globosa* is another important plant indicator used for screening cadmium sensitivity and is also used for estimating degree of cadmium contamination. Changes in the variety of species of phytoplankton, including *Euglena elastica, Phacus tortus*, and *Trachelonanas*, specify the contamination of marine ecosystems (Phillips and Rainbow, 1993; Jain et al., 2010).

12.5.3 ANIMAL INDICATORS

A sharp variation in the population of animals is indicative of detrimental changes caused due to pollution in the ecosystem. A change in factor-like population density may imply undesirable changes in the health of the ecosystem. A variation in population density may be as a result of the distorted relationship between populations and food sources; if food resources become scarce and cannot provide for the population demand reduction of population in question will follow (Phillips and Rainbow, 1993; Jain et al., 2010). Animal indicators are also employed for the detection of the amount of toxins present in the tissues of animals (Joanna, 2006; Khatri and Tyagi, 2015). Frogs are known for being important animal indicators designated to assess the quality of environment and changes taking place therein. They are mainly sensitive to the changes that take place in their freshwater and terrestrial habitats, making them as vital bio-monitors of ecological quality and change. Zooplanktons like Alonaguttata, Mesocyclopsedax, Cyclops, Aheyella are zone-based indicators of pollution (Jha and Barat, 2003; Zannatul and Muktadir, 2009; Jain et al., 2010; Hosmani, 2014). Various invertebrates also function as bio-indicators, these aquatic invertebrates are generally bottom feeders (also known as Benthos or macroinvertebrates), living near the bottom of aquatic ecosystems. These kinds of bio-indicators may be predominantly powerful indicators of watershed health as they are easy to distinguish in a lab, commonly live for more than one year, show limited mobility, and are good indicators of ecological conditions (Khatri and Tyagi, 2015).

12.6 HOW DO BIO-INDICATORS WORK?

Bio-indicators that are designed particularly to assess human exposure (e.g., through the food chain) can also be suggestive of the health of the organisms themselves (Burger et al., 1997), and more generally of the health of the ecosystem. Bio-indicators are specifically modified in physiological and behavioral ways to sense changes in environmental health. These specific changes happen to differ from organism to organism. Employing organisms as bio-indicators comprises many areas of science. Changes in behavior and population of a species can be sensed by scientists, but physiological changes must be detected using specials tests only. The technique of bio-assaying requires samples from organisms to detect changes in the environment. These tests may be used to ensure drinking water safety or to measure river health. In the future, as research identifies new ways to use microbes, these uses will expand to include testing of soil and air. Bioassays can be carried out in traditional ways or with new biotechnology-derived methods. These modern techniques include Fluorescence In-Situ Hybridization-FISH and DNA Microarray Technology.

12.7 SPECIES AS WATER QUALITY BIO-INDICATORS

Planktons, microbes, plants, and animals are utilized as bio-indicators to monitor the health of the natural ecosystem, for assessing environmental health and bio-geographic changes taking place in the environment. Figure 12.2 shows different natural species that have the potential to serve as bio-indicators for water quality.

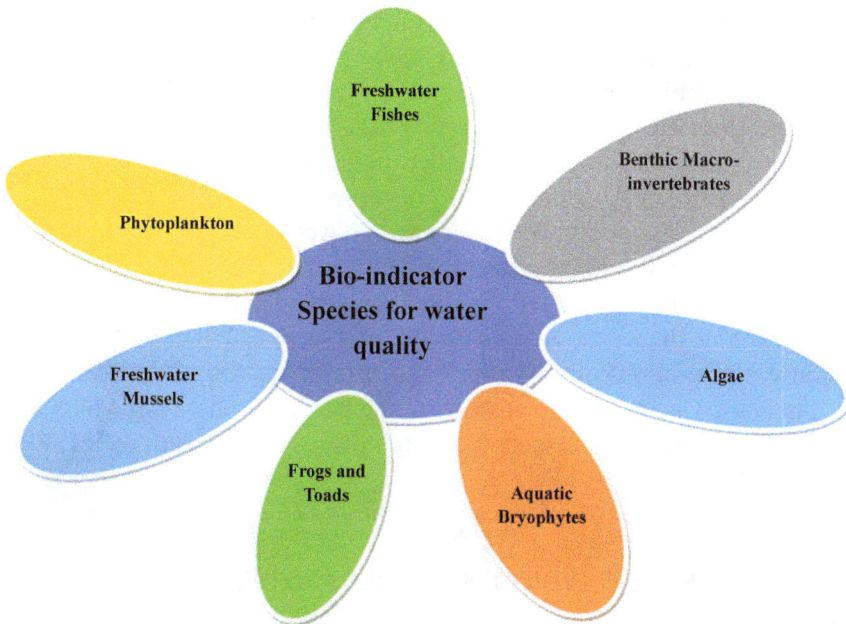

FIGURE 12.2   Different species of organism that function as biological indicators and help in monitoring the quality of water.

12.7.1 *FRESHWATER FISHES*

Fishes being a large taxonomic group possibly accounts for almost 50% of all vertebrate diversity, with an approximate number of 28,000–40,000 species

(Nelson, 2006). They are known to have occupied almost every aquatic habitat (Wootton, 1992), the freshwater as well as the saltwater. This universal presence allows the fishes to be nominated as one of the best ecological indicators of the aquatic ecosystems. Thus fishes have been considered since long by researchers as bio-indicators of water pollution due to their unique biological characters and their role in monitoring the health of freshwater ecosystems. Unlike many invertebrates, some of the characters among many which make them valuable as indicators can be mentioned as the following: the fishes are completely aquatic and spend all their lives in water, which makes them sensitive to even the slightest kinds of disturbances, like the hydrological alterations and impact of pollutants (Hogan and Vallance, 2005) thus fishes are excellent models existing in aquatic environments which are impacted by numerous disturbances and are closely analyzed as to how they respond to some stressors; mostly fish species have a lifespan of about 2–10 years and can reflect both current as well as long-term water quality parameters; almost all fishes comprise of diverse feeding habits, and are therefore able to assimilate ecological health over both the temporal as well as large spatial scales. In this way, the fishes are very much less susceptible to any natural microhabitat change other than smaller organisms, making them extremely useful for assessing regional and macro-habitat changes (Hogan and Vallance, 2005). Moreover, the fishes form a well-defined taxonomic group and are the largely studied group in aquatic environments in terms of their physiological and biological (morphological and anatomical) responses. Fishes provide diverse ecosystem "goods and services" such as the fisheries produce. More significantly, various fish species may directly affect human health as they are at the top position in the aquatic food chain, which is of much significance with respect to bio-monitoring using fish species (Zhou et al., 2008).

12.7.1.1   FISH: A BIO-INDICATOR SPECIES AT LOWER LEVELS OF ORGANIZATION

A variety of biomarkers can respond to toxic stress with different levels of specificity, and can also reveal the large-scale effects of a mixture of contaminants in organisms. Thus, some biomarkers respond to only one chemical or group of chemicals and are highly specific; however, the majority of biomarkers respond to environmental stress in general and are less specific (Lam, 2009). The most common effect of xenobiotics on fish is oxidative stress which includes a number of oxidative reactions that harm the health conditions of fish (van der Oost et al., 2003). Several such chemicals can result in amplified generation of free radicals, mainly oxygen free radicals, also recognized as reactive oxygen species (ROS). The ROS or their intermediates consist of hydrogen peroxide (H_2O_2), superoxide anion (O_2), and hydroxyl radical (OH). The increase in ROS production due to exposure to pollution can occur by a number of mechanisms, which include the uptake of organic xenobiotics and redox cycling metals, the introduction of oxyradical generating enzymes and the metabolism of xenobiotics to redox cycling derivates, such as quinones, (Livingstone, 2003). To assess the impact of pollutants and chemicals released into environment diverse kinds of biomarkers have been used in fishes whose last destination is aquatic ecosystems. For organophosphate and carbamate pesticides, inhibition of cholinesterases has been extensively used as a biomarker (García et al., 2000), among which, inhibition of acetyl-cholinesterase (AchE) is generally used as biomarkers of neurotoxicity.

12.7.1.2 FISH: A BIO-INDICATOR SPECIES AT HIGHER LEVELS OF ORGANIZATION

Bio-indicators have been used by several researchers at higher levels of organization (Nath Singh and Srivastava, 2010; Oertel and Salanki, 2003; Tejeda-Vera et al., 2007; Trujillo Jimenez et al., 2011). To recognize damage in sentinel organisms, individual level parameters like morphological changes are measured. Individual metrics comprise condition indices, such as the liver-somatic index, spleno-somatic index, condition factor, and visceral-somatic index. All these parameters thus indicate the overall health of the organism. Parameters of reproductive integrity consist of the fecundity, gonado-somatic index (GSI) and the prevalence of atretic oocytes (Barton et al., 2002; Tejeda-Vera et al., 2007). The metrics that can be used at the population level are size at age (growth indicator), size-frequency distribution (reproductive and mortality indicators), and sex ratio. These parameters indicate the structure of the population. At community level several ecologically significant responses have been used to assess the health of the fish community. Several factors like degradation and environmental stress usually result in changes in species richness, relative abundance, and their similarity that represents different components of the community. To assess species richness, it is of huge significance to use the number of native species, which in general decreases with the increased environmental degradation. For comparing communities diversity indexes are very useful.

The characteristics that may be usually found in a strained ecosystem, including various changes at community level: increase in the amount of "r" strategists; decrease in the size of organism; decrease in the lifespan of organism; shortening of food chains because of reduction in the energy flow at higher trophic levels and greater sensitivity of predators to stress. Thus, all the above-mentioned factors results in the decrease in species diversity and an increase in species dominance (Odum, 1985). To measure relative ecosystem health a number of biotic indices for fish communities have been developed. Karr and Dudley (1981) defined Biological Integrity as "the ability of supporting and maintaining a balanced, integrated, and adaptive community of organisms having species diversity, composition, and functional organization similar to that of natural habitat of the region," and they first proposed the biotic index for fish called the "index of biotic integrity" (IBI). It is to be noted that different physical, chemical, and biotic factors of the aquatic ecosystem may disturb this biotic integrity. Based on metrics or fish assemblage attributes, the IBI is assessed. The metrics should illustrate features, such as the taxonomy that is species richness and composition, trophic position of species, reproductive type that is oviparous and viviparous and tolerance features of the assemblage.

12.7.2   MACRO-INVERTEBRATES (BENTHIC)

Macro-invertebrates represent a group of those invertebrates that can be seen without the help of a magnifying glass or microscope. Aquatic macro-invertebrates spend whole or part of their life cycles in water and consist of many insects, mites, crustaceans, worms, and mollusks (Chessman, 2003). Macro-invertebrate assemblages contain an extensive range of trophic guilds, which have been described as per the adaptations for food procurement rather than the food eaten and are hence called functional feeding groups (FFG) (Cummins and Klug, 1979). FFG play a major ecological role since they show an important association between energy inputs and their transfer to other trophic levels through the food web (Hanson et al., 2010), take part in maintaining the balance between heterotrophs at the head of rivers, and autotrophy that depends on the primary production of the periphyton and macrophytes at downstream reaches of rivers. In this way, they are also in contact with contaminants entering the system, accumulating the contaminants and transferring them to other trophic levels. FFG form the basis of the river continuum concept developed by Vannote et al. (1980), which describes the structure and function of biological communities along a river system, and where the capability of these biological communities move towards efficient use of energy inputs through resource partitioning. Macro-invertebrates are universal as well as abundant in most of the streams; they occupy different habitats, and are present even in small streams where a limited fish fauna is usually supported. Macro-invertebrates also have restricted movement or their lifestyle is of sedentary type. Generally, macro-invertebrates have long life cycles of one year or more, which permits them to assimilate various environmental conditions over long periods of time (Mathuriau et al., 2011), so it is possible to monitor the effects of various contaminants along time. Hence, a macro-invertebrate assemblage represents a good picture of local environmental conditions. Macro-invertebrates show a wide tolerance spectrum to contaminants, ranging from highly sensitive groups to those that can withstand highly polluted conditions. Thus, the structure of the macro-invertebrate assemblage's can change significantly in response to environmental disturbances in predictable ways. Species richness and abundance of the macro-invertebrate assemblages go through a strong reduction in affected areas, and therefore more tolerant species predominate, whereas sensitive species of macro-invertebrates are present only in environments with least affected or un-impaired conditions. Thus, the structure and diversity of macro-invertebrates assemblages reflects the gradient of stream conditions from least impaired to highly impacted zones. These features form the basis for bio-monitoring using macro-invertebrates. With respect to ecological perspective of bio-monitoring, it is possible that some components of macro-invertebrate assemblages might not be present at the monitoring site during any time of the year. This will rely on the factors like life stages, seasonality, and the effects of natural and induced drift. Another factor that is to be taken into account is the nature of the substratum, because macro-invertebrates have different preferences for different substrate type. These variations do not necessarily depend on the effects caused by anthropogenic activities and may cause bias in bio-monitoring because of the absence of some macro-invertebrate groups, thus affecting those evaluation methods that are mainly based on scores of taxonomic groups.

12.7.3 ALGAE

In biological monitoring programs algae are used as an important constituent for assessing quality of water because of their short life cycle, nutrient requirement, and fast reproduction rate. They are used as valuable bio-indicators of environmental conditions as they respond quickly to water situations like water pollution by industrial and domestic waste discharge and effect the composition of species which are tolerant to these conditions (Gokce, 2016). As a valuable tool Algae were used for evaluating long-term changes in ecosystems such as eutrophication, management of water, changes in land use at watershed scale. In this sense, algae emerge as a valuable bio-indicator because they respond quickly to ecosystem changes thereby enabling a quick water quality assessment (Farrell et al., 2002). It is necessary to consider that the phytoplankton community changes rapidly with a change in water quality. Water quality which is a sum total of physical, chemical, and biological properties of water (European Environmental Agency, 1999). Phytoplankton biomass and many phytoplankton or algal species are used as indicators of freshwater pollution as they are very sensitive to changes in their environments (Reynolds, 1984). Depending on the type of pollution the algal biomass changes both qualitatively and quantitatively. In some cases new species colonize the lakes and in other cases local disappearance of other species occur (Gokce, 2016). There are two ways by means of which algal communities in aquatic ecosystem present the ecological water history, firstly via species differential revival rates and sensitivities to water substances and secondly through the accumulation and concentration of the substances in their cells (Law, 2011). Quality of water was defined by Wilhm using Shannon-Weaver diversity index under three classes (Table 12.1) showing higher H value for less polluted ecosystem and rich biodiversity and lower H value for more polluted ecosystem and poor biodiversity (Wilhm, 1975).

TABLE 12.1   Classes of Water Quality Determined by Using Shannon-Weaver Diversity Index (H)

Class	Water State	Shannon-Weaver (H)
I	Uncontaminated	>3
II	Moderately polluted	1–3
III	Highly polluted	<1

Algal pollution index (Table 12.2) value >20 determines higher organic contamination, index value from 15–19 depicts moderate organic pollution and the index value <15 shows extremely low or no organic contamination (Palmer, 1977).

TABLE 12.2 Algal Pollution Index Evaluates the Algal Species Tolerance to Organic Contamination and Determines Quality of Water

Algal Pollution Index	≤14	15–19	≥20
Water state	Low organic contamination	Moderate organic contamination	Higher organic contamination

12.7.4 EUGLENA GRACILIS

Although Euglena which is a motile freshwater photosynthetic flagellate is tolerant to acidity but it responds quickly to environmental toxicants such as organic or inorganic substances and heavy metals. Classic responses are a change in direction and inhibition of movement. Besides, this organism is very easy to handle and grow, which makes it a valuable tool for ecotoxicological assessments. One of the very valuable properties of this organism is the gravitactic orientation, which is very responsive to pollutants. The gravireceptors are affected by toxicants like organic or inorganic compounds and heavy metals. Thus the appearance of these substances is associated with the possibility of cell movement in water column. The extreme sensitivity of gravitactic alignment in *E. gracilis* has been reported for short-term tests (Ullah et al., 2013).

12.7.5 AQUATIC BRYOPHYTES

Bryophytes which are non-vascular vegetation present in aquatic ecosystems with high biomass and production are used to assess the ecological status. They are stress-tolerant species and number of species has a broad trophic range. Aquatic bryophytes are used as monitors to determine concentration of pollutants or are used as indicators to assess the presence or absence of pollutants (Burton, 1990; Taylor, 1990).

12.7.6 FROGS AND TOADS AS BIO-INDICATORS

Anurans which comprise of toads and frogs are used as bio-indicators of pollution studies (Simon et al., 2010). These absorb harmful chemicals through their skin and larval gill membrane and are very responsive to changes in their environments (Lambert, 1997). The ability of these organisms to detoxify pesticides which are absorbed or inhaled is very low which allows these chemicals especially organochlorines to accumulate in their bodies (Lambert, 1997). These organisms are used as model organisms for evaluating the effects of environmental toxicants that may cause the reduction of the amphibian population as they have previous skin that can easily absorb toxic pollutants (Lambert, 1997). The sustenance of ecosystem health depends on the consciousness, understanding, and management of ecological agents (Silvia et al., 2011). Frogs being vital environmental bio-indicators, have a chance of being affected by habitat changes, which in turn can affect both lower as well as higher trophic organisms.

12.7.7   FRESHWATER MUSSELS AS BIOLOGICAL INDICATORS

The quality of water and changes occurring in the aquatic environment have been documented by using several reliable indicators like freshwater mussels, which act as ecological repositories of the variations occurring in aquatic environment (Imlay, 1982; Neves, 1993; Naimo, 1995). Habitat modifications stemming from urbanization and industrialization impede the natural course of small brooks and rivers have had the most important effect on freshwater mussels (Bogan, 1993; Neves, 1993; Yeager, 1993). The physiochemical and biological characteristics of numerous water bodies have changed gradually from shallow habitats to vast ponds (Yeager, 1993; Hughes and Parmalee, 1999). The process of sedimentation is also known to have detrimental effects on freshwater mussel population. Sediment particles suspended in water column mainly soft and cohesive in nature are toxic for most of the mussel species and may affect their metabolism and overall growth processes (Marking and Bills, 1979).

12.7.8   PHYTOPLANKTON AS BIO-INDICATOR FOR WATER QUALITY

Since times immemorial there has been a close and intricate relationship between aquatic flora and pollution. Various aquatic organisms are known to be the biological indicators or indices of pollution and heavy metal contamination (Kolkwitz and Marrson, 1908). Different methods have been adopted to reveal organisms which are capable of monitoring the water quality (Knopp, 1954; Sladecek, 1973). Industrialization, urbanization, and other anthropogenic activities are the major causes of the release and deposition of harmful chemicals and other toxic pollutants in the environment (Ghorbanli et al., 2007; Hayes et al., 2007). The leaves of the plants growing on roadsides can be bio monitored to measure their exposure to harmful air pollutants, and to check their responses as stressor against them (Pandey et al., 2005; Sharma et al., 2007). In industrial areas with high levels of pollution the response from plants growing therein has been monitored biochemically and physiologically through proper techniques and methods (Joshi et al., 2009; Gupta et al., 2011).

12.7.9   DAPHNIA AS A POTENTIAL BIO-INDICATOR

Daphnia is used as a valuable bio-indicator in freshwater ecosystems for monitoring water quality due to its excellent features (Weber, 2002). Daphnia which is a freshwater crustacean fulfills all criteria of a bio-indicator and has been widely used to measure the ecological changes as well as effects of pollutants on freshwater ecosystems (Chen et al., 2012; Neves et al., 2015). This water flea also plays a vital role in building up monitoring criteria by environmental agencies (e.g., Environmental protection agency (EPA)), Organization for Economic Cooperation and Development (OECD) (Shaw et al., 2008). Daphnia have been used in ecotoxicological evaluation of water from many decades (Ambasht and Ambasht, 2003). In traditional bio-monitoring methods it involves toxicity assays such as acute or chronic toxicity tests to assess the toxicity as well as undesirable effects of toxicants with endpoints like growth, reproduction, and survival on Daphnia (Flaherty and Dodson, 2005; Heckmann et al., 2007). As per some reports Daphnia has been reported to be more tolerant than other biological organisms while Daphnia *magna* has the highest tolerance potential (Koivisto et al., 1992: Shaw et al., 2006).

12.7.10   SULFUR-OXIDIZING BACTERIA (SOB) AS A BIO-INDICATOR

In the last few years, sulfur-oxidizing bacteria (SOB) has become an important tool for the detection of heavy metals and other pollutants in water (Van Ginkel et al., 2011; Hassan et al., 2016). This SOB oxidizes elemental sulfur ($S°$) to sulfuric acid (H_2SO_4) under aerobic conditions in pollution-free waters (Madigam et al., 2009). The pH of water decreases

by production of hydrogen ions (H⁺) while as electrical conductivity
increases by SO_4^{2-} from this reaction (Hassan et al., 2012). However,
in the presence of harmful substances or chemicals the pH of water is
increased and EC is decreased because of inhibition of SOB activity (Oh
et al., 2011). Thus, changes in EC can be used to provide ecologically
appropriate indicator of toxicity that can support standard toxicity
tests using online bio-monitors (Van Ginkel et al., 2010; Hassan et al.,
2012, 2013). Therefore, SOBs are very valuable bio-indicators as they
can indicate a broad range of pollutants such as heavy metals, oxidized
chemicals, etc., (Gurung et al., 2011).

12.8 CONCLUSION

Bio-indicators can be a measure, an index of measures, or a model that
characterizes an ecosystem or one of its critical components. Bio-indicators
have proven to be reliable tools in monitoring or sensing the adverse effects
that industrial activity has on the global environment. This knowledge
will help in developing strategies and policies that will check or lower
such negative effects and make industry more sustainable. The role of
bio-indicators in sustainable development will help ensure that industry
leaves the smallest footprint possible on the environment. The many-
fold advantages of bio-indicators have outweighed their restrictions. The
bio-indicator is helpful, objective, straightforward, and reproducible.
Bio-indicators can be used at different levels, from the simple cellular
to the vast ecosystem scale, for assessing the changes taking place in a
particular biotic community. Planktonic monitors unite biological, physical,
chemical factors, and are utilized as an important part for evaluating health
status of water bodies. The conclusion can be drawn that bio-indication and
bio-monitoring have become promising tools for studying the impacts of
external factors on an ecosystem and its development and for differentiating
polluted and unpolluted areas.

ACKNOWLEDGMENT

Special thanks are due to Research Center, College of Pharmacy, King Saud
University, Riyadh, and Deanship of Scientific Research, King Saud University, Kingdom of Saudi Arabia.

KEYWORDS

- **aquatic bio-monitoring**
- **bio-indicator**
- **bio-monitoring**
- **sulfur-oxidizing bacteria**
- **toxicants**
- **water quality**

REFERENCES

(2017). *Protak Scientific "Biological Indicators."* Protak Scientific; United Kingdom: Protak Scientific. https://www.protakscientific.com/biologicalindicators (accessed on 27 October 2020).

Akbulut, A., (2003). The relationship between phytoplanktonic organisms and chlorophyll-a in Sultan Sazlığı. *Turk. J. Bot., 27,* 421–425.

Ambasht, R. S., & Ambasht, N. K., (2003). In: *Modern Trends in Applied Aquatic Ecology.* Springer Science & Business Media.

Aslam, M., Verma, D. K., Dhakerya, R., Rais, S., Alam, M., & Ansari, F. A., (2012). Bioindicator: A comparative study on uptake and accumulation of heavy metals in some plants leaves of M.G. Road, Agra City, India. *Res. J. Environ. Earth Sci., 4,* 1060–1070.

Bae, M. J., Kim, J. S., & Park, Y. S., (2012). Evaluation of changes in effluent quality from industrial complexes on the Korean nationwide scale using a self-organizing map. *Inter. J. Environ. Res. Pub. Health., 9,* 1182–1200.

Bartell, S. M., (2006). Biomarkers, bioindicators, and ecological risk assessment: A brief review and evaluation. *Environ. Bioindi., 1,* 60–73.

Barton, B. A., Morgan, J. D., & Vijayan, M. M., (2002). Physiological and condition-related indicators of environmental stress in fish. *Biological Indicators of Aquatic Ecosystem Stress,* 111–148.

Batiuk, R. A., Orth, R. J., Moore, K. A., Dennison, W. C., Stevenson, J. C., Staver, L. W., Carter, V., et al., (1992). *Chesapeake Bay Submerged Aquatic Vegetation Habitat Requirements and Restoration Targets: A Technical Synthesis* (pp. 166–169). Annapolis (MD), Chesapeake Bay Program 83/92.

Beyene, A., Addis, T., Kifle, D., Legesse, W., Kloos, H., & Triest, L., (2009). Comparative study of diatoms and macroinvertebrates as indicators of severe water pollution: Case study of the Kebena and Akaki rivers in Addis Ababa, Ethiopia. *Ecol. Indica., 9,* 381–392.

Bogan, A. E., (1993). Freshwater bivalve extinctions (Mollusca: Unionoida), A search for causes. *American Zoologist., 33,* 599–609.

Burger, J., Sanchez, J., Gibbons, J. W., & Gochfeld, M., (1998). Gender defenses in recreational use, environmental attitudes and perception of future land use at Savannah River Site. *Environmental and Behavior, 30,* 472–486.

Burton, M. A. S., (1990). Terrestrial and aquatic bryophytes as monitors of environmental contaminants in urban and industrial habitats. *Bot. J. Linn. Soci., 104*, 267–280.

Butterworth, F. M., Gunatilaka, A., & Gonsebatt, M. E., (2001). *Biomonitors and Biomarkers as Indicators of Environmental Change* (Vol. 2). Boston (MA), Springer Science & Business Media.

Cairns, J., & Pratt, J. R., (1993). A history of biological monitoring using benthic macroinvertebrates. *Freshwater Biomonitoring and Benthic Macroinvertebrates* (pp. 10–27).

Chen, L., Zhang, G., Zeng, Y., & Ren, Z., (2012). Influences of temperature, pH and turbidity on the behavioral responses of daphnia magna and Japanese Medaka (*Oryzias latipes*) in the biomonitor. *Proc. Environ. Sci., 13*, 80–86.

Chessman, B., (2003). *SIGNAL 2 - a Scoring System for Macro-Invertebrate (Water Bugs) in Australian Rivers.* Monitoring river heath initiative technical report no 31, Commonwealth of Australia, Canberra.

Cummins, K. W., & Klug, M. J., (1979). Feeding ecology of stream invertebrates. *Ann. Revi. Ecol. Sys., 10*, 147–172.

Demir, A. N., Fakioğlu, Ö., & Dural, B., (2014). Phytoplankton functional groups provide a quality assessment method by the Q assemblage index in Lake Mogan (Turkey). *Turk. J. Bot., 38*, 169–179.

Devlin, M., Bricker, S., & Painting, S., (2011). Comparison of five methods for assessing impacts of nutrient enrichment using estuarine case studies. *Biogeochemistry, 106*, 177–205.

Espinal-Carreon, T., Sedeno-Diaz, J. E., & Lopez-Lopez, E., (2013). Evaluacion de la calidad del aguaen la Laguna de Yuriria, Guanajuato, Mexico, mediantete cnicasmultivariadas: Un ana lisis de valoracion para dos e pocas 2005, 2009–2010. *Rev. Int. Contam. Ambie., 29*, 147–163.

European Environmental Agency, (1999). *Nutrients in European Ecosystems* (p. 155). Environmental Assessment Report No. 4. Copenhagen: European Environmental Agency.

Ferreira, W. R., Paiva, L. T., & Callisto, M., (2011). Development of a benthic multimetric index for biomonitoring of a neotropical watershed. *Braz. J. Biol., 71*, 15–25.

Flaherty, C. M., & Dodson, S. I., (2005). Effects of pharmaceuticals on Daphnia survival, growth, and reproduction. *Chemosphere, 61*, 200–207.

Flores, M. J. L., & Zafaralla, M. T., (2012). Macroinvertebrate composition, diversity and richness in relation to the water quality status of Mananga River, Cebu, Philippines. *Phili. Sci. Lett., 5*, 103–113.

Garcıa, L. M., Castro, B., Ribeiro, R., & Guilhermino, L., (2000). Characterization of cholinesterases from guppy (*Poecilia reticulata*) muscle and it's *in vitro* inhibition by environmental contaminants. *Biomarkers, 5*, 274–284.

Gerhardt, A., (1996). Behavioral early warning responses to polluted water, performance of *Gammarus pulex* L. (Crustacea) and *Hydropsyche angustipennis* (Curtis) (Insecta) to a complex industrial effluent. *Environ. Sci. Pollut. Res., 3*, 63–70.

Gerhardt, A., (2002). *Bioindicator Species and Their Use in Biomonitoring.* Environmental monitoring I. Encyclopedia of life support systems. UNESCO ed. Oxford (UK), Eolss Publisher.

Ghorbanil, M., Bakand, Z., Khaniki, B., & Bakand, S., (2007). Air pollution effects on the activity of antioxidant enzymes in *Nerium oleander* and *Robinia pseudoacacia* plants in Tehran. *Iran. J. Environ. Heal. Sci. Engi., 4*, 157–162.

Gökçe, D., (2016). Algae as an indicator of water quality. *Algae-Organisms for Imminent Biotechnology* (pp. 81–101). In Tech.

Guilhermino, L., Diamantino, T., Carolina, S. M., & Soares, A. M. V. M., (2000). Acute toxicity test with daphnia magna: An alternative to mammals in the prescreening of chemical toxicity? *Ecotoxicol. Environ. Safe, 46*, 357–362.

Gupta, S., Nayek, S., & Bhattacharya, P., (2011). Effect of airborne heavy metals on the biochemical signature of tree species in an industrial region, with an emphasis on anticipated performance index. *Chemi. Ecol., 27*, 381–392.

Gurung, A., Hassan, S. H. A., & Oh, S. E., (2011). Assessing acute toxicity of effluent from a textile industry and nearby river waters using sulfur-oxidizing bacteria in continuous mode. *Environ. Technol., 32*, 1597–1604.

Hanson, P., Springer, M., & Ramirez, A., (2010). Capítulo I Introduccion a losgrupos de macr invertebradosacuaticos. *Rev. Biol. Trop., 58*, 3–37.

Haribhau, M. G., & Rathod, R. G., (2012). Trace metals contamination of surface water samples in and around Akot City in Maharashtra, India. *Res. J. Recent Sci., 1*, 5–9.

Hassan, S. H. A., Van, G. S. W., & Oh, S. E., (2012). Detection of Cr^{6+} by the sulfur-oxidizing bacteria biosensor: Effect of different physical factors. *Environ. Sci. Technol., 46*, 7844–7848.

Hassan, S. H. A., Van, G. S. W., & Oh, S. E., (2013). Effect of organics and alkalinity on the sulfur-oxidizing bacteria (SOB) biosensor. *Chemosphere, 90*, 965–970.

Hassan, S. H. A., Van, G. S. W., Hussein, M. A. M., Abskharon, R., & Oh, S. E., (2016). Toxicity assessment using different bioassays and microbial biosensors. *Enviro. Internat., 92*, 106–118.

Hayes, A., Bakand, S., & Winder, C., (2007). Novel *in vitro* exposure techniques for toxicity testing and Biomonitoring of airborne contaminants. In: *Drug Testing In vitro-Achievements and Trends in Cell Culture Techniques* (pp. 103–124). Wiley-VCH, Berlin.

Heckmann, L. H., Callaghan, A., Hooper, H. L., Connon, R., Hutchinson, T. H., Maund, S. J., & Sibly, R. M., (2007). Chronic toxicity of ibuprofen to Daphnia magna: Effects on life history traits and population dynamics. *Toxicol. Lett., 172*, 137–145.

Hogan, A., & Vallance, T., (2005). *Rapid Assessment of Fish Biodiversity in Southern Gulf of Carpentaria Catchments*. Department of Primary Industries and Fisheries.

Holt, E. A., & Miller, S. W., (2011). Bioindicators: Using organisms to measure environmental impacts. *Nature Education Knowledge, 3*, 8.

Hosmani, S., (2014). Freshwater plankton ecology: A review. *J. Res. Manage. Technol., 3*, 1–10.

Hughes, M. H., & Parmalee, P. W., (1999). Prehistoric and modern freshwater mussel (mollusca: Bivalvia) faunas of the Tennessee River: Alabama, Kentucky, and Tennessee. *Regulated Rivers: Research and Management, 15*, 24–42.

Imlay, M. J., (1982). The use of shells of freshwater mussels in monitoring heavy metals and environmental stresses: A review. *Malacol. Revi., 15*, 1–14.

Jain, A., Singh, B. N., Singh, S. P., Singh, H. B., & Singh, S., (2010). Exploring biodiversity as bioindicators for water pollution. *National Conference on Biodiversity, Development and Poverty Alleviation*. Uttar Pradesh. Lucknow (India), Uttar Pradesh State Biodiversity Board.

Jha, P., & Barat, S., (2003). Hydrobiological study of Lake Mirik in Darjeeling, Himalayas. *J. Environ. Biol., 24*, 339–344.

Joanna, B., (2006). Bioindicators: Types, development, and use in ecological assessment and research. *Environ. Bioind., 1*, 22–39.

John, J., (2003). Bioassessment of health of aquatic systems by the use of diatoms. In: Ambasht, R. S., & Ambasht, N. K., (eds.), *Modern Trends in Applied Aquatic Ecology* (pp. 1–20). Kluwer Academic/Plenum Publishers, New York.

Joshi, N., Chauhan, A., & Joshi, P. C., (2009). Impact of industrial air pollutants on some biochemical parameters and yield in wheat and mustard plants. *Environmentalist, 29*, 398–404.

Karr, J. R., & Dudley, D. R., (1981). Ecological perspective on water quality goals. *Enviro. Manag.*, *5*, 55–68.

Karr, J. R., (1981). Assessment of biotic integrity using fish communities. *Fisheries*, *6*, 21–27.

Khatri, N., & Tyagi, S., (2015). Influences of natural and anthropogenic factors on surface and ground water quality in rural and urban areas. *Front. Life. Sci.*, *8*, 23–39.

Knopp, H., (1954). Ein neuer Weg zur Darstellung biologischer Gewasseruntersuchungen, erlautert an einem Gutellangsschnitt des Mains. Die. Wasserwirtschaft., *45*, 9–15.

Koivisto, S., Ketola, M., & Walls, M., (1992). Comparison of five cladoceran species in short- and long-term copper exposure. *Hydrobiol.*, *248*, 125–136.

Kolkwitz, R., & Marrson, M., (1908). Okologie der pflanzlichen Saprobien. *Ber. Deut. Bot. Ges.*, *26*, 505–519.

Lam, P. K., (2009). Use of biomarkers in environmental monitoring. *Ocean Coast. Manag.*, *52*, 348–354.

Lambert, M. R. K., (1997). Environmental effects of heavy spillage from a destroyed pesticide store near Hargeisa (Somaliland) assessed during the dry season, using reptiles and amphibians as bioindicators. *Arch. Environ. Contam. Toxicol.*, *32*, 80–93.

Landres, P. B., Verner, J., & Thomas, J. W., (1988). Ecological uses of vertebrate indicator species: A critique. *Conserv. Biol.*, *2*, 316–328.

Law, R. J., (2011). A review of the function and uses and factors affecting, stream phytobenthos. *Freshwater Rev.*, *4*, 135–166.

Li, L., Zheng, B., & Liu, L., (2010). Biomonitoring and bioindicators used for river ecosystems: Definitions, approaches and trends. *Procedia. Environ. Sci.*, *2*, 1510–1524.

Livingstone, D. R., (2003). Oxidative stress in aquatic organisms in relation to pollution and aquaculture. *Revuer. Med. Vet.*, *154*, 427–430.

Madigan, M. T., Martinko, J. M., Dunlap, P. V., & Clark, D. P., (2009). *Brock Biology of Microorganisms*. Pearson Education Inc., San Francisco.

Malik, D. S., & Bharti, U., (2012). Status of plankton diversity and biological productivity of Sahastradhara stream at Uttarakhand, India. *J. Appl. Natural. Sci.*, *4*, 96–103.

Markert, B., Kayser, G., Korhammer, S., & Oehlmann, J., (2000). Distribution and effects of trace substances in soils, plants and animals. *Tr. Met. Env.*, *4*, 3–31.

Marking, L. L., & Bills, T. D., (1979). Acute effects of silt and sand sedimentation on freshwater mussels. In: Rasmussen, J. R., (ed.), *Proceedings of the UMRCC Symposium on Upper Mississippi River Bivalve Mollusks* (pp. 204–211). Upper Mississippi River Conservation Committee, Rock Island, Illinois.

Mathuriau, C., Silva, N. M., Lyons, J., & Rivera, L. M. M., (2012). Fish and macroinvertebrates as freshwater ecosystem bioindicators in Mexico: Current state and perspectives. In: *Water Resources in Mexico* (pp. 251–261). Springer, Berlin, Heidelberg.

McGeoch, M. A., & Chown, S. L., (1998). Scaling up the value of bioindicators. *Trends Ecol. Evol.*, *13*, 46–47.

Meng, Q., Li, X., Feng, Q., & Cao, Z., (2008). *In the 2nd International Conference on Bioinformatics and Biomedical Engineering, (ICBBE 2008)* (pp. 4555–4558). (IEEE).

Metcalfe, J. L., (1989). Biological water quality assessment of running waters based on macroin-vertebrate communities: History and present status in Europe. *Environ. Pollut.*, *60*, 101–139.

Mishra, S., Barik, S. K., Ayyappan, S., & Mohapatra, B. C., (2000). Fish bioassays for evaluation of raw and bioremediated dairy effluent. *Biores. Technol.*, *72*, 213–218.

Morcillo, Y., Albalat, A., & Porte, C., (1999). Mussels as sentinels of organotin pollution: Bioaccumulation and effects of P450-mediated aromatase activity. *Environ. Toxicol. Chem.*, *18*, 1203–1208.

Naimo, T. J., (1995). A review of the effects of heavy metals on freshwater mussels. *Ecotoxicology, 4*, 341–362.

Nath, S. N., & Srivastava, A. K., (2010). Hematological parameters as bioindicators of insecticide exposure in teleosts. *Ecotoxicol., 19*, 838–854.

Nelson, J. S., (2006). *Fishes of the World* (4th edn., p. 559). John Wiley and Sons, New York.

Neves, M., Castro, B. B., Vidal, T., Vieira, R., Marques, J. C., Coutinho, J. A. P., Gonçalves, F., & Gonçalves, A. M. M., (2015). Biochemical and populational responses of an aquatic bioindicator species, *Daphnia longispina*, to a commercial formulation of a herbicide (Primextra® Gold TZ) and its active ingredient (S-metolachlor). *Ecol. Indic., 53*, 220–230.

Neves, R. J., (1993). A state-of-the-unionids address. In: *Conservation and Management of Freshwater Mussels* (pp. 1–10). Proceedings of a UMRCC symposium. Upper Mississippi River Conservation Committee, Rock Island.

Nkwoji, J. A., Igbo, J. K., Adeleye, A. O., Obienu, J. A., & Tony-Obiagwu, M. J., (2010). Implications of bioindicators in ecological health: Study of a coastal lagoon, Lagos, Nigeria. *Agric. Biol. J. Noth. Am., 1*, 683–689.

Nwajei, G. E., Obi-Iyeke, G. E., & Okwagi, P., (2012). Distribution of selected trace metal in fish parts from the River Nigeria. *Res. J. Recent Sc., 1*, 81–84.

O'Brien, D. J., Kaneene, J. B., & Poppenga, R. H., (1993). The use of mammals as sentinels for human exposure to toxic contaminants in the environment. *Environ. Health Persp., 99*, 351–368.

O'Farrell, I., Lombardo, R. J., Pinto, P. T., & Loez, C., (2002). The assessment of water quality in the lower Lujan River (Buenos Aires, Argentina), phytoplankton and algal bioassays. *Environ. Poll., 120*, 207–218.

O'uyang, Y., (2005). Evaluation of river water quality monitoring stations by principal component analysis. *Water Res., 39*, 2621–2635.

Odum, E. P., (1985). Trends expected in stressed ecosystems. *BioSc., 35*, 419–422.

Oertel, N., & Salanki, J., (2003). Biomonitoring and bioindicators in aquatic ecosystems. In: Ambasht, R. S., & Ambasht, N. K., (eds.), *Modern Trends in Applied Aquatic Ecology* (pp. 219–246). Kluwer Academic/Plenum Publishers, New York.

Oh, S. E., Hassan, S. H. A., & Van, G. S. W., (2011). A novel biosensor for detecting toxicity in water using sulfur-oxidizing bacteria. *Sensors and Actuators B: Chem., 154*, 17–21.

Palmer, C. M., (1977). *Algae and Water Pollution*. Available from the National Technical Information Service, Springfield VA 22161 as PB-287 128, Price codes: A 07 in paper copy, A 01 in microfiche; Report.

Pandey, S. K., Tripathi, B. D., Prajapati, S. K., Mishra, V. K., Upadhyay, A. R., Rai, P. K., & Sharma, A. P., (2005). Magnetic properties of vehicle derived particulates and amelioration by *Ficus* infectoria: A keystone species. *Ambio: J. Human Environ., 34*, 645–646.

Pereira, J. L., Antunes, S. C., Castro, B. B., Marques, C. R., Gonçalves, A. M., Gonçalves, F., & Pereira, R., (2009). Toxicity evaluation of three pesticides on non-target aquatic and soil organisms: Commercial formulation versus active ingredient. *Ecotoxicol., 18*, 455–463.

Phillips, D. J. H., & Rainbow, P. S., (1993). *Biomonitoring of Trace Aquatic Contaminants*. New York (NY), Elsev. Appl. Sc.

Posudin, Y., (2014). Bioindication. In: *Methods of Measuring Environmental Parameters* (pp. 145–146). John Wiley & Sons, Inc., Hoboken, NJ, USA. Proceeding of XXXIX-the Apimondia International Apicultural Congress, Dublin, Ireland.

Ranjeeta, C., (2012). Heavy metal analysis of water of Kaliasote Dam of Bhopal, MP, India. *Res. J. Recent Sci., 1*, 352–353.

Reece, P. F., & Richardson, J. S., (2000). Biomonitoring with the reference condition approach for the detection of aquatic ecosystems at risk. In: Darling, L. M., (ed.), *Proceedings of a Conference on the Biology and Management of Species and Habitats at Risk* (Vol. 2, pp. 549–552). BC Ministry of Environment, Lands and Parks, Victoria, BC and University College of the Cariboo, Kamloops, BC.

Reynolds, C. S., (1984). *The Ecology of Freshwater Phytoplankton* (p. 384). Cambridge: Cambridge University Press.

Sharma, A. P., Rai, P. K., & Tripathi, B. D., (2007). Magnetic biomonitoring of roadside tree leaves as a proxy of vehicular pollution. In: Lakshmi, L. V., (ed.), *Urban Planning and Environment: Strategies and Challenges* (pp. 326–331). Mc Millan Advanced Research Series.

Shaw, J. R., Dempsey, T. D., Chen, C. Y., Hamilton, J. W., & Folt, C. L., (2006). Comparative toxicity of cadmium, zinc, and mixtures of cadmium and zinc to daphnids. *Environ. Toxicol. Chem., 25*, 182–189.

Shaw, J. R., Pfrender, M. E., Eads, B. D., Klaper, R., Callaghan, A., Sibly, R. M., Colson, I., et al., (2008). Daphnia as an emerging model for toxicological genomics. *Adv. Experim. Biol., 2*, 165–328.

Silvia, C., Rosa, L., Souza, R. B. D., Giuliano, D., & Thiago, G., (2011). *Soil Contamination.* InTech. doi: 10.5772/25042. ISBN 978-953-307-647-8.

Simon, E., Braun, M., & Tóthmérész, B., (2010). Non-destructive method of frog (*Rana esculenta* L.) skeleton elemental analysis used during environmental assessment. *Water, Air, and Soil Pollution, 209*, 467–471.

Sladecek, V., (1973). System of water quality from the biological point of view. *Ergebnisse. Der. Limnol., 7*, 1–218.

Springer, M., (2010). Biomonitoreo acuatico. *Int. J. Trop. Biol., 58*, 53–59.

Srivastava, K. P., & Kumar, S. V., (2012). Impact of air pollution on pH of soil of Saran, Bihar, India. *Res. J. Recent Sci., 4*, 9–13.

Taylor, G., (1990). Bryophytes and heavy metals: A literature review. *Bot. J. Lin. Soc., 104*, 231–253.

Tejeda-Vera, R., López-López, E., & Sedeño-Díaz, J. E., (2007). Biomarkers and bioindicators of the health condition of *Ameca splendens* and *Goodea atripinnis* (Pisces: Goodeaidae) in the Ameca River, Mexico. *Environ. Int., 33*, 521–531.

Trujillo, J. P., Sedeno-Diaz, J. E., Camargo, J. A., & Lopez-Lopez, E., (2011). Assessing environmental conditions of the RıoChampoton (Mexico) using diverse indices and biomarkers in the fish *Astyanax aeneus* (Gu¨nther, 1860). *Ecol. Ind., 11*, 636–646.

Tsui, M. T., & Chu, L. M., (2003). Aquatic toxicity of glyphosate-based formulations: Comparison between different organisms and the effects of environmental factors. *Chemosphere, 52*, 1189–1197.

Ullah, A., Murad, W., Adnan, M., Ullah, W., & Häder, D. P., (2013). Gravitactic orientation of *Euglena gracilis*: A sensitive endpoint for ecotoxicological assessment of water pollutants. *Frontiers Environ. Sci., 1*, 4.

Uttah, E. C., Uttah, C., Akpan, P. A., Ikpeme, E. M., Ogbeche, J., & Usip, J. O., (2008). Bio-survey of plankton as indicators of water quality for recreational activities in Calabar River, Nigeria. *J. Appl. Sci. Environ. Manage., 12*, 35–42.

Valentine, J., Davis, S. R., Kirby, J. R., & Wilkinson, D. M., (2013). The use of testate amoebae in monitoring peatland restoration management: Case studies from North West England and Ireland. *Acta Protozool., 52*, 129–145.

Van, D. O. R., Beyer, J., & Vermeulen, N. P. E., (2003). Fish bioaccumulation and biomarkers in environmental risk assessment: A review. *Environ. Toxicol. Pharmacol., 13*, 57–149.

Van, G. S. W., Hassan, S. H. A., Ok, Y. S., Yang, J. E., Kim, Y. S., & Oh, S. E., (2011). Detecting oxidized contaminants in water using sulfur-oxidizing bacteria. *Environ. Sc. Technol., 45*, 3739–3745.

Vannote, R. L., WhyneMinshall, G., Cummis, K. W., & Sedell, J. R., (1980). The river continuum concept. *Can. J. Fish. Aquat. Sci., 37*, 130–137.

Weber, C. I. U. S., (2002). *Environmental Protection Agency Office of Water* (5th edn.). Methods for measuring the acute toxicity of effluents and receiving waters to freshwater and marine organisms. Diane Publishing Co., Washington, D.C.

Wilhm, J. L., (1975). Biological indicators of pollution. In: Whitton, B. A., (ed.), *River Ecology, Studies in Ecology* (Vol. 2, pp. 375–402). London: Blackwell Science Publ.

Wilkomirski, B., (2013). History of bioindication (*Historia bioindykacji*). *Monitoring Srodowiska Przyrodniczego, 14*, 137–142.

Wolf, P., Brischwein, M., Kleinhans, R., Demmel, F., Schwarzenberger, T., Pfister, C., & Wolf, B., (2013). Automated platform for sensor-based monitoring and controlled assays of living cells and tissues. *Biosen. Bioelectron., 50*, 111–117.

Wootton, R. J., (1992). *Fish Ecology* (p. 215). Chapman and Hall, New York.

Yeager, B. L., (1993). Dams: Impacts on warm water streams: Guidelines for evaluation. In: Bryan, C. F., & Rutherford, D. A., (eds.), *Warm Water Stream Committee* (pp. 57–113). Southern Division, American Fisheries Society, Little Rock, Arkansas.

Zannatul, F., & Muktadir, A. K. M., (2009). A review: Potentiality of zooplankton as bioindicator. *Am. J. Appl. Sci., 6*, 1815–1819.

Zhang, L. J., Ying, G. G., Chen, F., Zhao, J. L., Wang, L., & Fang, Y. X., (2012). Development and application of whole-sediment toxicity test using immobilized freshwater microalgae *Pseudokirchneriella* subcapitata. *Environ. Toxicol. and Chem., 31*, 377–386.

Zhou, Q., Zhang, J., Fu, J., Shi, J., & Jiang, G., (2008). Biomonitoring: An appealing tool for assessment of metal pollution in the aquatic ecosystem. *Ana. Chim. Acta, 606*, 135–150.

Index

For Product Safety Concerns and Information please contact our EU
representative GPSR@taylorandfrancis.com
Taylor & Francis Verlag GmbH, Kaufingerstraße 24, 80331 München, Germany

www.ingramcontent.com/pod-product-compliance
Lightning Source LLC
Chambersburg PA
CBHW060755220326
41598CB00022B/2442

* 9 7 8 1 7 7 4 6 3 8 8 3 5 *